W9-CGM-241

Design-Build Project Delivery

Managing the Building Process from Proposal through Construction

Sidney M. Levy

McGraw-Hill
New York Chicago San Francisco Lisbon London Madrid
Mexico City Milan New Delhi San Juan Seoul
Singapore Sydney Toronto

The McGraw-Hill Companies

Library of Congress Cataloging-in-Publication Data

Levy, Sidney M.
Design-build project delivery / Sidney M. Levy.
p. cm.
Includes index.
ISBN 0-07-146157-4
1. Building. 2. Construction industry—Management. I. Title.

TH145.L44 2006
690.068—dc22

2005056221

1 2 3 4 5 6 7 8 9 0 DOC/DOC 0 1 3 2 1 0 9 8 7 6

ISBN 0-07-146157-4

The sponsoring editor for this book was Cary Sullivan and the production supervisor was Richard C. Ruzycka. It was set in Century Schoolbook by International Typesetting and Composition. The art director for the cover was Handel Low.

Printed and bound by RR Donnelley.

This book was printed on recycled, acid-free paper containing a minimum of 50% recycled, de-inked fiber.

Contents

ABOUT THE AUTHOR

Sidney M. Levy is a construction consultant with more than 40 years' experience in the industry. The author of 15 previous books, including several devoted to international construction and the award-winning *Project Management in Construction*, Mr. Levy is the owner of a construction consulting firm in Baltimore, Maryland.

Preface

Design-build is a project delivery system with a record of reducing overall project costs and bringing capital facilities on-line more rapidly. But its true value may be in the cultural changes it effects in the process. Other design and build processes include third party relationships, owner contracts with design consultant and then owner contracts with builder, each party to the process having their own agenda and their own concerns, which frequently do not mesh.

Design-build offers a somewhat different approach. The collaborative effort that develops as owners work with design consultants and contractors from project genesis through design development and into construction creates an environment of trust as problems arise. These problems are then discussed, dealt with and resolved as a team. This working together toward a common goal—the successful completion of a construction project—may just be design-build's lasting legacy.

Other delivery systems had promised relief from the confrontational environment that frequently followed in the path of design-bid-build projects, where costs often exceeded budgets and sparked a round of redesign, rebid and rebudgeting. The construction manager approach afforded an owner the ability to bring a building professional into the project at an early stage of design development, utilizing their knowledge of local conditions, product availability, and a strong database of costs to avoid some of the problems of designing in a void. And this system has flourished in recent decades.

Design-build carries this owner-designer-builder interface much further by creating a single entity charged with transforming the client's program into an aesthetic and cost-effective reality.

The last 30 years or so of the twentieth century could be characterized as an era where construction and design-related disputes and claims were commonplace in both the private and public sectors. It became evident to all that things were not working the way they were conceived. Contractors blamed design consultants for producing incomplete construction documents and architects blamed contractors for "lowballing" bids and change order mania. Owners caught between the two professional groups argued that they were the real victims since they ultimately paid the price for defective drawings and opportunistic contractors.

Design-build began testing the waters of mainstream project delivery systems with promises of, if not eliminating, then drastically reducing the adversarial relationships that sprouted in previous decades while controlling the costs associated with the resultant disputes and claims. This process was not a new one, having its roots in the master builder's concept going all the way back to ancient Egypt and the Roman Empire, but it was a renewed effort to seamlessly take a project from design to completion with all parties working together and congratulating each other on a job well done, instead of facing one another in a court of law.

After all it was not until the mid-nineteenth century that the practice of architecture split away from the building trades and now, 200 years later, the industry is attempting to renew this notion of the master builder.

The design-build team focuses on achieving a common goal, and along the way each participant realizes that to do so, they will be required to assume some risk, allay professional prejudices, accept compromise when the occasion arises and be willing to listen, really listen, to others in the group, a hard thing to do in this business.

The design-build process provides this environment where the challenges that arise can be addressed and resolved in a nonconfrontational manner.

In our capitalistic society, the drive for greater profits also drives the effort to look for better ways to accomplish our individual and collective goals so that we become more efficient at what we do. Profit and creativity can live in the same house.

The advances in computer technology have radically changed our industry and more and more benefits will be seen in the years to come. The ability to create the virtual project on a computer screen has progressed from a laboratory exercise to a commercial product, and the confusing proliferation of software programs that can't communicate with each other are slowly being replaced by one universal interoperable system. Building technology is also changing, slowly at first, but more rapidly as contractors shed their reluctance to embrace these new products, components, and equipment.

As owners, architects, and builders learn to work together and respect each other's concerns and desires, the cultural differences of the past are slowly melting away.

Clients appreciate this "one-stop shopping" process where they, like their architect/engineer and builder counterparts, can work in an open environment pursuing their common goal—the more perfect project.

And design-build will be judged as one of the dominant forces in pushing capital delivery systems closer to that goal.

Design-Build Project Delivery—Managing the Building Process from Proposal through Construction explores this methodology from an owner's, architect's, and contractor's perspective and provides a roadmap for those contemplating a new way to create an old product.

Sidney M. Levy

Chapter

1

An Introduction to Design-Build

Design-build is an outgrowth of a project delivery system steeped in antiquity, dating as far back as the construction of the pyramids in 1596 B.C. It is also an industry-driven program to find a better project delivery system. The word *architect* in its Greek origin means *the work of a master carpenter*—so design-build firms may have plied the streets of ancient Athens.

The traditional design-bid-build project, in recent times, has become the design-bid–redesign-rebid and build project. Budgets prepared by either an owner's consultant or capital improvements team often fall short of the actual cost of construction, requiring expensive redesign, acceptance of less-than-value engineering suggestions, and delays in bringing the project on stream.

Searching for the optimum project delivery system goes on and on. The construction manager (CM) concept presented another alternative to the conventional design-bid-build method. Some sources trace the advent of the CM back to the State of New York's Wicks Law enacted in 1921 requiring four prime contracts (general construction, plumbing, electrical, HVAC) for public projects exceeding $50,000. With the absence of a central point of control and management, the creation of a CM fulfilled that need. As the CM concept matured, owners recognized the value of bringing the expertise of a general contractor into the picture during the design stage when their advice on constructability, costs, and knowledge of local markets could bring considerable value to the project. Hence the two-part CM contract where CM can be engaged, initially, during the design stage, and if their contribution during that stage is beneficial to the owner, they will be awarded the second part of the CM contract, one to provide management of construction services.

Design-build can be viewed as an evolutionary project delivery system, one that addresses many of the concerns owners have had, and will continue to have, as they ponder the way to achieve that perfect project.

The Search for a Better System

The construction industry, because of its vast outreach, touches everyone from homeowners to Fortune 500 companies and receives its fair share of publicity—some good and some not so good.

Articles appear from time to time in trade magazines and newspapers deriding design consultants for producing defective drawings that include errors and omissions, and owners complain about the lack of design accountability that ends up increasing the construction cost. They often ignore the fact that the designs of today's projects are much more complex than they were years ago, and yet owners demand production of these complex documents in a compressed time frame and at reduced cost.

These same magazines and newspapers, at other times, print articles about contractors rigging bids, working questionable deals with subcontractors, and producing shoddy work often resulting in job-site injuries or fatalities.

Articles point out unethical project owners who drum up excuses to avoid paying contractors or present them with an offer they can't refuse—*take 50% of your final payment or sue me.*

Even though these practices by architect/engineer, contractor, and owner may be only isolated cases, they tend to color the way in which the industry is viewed by the general public.

This ongoing drive to produce more cost-effective construction projects reached new highs, or some would say new lows in recent years, with a new approach—the reverse auction. An owner using a reverse auction would solicit bids on their Web site and post the bids they received for the particular proposed project.

Contractors, viewing those bids would have an opportunity to adjust their prospective bid to be somewhat or significantly lower than the low bid already posted. Often, no contractor prequalifications were required, and owners, by accepting low bids, would get exactly what they deserved—trouble.

If the theory behind reverse auctions wasn't enough of an ethical stretch, there were rumors that some owners were posting phony bids on their Web site to attract prices that had little to do with the legitimate cost of the project. The reverse auction has since died a quiet death.

Some other practices such as contractors *front-end loading* their schedule of values, the basis for their monthly requisitions, are frowned upon. Contractors can stretch the ethical envelope when *upfront* values of early operations are significantly higher than they should be. In effect, the builder would be able to requisition and get paid more money during the front end or start of construction than he or she was really entitled to receive.

In an industry so vital to the U.S. economy, good business practices and ethical business practices ought to be of prime concern to all parties to the construction process.

How the Construction Industry Is Perceived

In a 2004 study conducted by FMI (formerly known as Fails Management Institute) and the Construction Management Association of America (CMAA), the question of ethics in the industry was presented to architects, engineers, owners, construction managers, and general contractors.

More that 270 responses were received—8% from owners and architects, 23% from CMs, 29% from general contractors, 30% from subcontractors, and 10% from vendors. Ethical issues relating to design consultants, contractors and owners alike were revealed. The key concerns surfacing in this survey had to do with:

- A breakdown in trust and integrity
- Loss of reputation for the industry
- The need to provide a code of ethics and standards
- Creating an equitable bid process

About 84% of the respondents said that, in the past year, they had *experienced, encountered, or observed* industry related acts they considered unethical. Thirty-four percent said they had experienced multiple examples of unethical behavior.

Concerns voiced about owners were:

1. Owners authorizing work but failing to pay for it or being very late in their payments
2. Owners attempting to pass off their responsibilities to others
3. Owners lacking ethical behavior (for example, placing bogus bids on their reverse auction site)
4. Not enough dialogue between owners and the construction industry regarding expectations of both parties

Concerns voiced about architects and engineers were:

1. Owners stating that architects and engineers would do whatever was necessary to make them happy, often at the expense of the contractor
2. The need for architects to be fair and equitable in making decisions that affect contractors and the owner
3. Designers knowingly issuing drawings and other bid documents that are deficient

Concerns voiced about contractors:

1. Bid shopping
2. Change-order games
3. Payment games (receiving payment from an owner, but delaying payment to subcontractors or suppliers)
4. Claims games
5. Hiring unreliable subcontractors

The Case for Design-Build

While design-build does not directly focus on ethical issues, the very nature of its process can eliminate some of the concerns voiced by owner, design consultants, and contractors alike. The potential for change-order games that result in disputes and claims is drastically reduced in the design-build process. The involvement of architect, engineer, and contractor with the owner from design conception—through design development, through contract documents, construction, and commissioning—should provide an environment of trust simply by getting to know each other often through some heated discussions as well as through resolution of those same problems.

Trust can't be legislated or contracted but can be built by working together, and that is what is required and what the design-build process is all about.

The contractual and operations differences between non-design-build and design-build are displayed in Fig. 1.1.

The master builder approach

The design-build project delivery system employed in the United States today is a distillation of the precepts and practices behind a single source responsibility with a long and successful history. Master builders in the first half of the twentieth

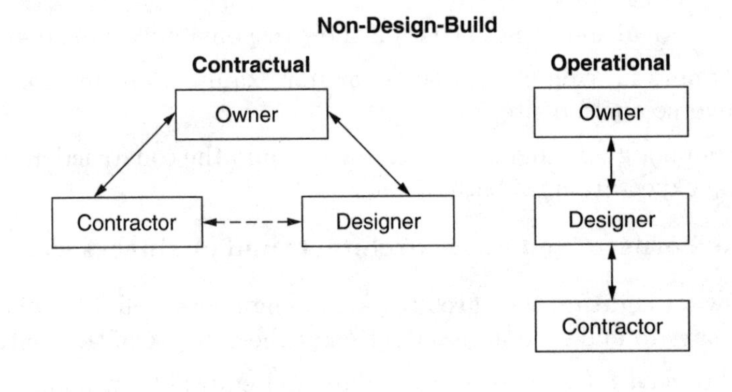

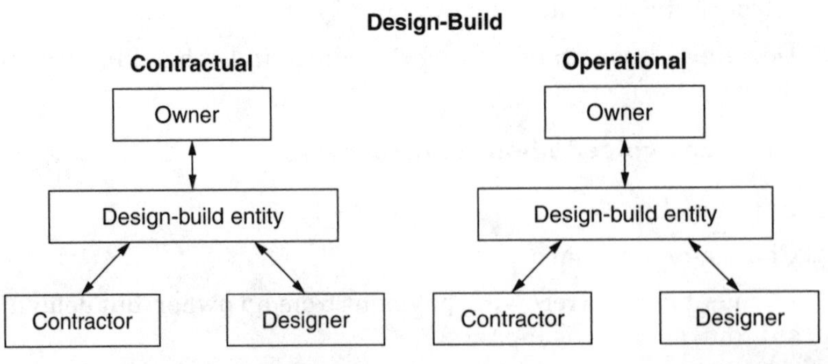

Figure 1.1 Contractual and operational differences between non-design-build and design-build projects. [*Courtesy: Legislate Analyst's Office (LAO), State of California.*]

century used their expertise and experience to offer clients a *package* that included a design to fit the owner's needs. Relying more on their extensive payroll of laborers, mechanics, skilled craftsman, plumbers, and electricians than third-party subcontractors, these master builders could effect design changes in the field often during an informal owner's walk-through inspection. No need for lengthy paper work, drawing revisions, or sketches—just do it.

Turnkey projects

Turnkey projects employed for many years in the process engineering, food, and pharmaceutical industries were utilized as a method of providing an owner with a complete facility, ready to operate and turn out product. All the owner had to do was turn the key to the front door. The contractor would be responsible not only for design and construction but also for equipping the facility to be ready to run. Sometimes this form of contract allowed the owner access to proprietary information not available otherwise. During this process, consulting engineers were usually hired by the owner to work with the turnkey contractor and act as the owner's representative not only through the design phase but also through construction and commissioning. These types of projects were frequently employed in cogeneration projects, refineries, and power plant construction, but have also found applicability in commercial and retail construction as well.

Build-operate-transfer—carrying design-build further

Another variation on the design-build process, known as *build-operate-transfer* (BOT), provides not only design and construction, but includes financing and operation of what is basically a concession-type project. These projects are associated with revenue-producing entities such as bridges, toll roads, and tunnels, where the toll rates are established by the BOT company in conjunction with the *owner*, usually a public agency. The revenue stream produced by the public's use of the facility generates the cash flow that will ultimately provide the BOT entity with a return on investment. Two of the virtues of BOT are (1) allowing construction of a public facility to be built with no increase in taxes or the need to float a bond issue, and (2) since the BOT entity will have to maintain the property for the concession period, usually 40 years, quality levels will be very high allowing maintenance costs to remain low. The Suez Canal, an early BOT project, under the supervision of Ferdinand de Lesseps in 1854, combined design, financing, construction, and operations for the canal for a period of 99 years. In more recent years this project delivery system was used to build the channel tunnel between Great Britain and France and, closer to home, the Dulles Toll Road in Virginia.

HUD and Government's Entrance into Design-Build

Back in the early 1970s, the Department of Housing and Urban Development (HUD), possibly unwittingly, advanced the government's venture into design-build when it initiated its Section 8 housing program. In order to generate sufficient, adequate housing for low-income senior citizens, the Section 8 program guaranteed developers near-market rental income for housing projects for the elderly. Adding accelerated depreciation to the pot, this program produced tens of thousands of clean, affordable apartments for the elderly, who were expected to pay a maximum of 25% of their income as rent—the government kicked in the rest. Section 8 allowed developers to design and build these, generally midrise, projects relying only on HUD's minimum property standards (MPS) as guidelines in the design. HUD's cost restrictions were based upon regionally adjusted *comparable costs*. Certain spatial requirements were mandated, but the basic design was left up to the developer who was relatively free to design the exterior of the building and various interior spaces not subject to HUD MPS. The developer in their proposal to the government presented a complete set of plans and specifications with associated costs for review and approval anticipating acceptance upon review so that a contract for construction would be awarded.

The United States Postal Service, a quasi-government agency, in the 1980s was also seeking a better way (translated—less expensive and less litigious) to construct their distribution centers. They tried design-build, and have been strong advocates of the system ever since. The federal government along with state governments has begun to utilize design-build to a greater extent, not only for roads and highways but also for vertical construction projects.

The Design-Build Advantage

Numerous studies over the years have been conducted to gauge reaction to this design-build approach. One such study in 1997 conducted by the Construction Industry Institute (CII) looked at 350 projects in order to compare construction management, design-bid-build, and design-build delivery systems. Of the projects compared, 20% were construction management , 35% design-bid-build, and 45% design-build. The projects ranged in size from 50,000 square feet to 2.5 million square feet, with costs of $30 to $2000 per square foot.

Design-bid-build was found to have the greatest median cost escalation at 4.84%, followed by the construction management approach at 3.34%, and design-build at 2.37%. Construction management and design-build projects had almost no delays; however, scheduling for design-bid-build projects grew by an average of 4.44%. In terms of speed of square feet construction per month, design-build produced the highest median activity at 9000 square feet plus, while design-bid-build was lowest at 4500 square feet per month.

Another study about this time, a survey of projects in 37 states conducted by Pennsylvania State University's College of Engineering, reached the following conclusions:

- Design-build project unit costs were 4½% less than CM-at-risk projects and 6% less than design-bid-build projects
- Design-build projects, measured in number of square feet constructed per month, was 7% faster than CM-at-risk projects and 12% faster than design-bid-build projects

- Factoring speed of design into the equation, the design-build method was 23% faster than a CM-at-risk system and 33% faster than the conventional design-bid-build project

Both studies validate the design-build process in terms of lower costs (in some instances) and a more rapid completion schedule.

The State of California, a trendsetter in many ways, wanted to look more closely at the design-build system as it applied to the public sector. A number of state laws on the books, some dating back to 1993, permit design-build projects in various highway and vertical construction projects. Seven laws statewide required local entities to report on their projects to the Legislative Analyst's Office (LAO), a nonpartisan office that provides fiscal and policy information to the state legislature. The LAO report presented in February 2005 compared two primary construction delivery systems—design-bid-build and design-build. One portion of their report is entitled "Contractual and Operational Differences between Non-Design-Build and Design-Build Projects" (Fig. 1.1). Figure 1.2 rather succinctly sums up the case for and against design-build in the public sector, which would appear to apply to the private sector as well. Figure 1.3 is a narrative view of design-bid-build and a comparison with design-build using a stipulated sum approach and design-build using construction management.

Some of these findings are not that surprising. Primarily in the private sector, the cycle of design-bid-build begins to bog down when initial bids received from contractors exceed the owner's budget. Generally a series of redesigns and repricing activities occur until budget and design are resolved. Quite often

Design-Bid-Build Versus Design-Build Advantages and Disadvantages	
Advantages	**Disadvantages**
Design-bid-build	
• Building is fully defined.	• Agency gets involved in conflicts and disputes.
• Competitive bidding results in lowest cost.	• Builder not involved in design process.
• Relative ease of assuring quality control.	• May be slower.
• Objective contract award.	• Price not certain until construction bid is received.
• Good access for small contractors.	• Agency may need more technical staff.
Design-build (stipulated price)	
• Price certainty.	• Limited assurance of quality control.
• Agency may avoid conflicts and disputes.	• Subjective contract award.
• Builder involved in design process.	• Limited access for small contractors.
• Faster project delivery.	
• Agency needs less technical staff.	

Figure 1.2 Construction delivery processes: pros and cons. [*Courtesy: Legislative Analyst's Office (LAO), State of California.*]

Design-Bid-Build

Under the design-bid-build system, the public agency first awards an architect/engineer contract to design the project based on subjective criteria of qualifications and experience of the architect/engineer. This contract generally accounts for a relatively small portion of the project's total costs—about 5% to 10%. After detailed project plans and drawings are completed, a contractor is selected to perform the construction work, which accounts for 90% to 95% of the project's costs. In almost all cases, contracts for construction work are awarded objectively based on competitive bidding.

Design-Build

With design-build, the public agency contracts with a general contractor to both design and build the project. The agency does not separately contract with an architect/engineer for design. That is the responsibility of the general contractor. The general contractor in turn subcontracts, through competitive bidding or otherwise, for an architect/engineer and various construction trade work. Design-build delivery methods have a number of variations, but most can be placed in one of two categories—*stipulated price* and *construction management.*

Stipulated price. With *stipulated price design-build* a public agency specifies how much it will pay for construction of a particular building. For example, the agency might provide only a programmatic description of the building it wants by specifying the size of the building, types of spaces, and perhaps some acceptable construction materials. The agency then asks competing firms to present proposals that illustrate a conceptual design and provide specifications for materials and building systems that it is willing to construct for the price stipulated by the agency.

Construction management. With *construction management design-build* the public agency awards a contract to a CM (frequently a construction firm, but sometimes an architect/engineer firm) on the basis of a fee. The CM designs the project and solicits bids from subcontractors and suppliers. The total of these bids plus the CM's fee determine the total price the agency pays for the buildings.

Figure 1.3 LAO narrative view of design-bid-build and design-bid. [*Courtesy: Legislative Analyst's Office (LAO), State of California.*]

during this process, value engineering proposals submitted by the contractor and reluctantly accepted by the design consultants result in a project of reduced costs as well as reduced value.

Theoretically, these events do not occur during the design-build process and its rapid growth in both the private and pubic sectors must be viewed as evidence that the system is working, and working effectively.

Public Sector Interest in Design-Build

As witnessed in the LAO's report from California, the growth of design-build by public agencies has been steadily increasing.

In an article in the *Journal of Management in Engineering*, published by the American Society of Civil Engineers (ASCE) in October 2004, the ASCE indicated that there were nearly $3 billion worth of water/wastewater projects either underway or in the bidding stage. During the year ending 2000, the article stated that projects worth about $37.2 billion were delivered using design-build, and, since 1994, design-build projects totaling $2.6 trillion had been approved by the Federal Highway Administration Special Experimental Projects program spread out over 25 states.

The acceptance of design-build in the public sector is evidenced further by the increase in state bills being introduced into state legislatures. As of June 2004, 159 bills relating to design-build were introduced in state legislatures across the country and a total of 34 bills passed in 13 states. Some of these 2004 rulings are:

California. Three laws passed in August 2004 allowing design-build for transit operators, transportation projects, and lease-back contracts for school districts.

Florida. A proposed design-build high-speed rail system previously approved has now been declared tax exempt.

Georgia. A new code allows design-build for buildings, bridges, and other projects not exceeding $10 million. A bill establishes new licensing arrangement for design-build contractors.

Maryland. A law permits counties to use design-build on public school projects.

Massachusetts. State allows CM-at-risk and design-build on public projects over $5 million.

New Mexico. Allows design-build and finance on public school projects.

Ohio. Permits design-build on a pilot project for a lodge and conference center at Geneva State Park in the state.

The Challenges of Design-Build

Institutional changes

One of the barriers facing the participants in design-build is an institutional one. The relationship between owner, architect, and contractor is radically changed. No longer is the architect the owner's agent, acting as a gatekeeper to the contractor. No longer does the contractor tend to view the architect/engineer as an adversary (if they ever did). No longer do the design consultants concern themselves with contractor selection, often anticipating an aggressive program of change-order requests depending upon which builder is selected.

One of the disadvantages of design-build voiced by skeptics is the lack of an owner's *gatekeeper* or ombudsman. But this problem can be averted by having the owner hire a professional as an owner's representative during the design development stage of a project and entrust him or her with construction services usually associated with an architect's construction services. And, of course, engaging a CM during construction would also provide an owner with a watchdog.

Trust becomes the operative word in this new relationship, and as we all know, institutional and cultural changes take place slowly.

Changing the mind-set of the individuals within the proposed design-build team may actually be one of the key elements for its success. In a recent book published by the American Institute of Architects discussing ethical issues relating to design-build, a common question among architects, *How do we keep the fox out of the henhouse?*, refers to the contractor part of the team. As contractors must avoid their focus on design deficiencies, so must designers change their view of the contractor as the fox (if that was ever the view of any design consultants).

Risk sharing

The concept of risk sharing changes somewhat in design-build. No longer can the contractor look to the owner and their design consultants for additional compensation due to inadequacies in the design documents. Contractors often complain that risk during construction is basically shifted by both owners and architects to them via obscure modifications to the standard contract for construction, and architects may perceive contractors as shifting risk back to them, citing incomplete or inconsistent design documents as their basis. And owners, oftentimes caught in the middle, complain that they bargained for a complete structure and that they have no responsibility for missing details or inconsistencies in the contract documents.

Embarking on a design-build venture requires all of us to reassess our concept of risk sharing.

Liability, bonding, licensing issues

Liability, bonding, and licensing issues can be a limiting factor in assembling a design-build team. Contractors routinely provide personal and property liability insurance certificates with multimillion dollar limits on their construction projects, something an architect/engineer firm is not frequently called upon to do. However, a contractor has very little, if any, knowledge of errors and omissions in insurance policies, something that a design consultant deals with frequently.

The same is true of bonding capacity, routinely tapped by contractors for public works projects and many privately funded projects. Contractors guard their good relationships with their bonding agent; substantial bonding limits are not created overnight but are the result of proven performance and are much coveted by contractors.

An architect/engineer firm's requirement for a bond—whether it be payment or performance or labor and material or maintenance is probably limited. And they have not had much experience in compiling all of the financial information required by sureties.

Requirements for contractor's licenses runs the gamut from having to submit detailed financial information, to exhibiting experience in their field of endeavor, to those states that have no requirements at all.

By contrast, the fields of architecture and engineering generally have rather strict requirements for licensing.

The Life Cycle of a Design-Build Process

Depending upon whether the project is in the public or private sector and the type of contract being considered, decisions will be made that influence the design-build life cycle (Fig. 1.4). Public sector work requires the preparation and dissemination of Requests For Proposals (RFP), and an evaluation and award process that may not be required for private sector work. A private owner desirous of selecting a design-build firm and deciding to negotiate a contract with them might skip the RFP/evaluation process entirely, but if they decide to advertise for bids they would follow this process much like a public agency.

The cycle of a design-build process begins with the program or project initiation and continues through to commissioning—a fairly typical process in any project's life, but the players and their roles are somewhat different.

Project initiation

This is the starting point where an owner planning the project must be able to define their needs and expectations. The owner must determine if they have qualified staff on board to begin to extract and define their program or whether they will require outside consultants to assist them. During this stage the owner needs to define their objective and consider budget and financial resources. Questions to be addressed are mentioned in Fig 1.5.

Project planning

If a design-build team is brought in at this early stage, they will begin to work with the owner's staff to develop the conceptual design and associated costs (Fig. 1.6).

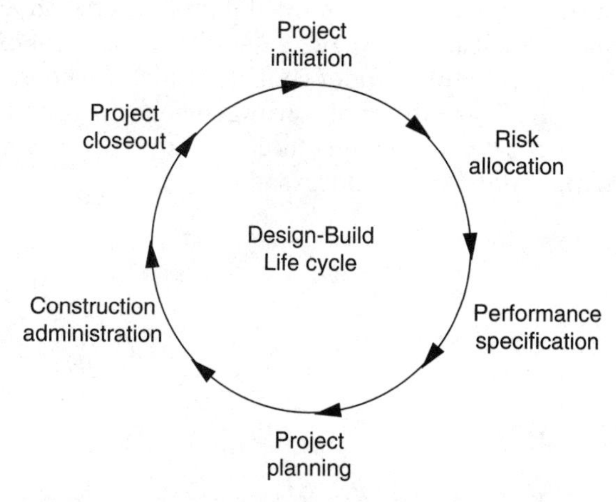

Figure 1.4 The design-build life cycle. [*Source: American Society of Civil Engineers (ASCE).*]

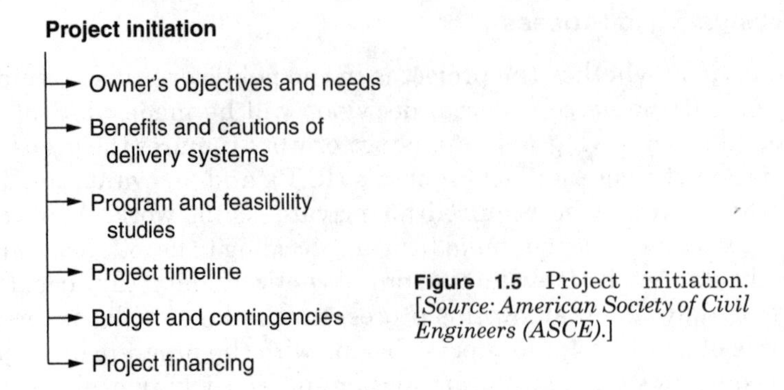

Figure 1.5 Project initiation. [*Source: American Society of Civil Engineers (ASCE).*]

Addressing schedule and costs at this stage will be necessary so that the owner can begin to consider various life-cycle options and evaluate proposed building systems and components. Scheduling alone can become a key issue. The difference in time and cost of one structural system over another can hinge on the time of year that is anticipated for commencement of the project. In cold climates, a cast-in-place concrete structure may not be cost-effective if construction is due to start in December.

Risk allocation

Risk sharing is an integral part of the design-build process and invites discussions involving insurance, what limits are required, and who is to furnish the necessary policies.

How will risk be shared between design-builder and owner? Questions will arise between members of the design-build team: who will furnish general liability insurance, errors and omissions insurance, and so forth? Are payment and performance bonds required, and if so, who will supply them? In a contractor-led design-build team, acquisition of a bond is usually not a problem, but when an architect-led design-build team is under consideration, the bonding arrangement may not be so simple. The subject of contingencies will arise at this stage of the project, and the need for an owner- and/or a contractor-controlled contingency will surface and ought to be addressed (Fig. 1.7).

Figure 1.6 Project planning. [*Source: American Society of Civil Engineers (ASCE).*]

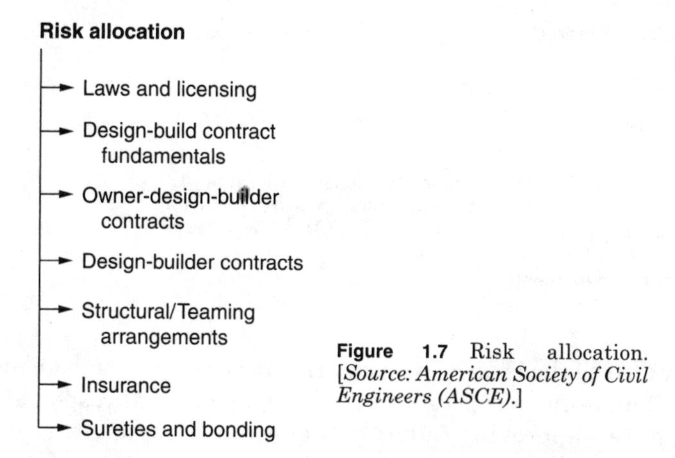

Figure 1.7 Risk allocation. [*Source: American Society of Civil Engineers (ASCE).*]

Project planning—formulation of the performance specifications

This phase of the project, which in some cases may precede the risk-allocation phase, focuses on methods by which the owner will solicit and award a contract for design-build. In the public sector the conventional two-phase process will probably be utilized whereby a RFP is prepared for each phase, a notice of solicitation of bids published, and a bid receipt and evaluation procedure established. The rules of engagement in the private sector can be more relaxed; although the same steps may be used, they may not be formalized (Fig. 1.8).

Contract award and construction administration

Prior to final contract review and execution, the owner and the design-build team should consider several activities related to the construction process and form a construction administration exhibit to that contract. Among items to consider are the process of submitting and documenting payment requests, change-order preparation and related fees, and how quality control and quality assurance methods will be employed and documented (Fig. 1.9).

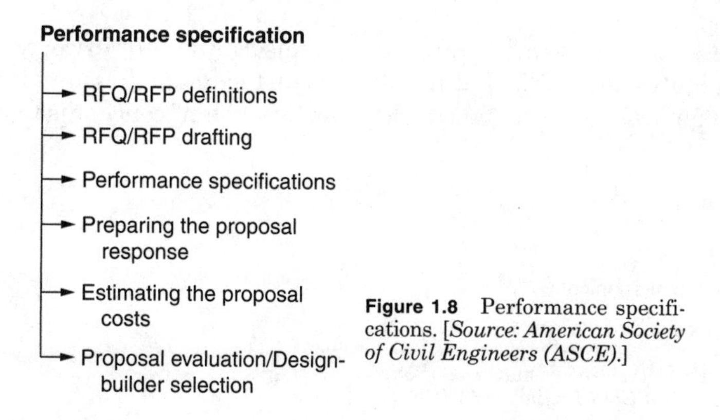

Figure 1.8 Performance specifications. [*Source: American Society of Civil Engineers (ASCE).*]

Figure 1.9 Construction administration topics. [*Source: American Society of Civil Engineers (ASCE).*]

There may also be the issue of whether the owner will engage an owner's representative. If an owner's representative is engaged, what responsibilities will this person have—approving monthly requests for payment, reviewing and approving change orders, ability to speak for the owner, and make decisions on matters involving time and money?

Closeout and commissioning

Prior to the start of construction, thought must be given to the project's closeout and commissioning. The more complex the building and its systems, the more importance will be placed on the commissioning process.

At what point does the design-build team consider its contractual obligation fulfilled (Fig. 1.10)? The first trigger occurs when a certificate of occupancy is obtained and all warranties and guarantees are provided to the owner.

Sometimes the commissioning process is quick and complete and at other times it is painstakingly slow. How the design-build team handles the commissioning process will determine what grades they are given by the owner. But what about statute liability issues that may remain the responsibility of the design-build team for the length of time required by law, structural failures, and the like? Project closeouts performed professionally and promptly are what all design-build teams strive for.

The Team

The expression "the team" is often used indiscriminately in correspondence, meeting minutes, and RFIs, but too often as just *window dressing*. It is not how often the expression is used, but rather how this "team" concept is truly put into

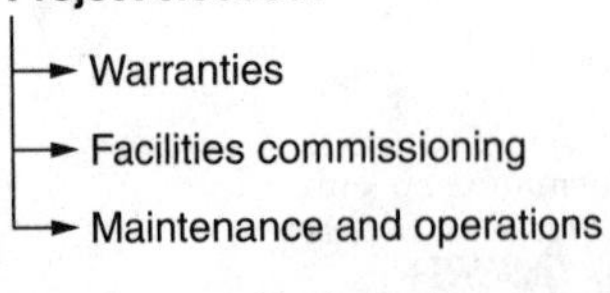

Project closeout

- Warranties
- Facilities commissioning
- Maintenance and operations

Figure 1.10 Project closeout matters. [*Source: American Society of Civil Engineers (ASCE).*]

practice that is important. The give-and-take of a collaborative effort does not come naturally in a competitive environment like the building business, but this team building is a process that requires an effort by all participants to make it work.

This collaborative approach is one of the virtues of design-build, a primary obligation to create a team— owner/designer/contractor—that will work together through the entire project to achieve a common goal.

Not until owner, design consultants, general contractor, and subcontractor are requested, individually, to express the goal they hope to achieve in the project does this *common goal* concept begin to materialize. A lesson learned during the writer's first exposure to partnering drove home an important point. At the initial partnering meeting with the Connecticut Department of Transportation, the owner's representative, the design architect and engineer, the general contractor, and various subcontractors were all requested by the partnering facilitator to prepare a list of goals they hoped to achieve. When the items on each list were incorporated in a single one, it became evident that each seemingly disparate entity expressed the same goals: complete the project with no claims or disputes, receive prompt payment, avoid change orders, and make a profit.

Each participant realized that they actually shared a common goal, which was achievable by their collaborative efforts.

The Changing Industry

The information technology era has brought with it many unanticipated benefits. The ability to transfer documents, photographs, drawings, and sketches instantaneously has transformed the construction industry in much the same manner that computer-assisted design has transformed the design industry. The transfer of information has been responsible for a slow but steady increase in productivity. The ability of an individual within a design or construction firm to deal with complex problems more quickly and with more accuracy is proven every day as requests for information and responses to those requests speed back and forth via copper or fiber optic highways.

But what has remained constant is a desire to work productively, profitably, and without conflict. The design and construction industries are basically service industries and the desire of the clients they serve has not changed as witnessed in another portion of the FMI/CMAA survey below devoted to finding out what owner's concerns in a project really are:

- Issues of coordination, collaboration, and communication continue to challenge owners and are the source of unnecessary confusion on projects.
- Find a way to deal with the leading cause of overruns, incomplete drawings, poor preplanning, and the increase in cost of materials and equipment.
- Not enough time is devoted to the predesign stage of the project.
- Owners must exhibit more control over scope and prevent "design creep."

- Seventy percent of the respondents to this survey said they have experienced a decline in the quality of the design drawings, and that architects need to be more responsible for completing a quality design to avoid change orders.
- Architects need to be more responsible for completing the design to avoid the proliferation of RFIs that seem to plague most projects.
- Owners expect their CMs to provide leadership in managing the projects from beginning to end.

The attraction of design-build from the owner, architect, engineer, and contractor viewpoints is that it allows an opportunity to work together in a more congenial environment that will ultimately provide a less stressful, more productive, and more profitable method of doing business.

The arena for design-build is growing rapidly. Fielding a professionally managed team and becoming a participant in this game is a worthy goal.

The following chapters will, hopefully, point the way to achieving this goal.

Chapter

2

Traveling the Path to Design-Build

The process of designing and constructing a project has changed considerably in the past century, but always with an eye to creating a more efficient and cost-effective product.

In Great Britain, design-build is referred to as a "packaged" project, an apt term because an owner no longer buys a service, but buys a package—a product.

Design-bid-build, a mainstay project delivery system in the public sector for decades, was a way in which the public was assured of obtaining best value by virtue of its low-bidder award process. In theory that concept worked, but in practice, not always. In many instances, prequalification of bidders was minimal and the ability to provide a bond was often viewed as assurance that a bidder would perform adequately. Quite often the public agency's budget was inadequate, requiring redesign and rebid after the initial round of proposals were received. The added costs for redesign and the impact of inflation on construction costs exacerbated the problem, frequently leading to a series of *value engineering* options that often reduce both price and quality. Then there were the unscrupulous contractors who took advantage of the ambiguities in the bid documents to "low ball" their bid, knowing that if awarded the project, they would unleash a flood of change orders to enhance their profit. Unlike work in the private sector, where an owner would have nothing to do with an unethical contractor once they were rid of them, unless a contractor was "blackballed" by a public agency or was unable to obtain a bond, they were free to continue their assault on publicly funded projects. The proliferation of claims consultants and construction litigation from the 1970s through the 1990s was a clear signal that something was wrong and needed to be changed. Owners battled with contractors who pointed fingers at the design consultants, while lawyers salivated in the background. The shortcomings to the design-bid-build process were evident even when the system proceeded reasonably well. The interest in design-build that began to take stage front began in the 1980s.

The top 100 construction companies tracked by *Engineering News Record* magazine reveal the change in their product mix over the past several years. In 1998, there were 14% more design-build projects than CM-at-risk projects; in 1999, there were 31% more design-build projects than CM-at-risk; and in the year 2000, there were 28% more design-build projects than CM-at-risk. Many experts have forecast that by 2010, design-build will represent 50% of the market in the continuous search for a better way.

Some interim procedures were put in place by owners to dilute the tensions that frequently existed between contractors and design consultants in other forms of project delivery systems.

Partnering

In an attempt to put an end to some of the deceptive practices by a few contractors, several government agencies instituted a program called *partnering*, at first voluntary, but later mandated. A facilitator was hired by the public agency to implement and monitor this process. The owner's representatives and their design team would be invited to meet with the general contractor and their subcontractors for the purpose of getting to know each other prior to the start of construction. The facilitator would state the goals of the partnering concept—banding together as a team, recognizing each participant's individual and collective goal, and working together to achieve those collective goals. These tenets would be formalized into a partnering agreement in which each participant would agree to reduce or eliminate any adversarial relationships with other participants, reduce or eliminate change orders, cooperate to complete the project on time and with the highest quality levels, and to allow participants to achieve their profit goals. One of the lessons learned by those attending a partnering session was an eye-opener—all of the participants had the same goals. At the beginning of the partnering program, the facilitator would ask each attendee—owner, design consultants, general contractor, or CM, and all attending subcontractors—to write a list of their expectations. When all of these individual lists were consolidated into one on a blackboard, there would be some gasps—everyone wanted the same thing—provide quality work, reduce change orders, avoid disputes and claims, be paid promptly, and complete their work on or before schedule. So although each attendee was a different part of the puzzle, they all wanted to participate in the completion of the puzzle.

Unfortunately, one cannot legislate goodwill or cooperation, and while the process, in some cases, worked for a while, unless rejuvenated by subsequent facilitator meetings, partnering degenerated into the more familiar grumbling and squabbling.

But for those organizations that participated in the partnering process, it did make them aware that all parties to the construction process ultimately pursue the same goals—fairness in treatment, payment for services provided, and the desire to do a good job. Maybe they might carry these virtues over to their next project.

Dispute Resolution Measures

The American Institute of Architects (AIA) recognized the need to address dispute resolution when they updated their standard form of contract in 1997. Article 4 of AIA document A201—General Conditions of the Contract for Construction—deals with dispute resolution. The latest version of AIA A201 stipulates that mediation is the first step to be taken to resolve a dispute. If unsuccessful, arbitration is the next step to be taken and if all else fails, litigation should be considered a final step in resolving a dispute.

The Associated General Contractors of America (AGC) in their AGC document No. 200 includes a list of rather precise provisions to resolve disputes:

1. Create a Dispute Resolution Board (DRB) to be composed of one member selected by the owner, one member selected by the contractor, and a third member selected by the two owner-contractor selected members. This board will meet periodically throughout the length of the construction project to track the construction process, and when called upon, will make advisory recommendations to avoid or settle any potential disputes or claims that have arisen.
2. Establish a procedure to invoke arbitration to be conducted in accordance with the Construction Industry Rules of the American Arbitration Association (AAA).
3. Conduct a minitrial where top management from the owner's side and the contractor's organizations will submit their individual positions to a mutually acceptable individual who will make a nonbinding recommendation to the parties (a process very similar to mediation).
4. Go to binding arbitration pursuant to the Construction Industry Rules of the AAA.
5. And as a final resort, proceed to litigation.

The Cost-Plus-Guaranteed Maximum Price Contract

This form of contract is frequently used when negotiating a contract between owner and builder, generally before the design documents have been fully completed. It allows for an early start to the project with some safeguard to the owner for total project cost.

The cost-plus-guaranteed maximum price type contract, while solving some of the ills of a lump sum or stipulated sum agreement may be satisfactory in some instances and not so in others. Contractors including significant contingencies, after preparing estimates based upon 60% to 75% complete drawings, may have increased the Guaranteed Maximum Price (GMP) higher than actually warranted affording them a rather safe cushion to ensure that they would not exceed the GMP. Questions about interpretation of what should or should not have been included in the contractor's estimate to cover the remaining 40% to 25% of the design will surface many times during the project and can become objects of mistrust among the project's participants. Questions regarding which costs should rightfully be charged to the project will also be raised, not only during the requisitioning process, but also when the contractor submits their final cost report and cost analysis. The sharing of savings as called for in the contract may be insignificant causing the owner to wonder why the contractor's requests during construction to substitute a certain material or piece of equipment did not represent higher cost savings. The GMP-type construction contract is effective when there is good communication between owner, architect/engineer, and contractor, and the atmosphere of trust established in the beginning stages of the project prevails up to the end.

Fast tracking

A derivation on the GMP contract concept that gained popularity in the 1980s and 1990s was called *fast track* and was later changed to *flash track* as owners demanded more rapid completion of schedules. Under the fast track system the designer would complete various drawings to allow the contractor to order some materials via the use of a letter of intent, before receiving a fully executed contract from the owner. For example, the first drawings the architect would prepare would be for foundations and superstructure. This would allow the general contractor to order shop drawings for reinforcing steel for the foundations and possibly shop drawings for the structure.

Pursuant to the issuance of a formal contract, the owner would issue a letter of intent that spelled out the specific scope of work, with associated costs, that the contractor was authorized to proceed with during this interim time frame. If, for any reason, no formal contract was issued, the contractor would be reimbursed by the owner for all work completed or in progress conforming to the stipulations in the letter of intent. If no such stop work order was issued, these preliminary activities would be folded into the scope and cost of work in the forthcoming contract. Speeding up this fast track concept morphed into an even more rapid process called *flash track*. This project delivery system is particularly effective when one of the owner's prime concerns is completing the project rapidly so that the revenue stream could begin quickly—say as an office developer who has preleased 75% of the office space or a condominium project that presold a substantial number of units.

The Construction Manager

Whether New York's 1921 Wicks Law was, in fact, the rationale behind the movement to construction management, the concept gained favor among owners of large projects and interest spread further downward to owners of midsize projects.

CM's value as estimator

Many problems associated with a project's genesis can be traced back to the quality of the budget assembled by the owner. Although there is a lot of building database information available on the Internet or in cost guides or from design consultants with a history of costs from previous projects, current cost data for the locale in which a project will be built is best obtained from a contractor working in that area and on similar projects. Unless a general contractor had been selected by an owner to work with the design consultants as design develops, prior to the widespread use of construction managers (CMs) hard-cost data was hard to come by. The two-part CM contract, where one part applies to preconstruction services and the other to construction services, allows the owner to tap into a well-documented database of construction costs, but delay further contractor commitments to a later date, if they so desire.

In the preconstruction phase, the CM approach using their experience in costs, constructability, and material and local labor availability provides the owner with the necessary construction expertise while the owner's architect and engineer work through the design development process.

CM agency and CM-at-risk

The CM agency approach engages the construction expert as the owner's agent and conducts all such matters as such. The CM-at-risk changes this relationship to some degree.

When the concept of CM-at-risk gained popularity, it also gained critics. It appeared that the CM may now have two masters—the project owner and the CM's own interests in preserving the contract sum since they would be *at risk* for all costs exceeding that contract sum. Would some of the CM-at-risk's decisions be based solely on what was best for the owner if they resulted in decreased fees or even total loss of their fee? Any CM, whether they be an agency type or an at-risk type, if they plan to remain in business for any length of time, will certainly keep the owner's interests ahead of their own in order to develop or maintain a sterling reputation. This is probably the most effective brake on the at-risk approach and should mollify critics.

Some owners of smaller projects wishing to cash in on the benefits of using a construction manager on their project, often hired small local contractors who were ill prepared to meet the standards of the profession and CM received some undeserved black marks.

Today CM is a force in the execution of construction projects and its use is growing as the concept fits perfectly into the design-build process as owner's representatives.

The Program Manager

A program manager (PM) provides management services spanning a wider spectrum of an owner's expansion program which, in some instances, can be somewhat analogous to managing a turnkey project. The PM may engage a wide range of consultants to meet many of the program goals required by the owner. The design and construction of a sports facility is a good example of the rationale for hiring a PM. Along with the design and construction of the facility itself, there are many other issues to be dealt with. Negotiating contracts with food concessionaires and coordinating their spatial requirements and utilities needs would be an important part of the program manager's responsibility. Negotiating fees with companies wishing to furnish those large electronic signs so prevalent in sports stadiums today, providing structural supports for signage, coordinating electrical requirements between vendor and electrical engineer, establishing and confirming scheduling commitments would all be a part of the PM's duties. Coordinating the owner's audio and visual communication systems, including provisions for closed circuit and network television often comes under the aegis of the PM and all under an immovable completion date—opening day.

PMs may assist an owner in selecting finance sources, figuring out innovative ways to fund the project(s) and engaging in a variety of functions involving time and money as they relate to the owner's program.

The Design-Build Process—Searching for the Holy Grail

Design-build is the latest stage in the evolutionary process of designing and building a more perfect construction project by either a negotiated or competitive bid process.

Negotiated design-build projects

Negotiated work in the private sector is basically limitless, but in the public sector there are various legal limits and impediments placed upon some state and federal agencies seeking to negotiate design-build work.

There are significant differences between negotiating a contract where the owner has engaged an architect to develop the project's program prior to inviting contractors to submit bids and negotiating a design-build contract for design and construction.

Part of the design-builder's task is explaining how this project delivery system is somewhat different from the more conventional design-bid-build process.

Design-build requires a new perspective on prequalification. Not only will an owner be reviewing the qualifications and experience of the designer and the builder, they will also need to determine if the team being presented has a history of successfully working together on previous, similar projects or separately as team members on similar scope design-build projects. How innovative has the team been on these previous projects and have they completed these projects within the initial schedule and budget framework?

A clear definition of the project program is essential for success. Does the owner have sufficient, qualified staff available to define and develop the project's program when working together with the design-build team during the conceptual phase? Failure of design-build projects can often be traced back to an ill-defined or poorly defined owner-presented program. And, conversely, many successful design-build projects have, as their beginning, a well-articulated program of owner needs and expectations.

Performance specifications for essential services and design will be needed. Commencing with the structural design, the owner's staff must present enough information for the development of spatial requirements, live loads, power, data communication, and lighting requirements in general and in specific areas, expectations for heating and cooling levels—all in terms of *performance* requirements that will allow the design-build team some flexibility in their approach to the project.

The owner's representative. One concern owners have is whether there are checks and balances during the design and construction phases of the project. The hiring of an experienced owner's representative, if there are no professionals already on staff, will provide the owner with assurance that their interests are being protected. The sooner this owner's representative can be brought on board, the better—for all participants in the project. The engagement of an owner's representative during the prequalification and selection process should be encouraged. The growing use of CMs in design-build is testament to the recognition of this need.

The contract for design-build work. The basic contracts for design-build work are sufficiently different from standard lump sum or GMP contracts that time must be devoted to their preparation, explanation, review, and acceptance. A subsequent chapter in this book discusses various issues relating to contracts.

Competitively bid design-build work

When a design-build project is being considered and the decision is made to seek competitive bids, the selection and evaluation process is substantially different from a design-bid-build project. After all, three important evaluations are to be made, one involving acceptance of design, one involving scope of work, and one involving total costs. Owners in the private sector might look at some of the evaluation procedures adopted by public sector agencies in creating their Requests For Proposal (RFP).

There are various ways in which to approach these types of proposals to help evaluate all bids; first of all, will there be a selected list of prospective bidders or a general invitation to submit a proposal after which a short list will be prepared? Let's assume for this discussion that the owner has short-listed the bidders and will now attempt to evaluate those responses. One of several methods can be used as described below.

Weighted criteria method. The owner will establish a point system for evaluating proposals, assigning various points for qualitative issues:

- How well does the design meet my program?
- How well do the systems fit my program?
- How much does cost enter into my selection decision?

TABLE 2.1 A Typical Point System Approach

Proposer	Design (25 points)	Systems approach (20 points)	Price (40 points)	Schedule (15 points)	Total score (100 points)
DB Firm A	25	15	30	15	85
DB Firm B	20	15	35	10	80
DB Firm C	15	20	25	10	70

In combination with an interview, each proposer's submission will be evaluated by the point system, and theoretically awarded on that basis. A typical point system approach is illustrated in Table 2.1.

In Table 2.1, Bidder C is clearly eliminated, Bidders A and B are fairly close in rating and an interview with each may be in order to make the final evaluation and award.

Adjusted low bid. This system is a variation on the weighted system inasmuch as it takes into account other subjective evaluations. Each proposer is interviewed and graded on their oral presentation on, say, a score of 0 to 100. These grades are expressed as a decimal, using the point system example above—a score of 85 becomes .85, a score of 80 becomes .80, a score of 70 becomes .70. When each bidder's envelope containing their project cost is opened it will be adjusted by dividing the project cost by the oral presentation rating expressed as a percentage.

As an example, let's look at the previous bidders who scored 85, 80, and 70 on their oral presentations and how the bids would be correspondingly adjusted using this method. See Table 2.2.

By using the weighted method for evaluating the oral presentation portion of the three bidders, Bidder A is the apparent winner based upon their slightly higher bid price ($50,000 more than Bidder B) but their oral presentation was superior to both Bidders B and C. Bidder C's significantly lower price ($200,000) was offset by their less-effective oral presentation, and they ran a close second to Bidder B. It is very difficult to objectively rate any series of bids, whether negotiated or competitively bid because even if accompanied by a detailed exhibit of qualifications, exclusions and inclusions of certain scope issues will frequently escape both presenter and reviewer. In the design–build process, objectivity is thrown a further curve ball because of the part that aesthetics in the eyes of the beholder plays in the selection process.

TABLE 2.2 Adjusted Low Bid Approach

Bidder	Oral presentation score	Project cost	Adjusted low bid*
Bidder A	85	$1,200,000	$1,411,764
Bidder B	80	$1,150,000	$1,437,500
Bidder C	70	$1,000,000	$1,428,571

*The award, obviously, would be based upon the actual price submitted not the adjusted price which was only developed to assist in the selection process.

At least these two rating systems will have some semblance of reducing some subjectivity to objectivity.

The equivalent design and low bid approach. This award procedure utilizes the same short list bidders approach in which sealed bids are submitted by the bidders and oral presentations are made by each of them; however, the owner critiques each proposal and allows the bidder time in which to reply to their critique.

The revised submittals are to include not only responses to the owner's comments but any adjustments in project costs resulting from those comments and the required response. The base bid and revised bids are compared and evaluated. If the nature of the critique was to furnish all bidders with more standardized design and performance criteria, then the final award may be based solely upon the most competitive price.

A fixed price–best design approach. Using the short-listed bidders competition approach, the fixed price–best design approach permits the owner to establish a fixed price for the project with only design requiring a subjective review. Oral presentations allow all responders to explain their design criteria so the owner may make a qualitative evaluation to ascertain compliance with their program requirements. The best design for the given budget will receive the award.

Design-build has its advantages and its disadvantages

On the plus side, design-build has a proven track record of success in delivering a project quickly and often at less cost than the design-bid-build or CM approach.

But there are disadvantages to the process as previously discussed.

- An owner that does not have staff to adequately develop a program will have difficulty defining and presenting their needs to the design-build team.
- The process involved in design-build may bypass the competitive bidding process, possibly not affording the owner the best price.
- Unless the owner has an experienced person on staff to interact with the design-build team, they may need to hire a professional owner's representative, adding cost to the total project.
- In some areas, legislation or licensing laws exist that won't allow the bundling of design and construction services into one firm.

The Bridging Approach to Design-Build

Bridging is a process where design-build can be approached obliquely rather than head on. Sometimes referred to as design-design-build, it is a process whereby an owner contracts with a design professional to create a set of partial design documents that will be used to solicit bids in the marketplace. The owner can, in effect, test the marketplace, limiting their financial exposure and obtain more definitive pricing information by presenting a basic design to bidders. By the issuance of partial design documents an owner can invite suggestions to change the design or allow submission of value engineering proposals at a stage that would not require significant redesign costs. A bridging consultant can work closely with the owner to prepare not only design development drawings, but provide preliminary budget numbers and design and construction schedules.

Going forward, the owner has the option of engaging the "bridging" architect to complete their initial design concept, developing a new one based upon ideas developed during the bridging exercise, or contracting with a different architectural firm to pursue an alternative design.

This original or modified bridging design can then be used in several ways; it can be incorporated in an RFP as a concept to be further developed by the bidder or as an invitation to critique, modify, or submit an entirely new plan for consideration.

TABLE 2.3 Bridging Responsibilities

Owner	Bridging consultant	Design-build firm
1. Defines scope of project 2. Selects bridging consultant 3. Develops budget	1. Prepares program 2. Develops conceptual drawings (approximately 30%) 3. Prepares RFP for design-build firms	1. Architect/engineer prepares contract documents 2. Contractor receives subbids and constructs facility

An architect assuming the role of bridging architect must consider certain liability issues prior to accepting this type of commission. By developing a preliminary design that may be further developed by another architectural firm, if design errors occur as the original design is enhanced, who has the liability? Who owns that design—the owner once it is paid for or the architect under license? Who is the architect of record? Does the other architect engaged to complete a bridging architect's design get the credit or negative comments that may rightfully belong to the original designer?

Rule 4.201 of the American Institute of Architects Code of Ethics and Professional Conduct pertains to "credit for design" and states that credit for work performed by a member is to be recognized as such and other participants in a project are to be given their proper share of credit. An owner considering engaging an architect to produce a bridging design should be cognizant of this question of credit and responsibility for design and include appropriate language in the bridging contract to deal with these issues.

How Effective Is Design-Build?

The study conducted by the Pennsylvania State University described in the previous chapter concluded that the design-build process had many advantages over design-bid-build.

In 2002, a report prepared for The National Institute of Science and Technology (NIST) and the Construction Industry Institute (CII), in conjunction with ongoing research by the Building and Fire Research Laboratory (BFRL) looked at the impact of project delivery systems on project outcomes. This extensive report entitled *Measuring the Impacts of the Delivery System on Project Performance—Design-Build and Design-Bid-Build* runs to more than 100 pages and consists of a review of more than 1000 projects containing information submitted by owners and contractors.

Owners reported that 75% of their projects were design-bid-build, while contractors were more evenly split—56% of their projects were design-bid-build and about 44% employed design-build.

Owner-submitted project data revealed that design-build was more prevalent in projects with high value. Only 18% of all projects using design-build were less than $15 million in value, 25% were valued between $15 million and $50 million, and 47% exceeded $50 million. Although not stated in the study, the prevalence of design-build in high value projects may be due to the fact that these owners are more sophisticated and have staff capable of developing the detailed programs so necessary to create a successful design-build project.

There are many pages of charts and graphs but the abstract in the report is rather succinct:

- Design-build projects are four times larger than design-bid-build projects in terms of project cost.
- Public sector projects made less use of design-build project delivery systems than private sector projects. (This was probably due to the fact that only with the passage of the Clinger-Cohen Amendment of 1996, were federal agencies allowed to use design-build for public buildings.)
- Industrial projects made greater use of design-build than building projects in the residential or commercial sector.
- Overall, owner-submitted design-build projects outperformed design-bid-build projects in cost, schedule, changes, and rework.

Figure 2.1 displays the average value of owner and contractor submitted data on design-build versus design-bid-build projects in the various sectors of construction activity.

NIST, by using the CII benchmarking and metrics database system, were able to measure the impact that design-build and design-bid-build have on selected performance outcomes and practices such as cost and schedule on projects valued less that $15 million, between $15 and $50 million, and over $50 million. They also evaluated design-build industrial projects and design-build addition and modernization projects.

Figure 2.2 contains the NIST purpose and scope statement, their method of collecting data, an explanation of their analysis and their benchmarking and metrics questionnaire contents.

Their summary findings are recapped in Figs. 2.3, 2.4, and 2.5.

Average Project Cost—Owner DB and DBB Projects

Category	Owner DB Projects ($ millions)	Owner DBB Projects ($ millions)
Public	69.5	21.0
Private	81.7	23.4
Domestic	44.8	22.8
International	165.2	22.0
Buildings	52.3	15.6
Industrial	84.0	26.4
<$15 Million	7.9	6.0
$15-$50 Million	29.9	26.9
>$50 Million	216.2	98.7
Addition	84.8	16.4
Grass Roots	84.8	31.5
Modernization	71.9	21.5
All Owners	80.5	22.7

Owner-submitted DB projects tended to be much larger in all of the subsets analyzed. The only exception to this trend occurred when projects were subsetted by project size. DB and DBB projects in the less than $15 million and the $15 to $50 million cost ranges were similar in size. Overall, owner-submitted DB projects were over three and one-half times larger than DBB projects.

Average Project Cost—Contractor DB and DBB Projects

Category	Contractor DB Projects ($ millions)	Contractor DBB Projects ($ millions)
Domestic	62.7	21.9
International	225.1	41.7
Buildings	20.1	15.9
Industrial	108.0	24.9
<$15 Million	9.7	4.9
$15-$50 Million	29.2	27.9
>$50 Million	202.9	150.0
Addition	86.6	22.8
Grass Roots	126.1	41.3
Modernization	80.4	10.5
All Contractors	104.6	24.1

Figure 2.1 Average value of owner and contractor DB and DBB submitted projects. (*Source: U.S. Department of Commerce, National Institute of Standards and Technology.*)

Using the Construction Industry Institute (CII) Benchmarking and Metrics (BM&M) database, this study seeks to measure the impact that the use of these delivery systems has on selected performance outcomes and practice use. The database currently comprises over 1,000 projects submitted by both owners and contractors and represents actual project experience systematically collected since 1996. While the type of information collected has remained relatively the same over this time period, changes have been made in specific areas of questionnaire content and format to accommodate new developments resulting from CII research and to enhance the user interface. Seven versions of the questionnaire have been produced. Each version of the questionnaire collected data on the five following performance metrics: cost, schedule, safety, changes, and rework. Practice use metrics have also been collected in each questionnaire version, but the number of practices measured has expanded over time. Version 1.0 gathered data on four practices and versions 2.0 through 4.0 gathered information on six. Version 5.0 collected data on eight practices; and versions 6.0 and 7.0 included nine practices. Productivity metrics were included in versions 6.0 and 7.0. Table 1.1 shows the major components of each version of the BM&M questionnaire.

Table 1.1 Benchmarking & Metrics Questionnaire Contents by Version

	Version						
	1.0	2.0	3.0	4.0	5.0	6.0	7.0
Performance Metrics							
Cost	✓	✓	✓	✓	✓	✓	✓
Schedule	✓	✓	✓	✓	✓	✓	✓
Safety	✓	✓	✓	✓	✓	✓	✓
Changes	✓	✓	✓	✓	✓	✓	✓
Rework	✓	✓	✓	✓	✓	✓	✓
Productivity						✓	✓
Practice Use Metrics							
Pre-project Planning	✓	✓	✓	✓	✓	✓	✓
Constructability	✓	✓	✓	✓	✓	✓	✓
Team Building	✓	✓	✓	✓	✓	✓	✓
Zero Accident Techniques	✓	✓	✓	✓	✓	✓	✓
Project Change Management		✓	✓	✓	✓	✓	✓
Design/Information Technology*		✓	✓	✓	✓	✓	✓
Materials Management					✓	✓	✓
Planning for Startup					✓	✓	✓
Quality Management						✓	✓

This was redesigned and renamed Automation and Integration in Version 7.0.

Figure 2.2 The NIST design-build/design-bid-build study-purpose and scope statement. (*Source: U.S. Department of Commerce, National Institute of Standards and Technology.*)

For the purposes of this study, only Versions 2.0 through 6.0 of the questionnaire were used since these contained the most complete set of data on the practices analyzed. Data from both domestic and international projects were included.

The resulting analytic dataset was divided into four categories: owner DB projects, owner DBB projects, contractor DB projects, and contractor DBB projects. The categorization was determined by analyzing the Project Participants section of the BM&M questionnaire. In this section, respondents were asked to indicate the functions performed by each company participating in the project and the approximate percentage of the function that each company performed. Owner projects were defined as DB if the same company performed over 50% of both the design and construction functions; otherwise, owner projects were defined as DBB. Note that for purposes of this analysis, projects that would be considered to be EPC (Engineer, Procure, and Construct) were included in the DB category. Like owner-submitted projects, contractor-submitted projects were categorized as DB if the same company performed the majority of the design and construction functions based on the percentages of the functions performed. Contractor projects were categorized as DBB if the company performed either of the following: (1) the design function only, (2) the construction function only, (3) greater than 50% of the design and less than 50% of the construction, or (4) greater than 50% of the construction and less than 50% of the design. Among owner and contractor-submitted projects, there was a relatively small number of projects that were difficult to classify due to missing or incomplete data. A secondary set of decision rules was developed for these projects using available data, such as, the amount of design work completed at the start of construction. Projects that could not be classified by these rules were excluded from the analysis. The resulting analytic data set comprised 326 owner projects and 291 contractor projects.

The five performance outcomes (cost, schedule, safety, changes, and rework) and the following practices, preproject planning, constructability, project change management, design/information technology (D/IT), team building, and zero accidents, were compared between owner DB and owner DBB projects, and contractor DB and contractor DBB projects. The practices analyzed were limited to the above six because it is for these that the most data are available. Minimal amounts of data are currently available for the other practices, rendering analysis of these impractical. Special emphasis was also placed on analyzing how safety performance was affected by fast tracking versus non-fast tracking, and by adherence to planned construction duration.

Figure 2.2 *(Continued)*

Quality as a Concern

One concern voiced by design-build team members is how to develop quality standards during the design and the construction stages. Quality issues during the design stage include: reduction of errors and omissions, coordination of drawings, and avoiding any conflicts between one design discipline and another. One of the major problems facing a design team, especially where civil, structural, and MEP (mechanical, electrical, and plumbing) design is subcontracted by the architect, is assuring that all systems fit within their prescribed place and space. In the conventional design-bid-build process, owners would be faced with options of lowering ceiling heights halfway through construction because someone failed to verify that ductwork, or fire protection mains would not fit into the space allotted to them. Owners would probably have been required to pay for any extra work to ensure that everything was fit. But in the design-build

COST: Owner-submitted DB projects had better performance in all but 1 out of the 5 cost-related metrics analyzed. Contractor-submitted DB projects had better performance in only 1 out of the three cost-related metrics.

SCHEDULE: Owner-submitted DB projects performed significantly better in 3 out of the 9 schedule metrics analyzed. Contractor-submitted DB projects performed significantly worse in 3 out of the 4 metrics analyzed.

SAFETY: Safety performance was mixed for both owner-submitted and contractor-submitted DB and DBB projects.

CHANGES: Owner-submitted DB projects performed significantly better in the change cost and change schedule metrics. Contractor-submitted DB projects performed significantly better only in the change cost factor.

REWORK: Owner-submitted DB projects performed significantly better in rework. Contractor-submitted DB projects outperformed DBB projects, but there were no significant differences between the two.

PRACTICE USE: Owner-submitted DB projects performed significantly better in 5 out of the 6 practices analyzed. Contractor-submitted DB projects performed significantly better in 1 out of the 6 practices.

Figure 2.3 Summary of cost, schedule, changes, safety, rework performance valuations of owner/contractor DB and DBB projects. (*Source: U.S. Department of Commerce, National Institute of Standards and Technology.*)

mode, these types of problems would be placed at the feet of the design-build team to resolve to the full satisfaction of the owner and at no additional cost to the owner.

During construction the historical role of the architect as the owner's watchdog was to monitor compliance with the quality standards as expressed in the contract documents. That this role may become blurred in the process of combining of design and construction will always be of major concern to any design-build team. What needs to be put into place to assure the owner that high quality standards will be incorporated into the design and monitored during construction?

One of the first assurances will rest with the desire of the design-build team to maintain their reputation as a quality design-builder; or for a first time design-build team, a need to establish a reputation for quality work. Since the designers and contractors will be working in a collaborative mode during the design and the construction phase of the project, many of the quality issues that would normally

COST: Cost performance was mixed for owner-submitted projects. Contractor-submitted DB projects performed better at the lowest and highest cost ranges.

SAFETY: Owner-submitted DB projects performed worse at the lower two cost ranges. Contractor-submitted DB projects performed somewhat better at the lowest and highest cost ranges and worse at the middle range.

SCHEDULE: Owner-submitted DB projects performed better in all cost ranges. Contractor-submitted DB projects performed worse at the lower two cost ranges.

CHANGES: Owner-submitted DB projects performed better at all cost ranges. Contractor-submitted DB projects performed better at the lower two cost ranges.

REWORK: Owner-submitted DB projects performed better at the lowest and highest cost ranges. Contractor-submitted DB projects performed better in the field rework cost factor.

PRACTICE USE: Owner-submitted DB projects had better performance at all cost ranges. Contractor-submitted DB projects performed somewhat better at the lowest cost range, and contractor-submitted DB projects performed worse at the two highest cost ranges.

Figure 2.4 Summary of overall performance for projects valued from $15 to $50 million. (*Source: U.S. Department of Commerce, National Institute of Standards and Technology.*)

surface during construction may possibly be discovered, addressed, and resolved by the collaborative effort of the team of contractor, civil, structural and MEP engineers. This collective approach to quality differs from that of the conventional design-bid-build approach because the contractor is able to input their quality-related experiences during the formative stages of the project and not after the design has been completed and released for bid. When qualified subcontractors are brought on board by the design-build team early on, their review of the design development documents and their input will also provide another quality check. Site visits to ensure compliance with their design and to respond to queries from the contractor's superintendent or subcontractors will substantially diminish the need for requests for information. In conventional design-bid-build, unless the owner elects to have a full blown construction services contract with the architect, visits by the various design disciplines may only occur during monthly visits when the contractor has sent the requisition for payment to the owner, or when called to the site to avoid or correct problems. The team of design-contractor and their consultants will be available any time a problem arises and will make periodic site visits to check on quality issues and compliance with the design documents—with no additional cost to the owner.

COST: Owner-submitted DB projects had better performance in additions and modernizations. Contractor-submitted DB projects had mixed results for cost-related metrics.

SCHEDULE: Owner-submitted DB projects had better performance in all project nature categories. Contractor-submitted DB projects had worse performance in all project nature categories.

SAFETY: Owner-submitted DB projects had worse performance for additions and better performance for grass roots and modernizations. Contractor-submitted DB projects had mixed results for additions and worse performance for grass roots and modernization projects.

CHANGES: Owner- and contractor-submitted DB projects had better performance in all project nature categories.

REWORK: Owner-submitted DB projects had better performance in grass roots and modernizations. Results were mixed for contractor-submitted projects.

PRACTICE USE: Owner-submitted DB projects had better performance in all project nature categories. Contractor-submitted DB projects had better performance in additions and worse performance in grass roots. Performance was mixed in modernizations.

Figure 2.5 Valuation of modernization and addition design-build projects. (*Source: U.S. Department of Commerce, National Institute of Standards and Technology.*)

The owner's quality responsibilities

Quality issues should be addressed in the owner's RFP if the project is to be competitively bid. If quality requirements are not readily available or definable by the owner, the owner may wish to retain a consultant to assist in their preparation. If the bridging concept is employed, that architect should focus on providing quality requirements and/or expectations along with their conceptual design. When a project is to be negotiated, it serves the design-build team well to assist the owner in establishing quality levels so that there is a clear understanding of the expected standards of work.

If an RFP is issued by the owner soliciting proposals in a competitive environment, a design-build team responding could point out the need to include a quality program and even include a rough outline of a program in their response. This will not only put all bidders on a more equal playing field, but may also earn the bidder some points that might lead to a negotiated contract.

The ASCE 2004 study. The American Society of Civil Engineers (ASCE), concerned about quality levels in design-build, had two of its members research the question of quality. The researchers published an article in the *Journal of Management* *in Engineering* in 2004 and the study noted that in the conventional design-bid-build project, the owner established quality levels via a set of plans and specifications and also established the time frame for construction via a milestone schedule. The cost of the project, as opposed to the budget, was determined by the competitive bid process. This approach differs somewhat from design-build where the owner may fix the cost of the project, but the level of quality and, frequently the time frame for design and construction, are based upon competitively bid proposals or by negotiation.

The question of establishing the level of quality acceptable to an owner therefore becomes a function of proper instructions to the project bidders—either in the proposal for design and/or the proposal for construction.

Defining quality. The ASCE team defined quality as *the totality of features and characteristics of a product or service that bears on the ability to satisfy given needs.* A definition established by the American Society for Quality (ASQ) focused on the ASQ phrase, *satisfying given needs.*

ASQ defined varying types of quality as:

1. *Relative quality*—a loose comparison between product features and characteristics.
2. *Product-based-quality*—a precise and measurable variable, and differences in quality reflect differences in some products, namely,
   a. *User-based:* fitness for intended use
   b. *Manufacturing based:* conformance to specifications
   c. *Value-based:* conformance to an acceptable cost

The ASCE survey team, in preparing their report, reviewed 78 RFPs for public design-build projects advertised between 1997 and 2002, totaling $3.0 billion in value. In most cases the owners required the bidder to prepare a firm fixed-price value on a project that was yet to be fully designed. In the conventional design-bid-build project, the plans and specifications containing quality levels would have been available to bidders, and, on awarding, these quality levels become contractor obligations.

Conversely in the design-build contract award cycle, unless quality standards are included in the RFP, the cost, schedule, and levels of quality become the basis for the competition among bidders and evaluation by the owner, as reported by ASCE.

There are six approaches for owners to articulate quality levels:

1. *By qualification.* The owner's RFP would include specific requirements to establish the design-build firm's qualifications to include successful experience in similar projects and the qualifications of various individuals who would be responsible for design and construction.
2. *By evaluated program.* This form of RFP would contain a requirement for the bidder to present a detailed proposal of their Quality Management (QM) program so that the owner could evaluate it along with those furnished by the competition.

3. *By specified program.* The bidder would be required to submit a detailed QM program in response to an owner-specified program. The owner would then be able to review compliance with the program when analyzing bidder's responses.
4. *By performance criteria.* The RFP would reply to the owner-furnished technical performance data to be reviewed and compared with responses from the other bidders.
5. *By specification.* The RFP would require the respondents to submit detailed technical solutions to the owner's technical specifications allowing the owner to verify compliance with their requirements during the design submittal review process.
6. *By warranty.* The owner's RFP would include specific performance warranty requirements or perhaps a maintenance bond or bonds.

After reviewing all 78 RFPs, various approaches to quality were tabulated by the ASCE researchers for both horizontal projects such as roads, bridges, and tunnels, and vertical construction such as schools, libraries, and institutional facilities (Table 2.4).

So one can see that vertical (building) construction projects established quality levels first and foremost by qualification of the design-build bidder.

This concept was reinforced during an interview with Mr. Bob Fraga, a senior manager at United States Postal Services in Arlington, Virginia, in May 2005 who said that much of their success with design-build projects can be attributed to their very intense prequalification program.

When analyzed by project type, the breakdown for preconstruction and post-construction award processes is given provided in Tables 2.5 and 2.6.

Table 2.5 shows that commercial owners and residential owners require a design-build team to submit a detailed qualification statement as part of a preaward review process. Once again the need to scrutinize bids in a prequalification or short list process is stressed.

How should quality management issues be addressed by owners

The ASCE study reviewed all of the 78 RFPs to see how owners were dealing with QM issues.

No distinction between quality control (QC) and quality assurance (QA) was made as each owner had their own definition and interpretation of what constitutes

TABLE 2.4 Approaches to Quality as Differentiated by Vertical and Horizontal Type Projects

Project type	Qualification	Evaluation	Specified program	Performance	Specification	Warranty
Horizontal	7	16	5	2	1	1
Vertical	27	9	4	2	2	3

TABLE 2.5 Preaward Quality Management Requirements by Project Type

Project type [total nos. in ( )]	Design quality in evaluated plan	Construction quality in evaluated plan	Quality plan in evaluated plan	Quality qualifications in evaluated plan
Residential (18)	3	2	7	12
Commercial (18)	6	7	7	14
Industrial (8)	1	2	4	4
Other types (3)	0	0	2	2

quality standards and the method to assure that those quality standards had been achieved. But certain benchmarks standards were established by owners.

Design QM. A plan would be required of the design-build team. Specifically, each respondent would be required to submit a plan for evaluation that included its proposed approach to establishing and managing design quality.

Construction QM. A plan would be required of the bidders whereby each respondent would be required to submit a plan that detailed their approach to controlling quality in the construction phase of the project.

A team QM plan. The RFP would require the design-builders to submit a plan, for evaluation and comparison with the other respondents, to describe the approach for managing quality without being specific to either design or construction. This would be comparable to a quality plan (TQM) plan stating the company's total management approach to quality.

Quality specifications required. A request in the RFP for the qualifications of the key personnel to be assigned to the project if an award were issued. Past performance of the design-build entity and its components would also be required.

Design QM plan after award. If an award is made, the bidder will be required to submit a plan for approval that presents their approach to managing quality of design.

Construction QM plan after award. If an award is made to the bidder, according to the requirements in the RFP, a plan, submitted for approval, would be required to show the proposed approach to manage quality in construction.

The conclusions arrived at in this survey were that in design-build, the owner has an opportunity to ensure that the project under consideration will

TABLE 2.6 Postaward Quality Management Requirements by Project Type

Project type [total nos. in ( )]	Design quality plan required	Construction quality plan required
Residential (18)	4	17
Commercial (18)	7	14
Industrial (8)	0	7
Other types (3)	0	1

achieve the requisite quality of levels, if the proper steps are taken in the preparation of the RFP. There are at least six different approaches to QM that can be incorporated into the RFP, quality by qualification, by evaluated program, by specified program, by performance criteria, by specifications, and by warranty.

The recommendations by these ASCE researchers to owners were as follows:

1. Include QM requirements for design and construction in the proposal.
2. Request quality-specific qualifications for members of the design and construction team.
3. Owners should establish the project's quality management system before award and in the bid proposal ask each respondent to accept and suggest changes to enhance the owner's QM program. Evaluation of the RFP proposals would therefore include the QM program.

The Austin Company of Cleveland, Ohio—A Case Study in Design-Build Evolution

An advertisement by a design-build firm in a Cleveland, Ohio, newspaper proclaimed their ability to provide a "square deal way of planning, erecting, equipping, and maintaining buildings." The article further states "It makes you (referring to the owner) your own architect, engineer, and builder, plus our specialized knowledge, experience, and facilities."

This ad was not a recent one, but dates back to 1907 when the Austin Company of Cleveland, Ohio, announced their integrated approach to project delivery systems to the public whereby they could provide a single source for design and construction.

One of the earliest practitioners of the design-build method in the United States, the Austin Company grew from its early beginnings in Cleveland in the late 1870s to a $640 million powerhouse today, still engaged in "planning, erecting, and equipping" buildings for clients around the world.

It all began with Samuel Austin and his *Austin Method*. Mr. Austin, a carpenter by trade, came to America from England after he had read advertisements for workers to rebuild Chicago after the disastrous fire of 1872. Between 1873 and 1879 Samuel traveled back and forth between the United States and England as work shifted from country to country during the Great Depression of that era. Samuel Austin worked for $1.50 per day, when he worked, which was no more than 3 to 4 days a week, not much for a recently married man. He started his own building business in Cleveland in 1879 with the basic philosophy that good materials and the best workmanship would be the way to success. When his son Wilbur graduated from Case (later to be known as Case Western Reserve University) in 1899, he served a two-year apprenticeship in Cleveland and Europe. With the addition of this young engineer the firm, Samuel Austin & Son, was born, and so was the *Austin Method*.

The Panic of 1907 was hard on the new company and not much work was coming in, but the invention of electric lighting and the incandescent lamp created a need for new factories to manufacture this amazing new product. The Austin Company began to construct a number of large projects for the National Electric Lamp Association (later to become General Electric) and Austin-engineered and Austin-built electric light bulb factories sprang up in Ohio, Rhode Island, Missouri, Minnesota, and California, giving the firm a national exposure and the beginnings of a strong design-build capability.

A sales brochure entitled *The Austin Book of Buildings* published in 1925 illustrated their approach to design-build.

They had previously purchased a steel fabricator, Bliss Mill, that designed and fabricated a standard or modified standard structural steel frame that would allow Austin to quote on and furnish a building's structural steel system in very short order. The *Book of Buildings* included various types of "standard" or "standard modified" factory-type buildings they advertised that they could build, in some cases, in 30 days (Figs. 2.6, 2.7, and 2.8).

They also stated in the brochure that they could provide building design and construction specifications for other types of buildings such as a multistoried office building, referred to as Austin No. 8 Type Building (Fig. 2.9). In a rather novel approach, at that time, they provided potential clients with cost information in the back of the book. Building component costs were indicated on a sliding scale, 100% being the most expensive, and rating others downward from there. So a client could look through floor-construction details and costs (Fig. 2.10), as an example, and determine that a concrete slab, hot-mopped with coal tar pitch topped with a wood. *Bloxonend* flooring system was the most expensive while the wood subfloor and maple surfaced flooring was the least expensive. Wall construction was treated in much the same way (Fig. 2.11) with a 13-inch common hard brick weighing in at 100% and corrugated iron priced at 15% to 21%. Austin also provided prospective clients with a primer on insurance (Fig. 2.12) advising the customer that the increased cost of certain types of construction may result in decreased costs of insurance. And they also made their customers aware of the costs of various soil-bearing capacities and how these capacities affect building costs (Fig. 2.13).

Their ability to design and construct complex projects resulted in their outgrowing their small rented offices and they soon moved into their own building on Euclid Avenue and Nobel Road where they remained until 1953. Along with the move in 1913, they formalized the *Austin Method* by publishing a sixty page book illustrating their accomplishments and stating the policy of this somewhat unique process.

> The Austin Method placed building on a square deal basis. With mutual confidence, the interests of the owner and builder become identical. The owner guaranteed the builder fair pay for his services; and the builder guaranteed the owner a fair return on his money.
>
> The Austin Method has been in operation since 1901. It has controlled the erection of buildings from Rhode Island to California.

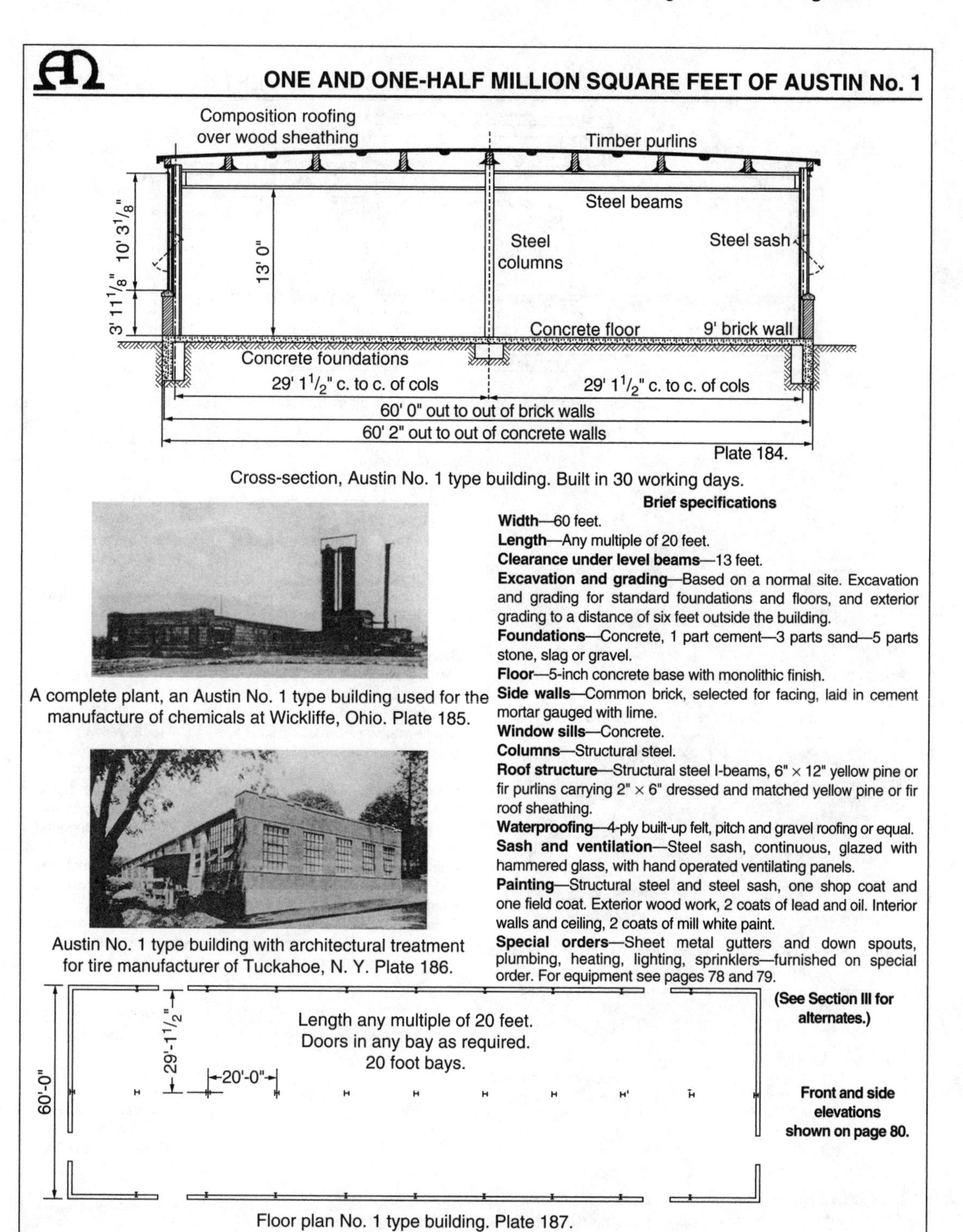

Figure 2.6 Austin's No. 1 Type Building profile. (*Courtesy: The Austin Company, Cleveland, Ohio.*)

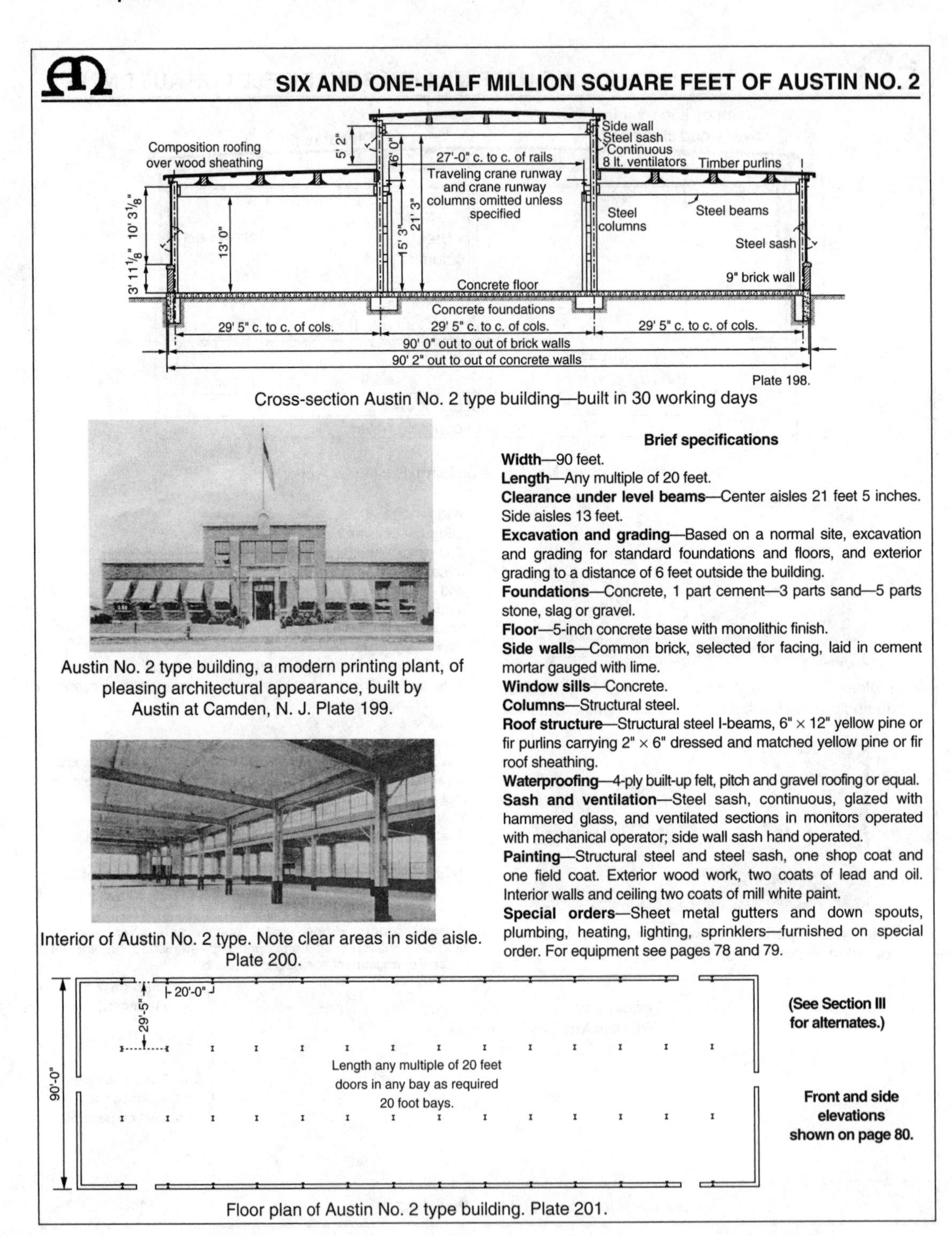

SIX AND ONE-HALF MILLION SQUARE FEET OF AUSTIN NO. 2

Cross-section Austin No. 2 type building—built in 30 working days

Austin No. 2 type building, a modern printing plant, of pleasing architectural appearance, built by Austin at Camden, N. J. Plate 199.

Brief specifications

Width—90 feet.
Length—Any multiple of 20 feet.
Clearance under level beams—Center aisles 21 feet 5 inches. Side aisles 13 feet.
Excavation and grading—Based on a normal site, excavation and grading for standard foundations and floors, and exterior grading to a distance of 6 feet outside the building.
Foundations—Concrete, 1 part cement—3 parts sand—5 parts stone, slag or gravel.
Floor—5-inch concrete base with monolithic finish.
Side walls—Common brick, selected for facing, laid in cement mortar gauged with lime.
Window sills—Concrete.
Columns—Structural steel.
Roof structure—Structural steel I-beams, 6" × 12" yellow pine or fir purlins carrying 2" × 6" dressed and matched yellow pine or fir roof sheathing.
Waterproofing—4-ply built-up felt, pitch and gravel roofing or equal.
Sash and ventilation—Steel sash, continuous, glazed with hammered glass, and ventilated sections in monitors operated with mechanical operator; side wall sash hand operated.
Painting—Structural steel and steel sash, one shop coat and one field coat. Exterior wood work, two coats of lead and oil. Interior walls and ceiling two coats of mill white paint.
Special orders—Sheet metal gutters and down spouts, plumbing, heating, lighting, sprinklers—furnished on special order. For equipment see pages 78 and 79.

Interior of Austin No. 2 type. Note clear areas in side aisle. Plate 200.

Floor plan of Austin No. 2 type building. Plate 201.

Figure 2.7 Austin's No. 2 Type Building profile. (*Courtesy: The Austin Company, Cleveland, Ohio.*)

TWO AND ONE-HALF MILLION SQUARE FEET OF AUSTIN NO. 4

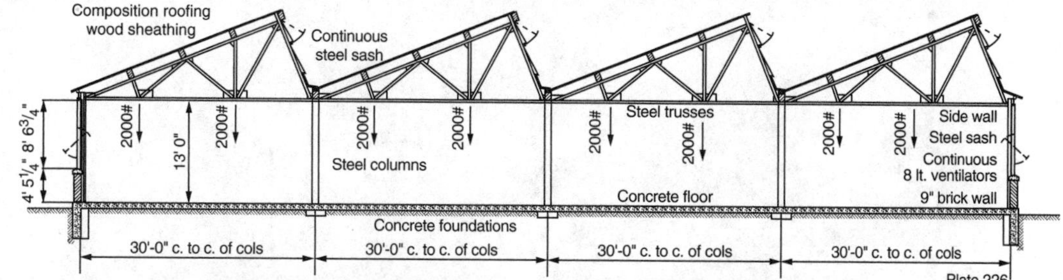

Composition roofing wood sheathing
Continuous steel sash
Steel trusses
Steel columns
Side wall Steel sash
Continuous 8 lt. ventilators
9" brick wall
Concrete floor
Concrete foundations
30'-0" c. to c. of cols
Plate 226

Cross-section Austin No. 4 type building—built in 60 working days

Austin No. 4 type building for large book publisher, Bloomfield, N. J. note evenly distributed daylighting. Plate 227.

Austin No. 4 type building for textile manufacturing at Cumberland, Md. Plate 228.

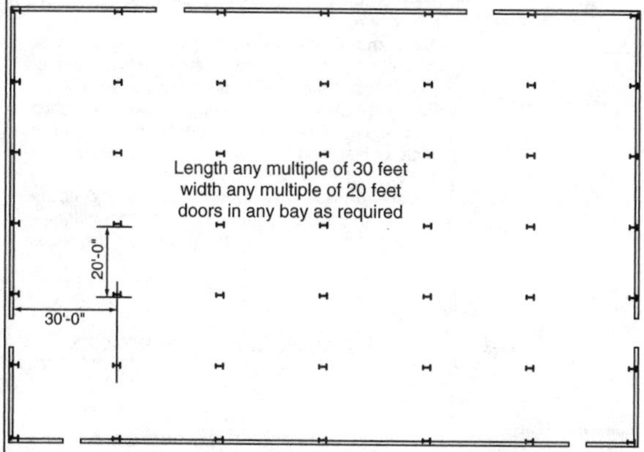

Brief specifications

Width—Any multiple of 20 feet.

Length—Any multiple of 30 feet.

Clearance under trusses—13 feet.

Excavation and Grading—Based on a normal site, excavation and grading for standard foundations and floors, and exterior grading to a distance of 6 feet outside the building.

Foundations—Concrete, 1 part cement—3 parts sand—5 parts stone, slag or gravel.

Floor—5-inch concrete base with monolithic finish.

Side walls—Common brick, selected for facing, laid in mortar gauged with lime.

Window sills—Concrete.

Columns—Structural steel.

Roof structure—Structural steel trusses with level bottom chord. 6" × 12" yellow pine or fir purlins carrying 2" × 6" dressed and matched yellow pine or fir roof sheathing.

Waterproofing—4-ply built-up felt and asphalt roofing or equal.

Sash and ventilation—Side wall steel sash, continuous, glazed with hammered glass, and ventilated sections hand operated. Upper row of sawtooth sash, 4 feet deep, hinged at top and mechanically operated; lower row, 4 feet deep fixed.

Painting—Structural steel and steel sash, one shop and one field coat. Exterior wood work, two coats of lead and oil. Interior walls and ceiling, two coats of mill white paint.

Special orders—Sheet metal gutters and down spouts, plumbing, heating, lighting, sprinklers—furnished on special order. For equipment see pages 78 and 79.

(See Section III for alternates.)

Length any multiple of 30 feet
width any multiple of 20 feet
doors in any bay as required

Floor plan Austin No. 4 type building. Plate 229.

Front and side elevations shown on page 80.

Figure 2.8 Austin's No. 4 Type Building profile. (*Courtesy: The Austin Company, Cleveland, Ohio.*)

FOUR AND ONE-HALF MILLION SQUARE FEET OF AUSTIN NO. 8

Austin No. 8 type building at Minneapolis with special architectural treatment, for large manufacturer of electrical equipment. Plate 249.

Austin No. 8 type building

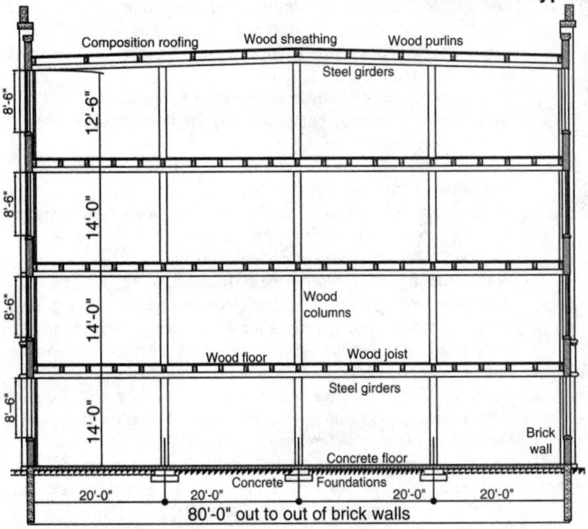

Cross-section Austin No. 8 type building. Plate 250.

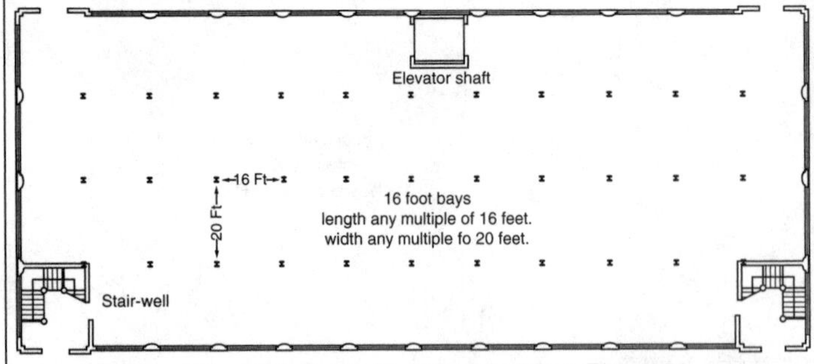

Floor plan, Austin No. 8 type building. Plate 251.

Brief specifications

Size—Any size in panels of 16 by 20 feet.

Clearance—12 feet clear with floor heights normally 14 feet. (See cross-section.)

Excavation and grading—Based on a normal site, excavation and grading for standard foundations and floors, and exterior grading to a distance of 6 feet outside the building.

Foundations—Concrete, 1 part cement—3 parts sand—5 parts stone, slag or gravel.

Floor—Steel girders, yellow pine or fir floor beams and sub-floor with maple finish, designed for 125 lbs. live load. For heavy machine shop or heavy manufacturing and warehouse purposes a live load capacity of 200 lbs. or more should be specified.

Stairs—Wood; steel at additional cost.

Side walls—Common brick selected for facing, laid in cement mortar gauged with lime; special face brick as required at extra cost.

Window sills and coping—Concrete.

Columns—Wood, standard. Steel, additional.

Roof structure—Structural steel girders, yellow pine or fir roof sheathing on yellow pine or fir purlins.

Waterproofing—4-ply built-up felt, pitch and gravel roofing or equal.

Sash and ventilation—Steel sash between brick pilasters, glazed with hammered glass, clear or wired at additional cost. Ventilation as ordered.

Painting—Structural steel and steel sash, one shop coat and one field coat. Exterior woodwork two coats of lead and oil. Interior walls and ceiling, two coats of mill white paint. (Colored for dado.)

Special orders—Sheet metal gutters and down-spouts, plumbing, heating, lighting, sprinklers—furnished on special order. For equipment see pages 78 and 79.

(See section III for alternates.)

Elevation shown on page 80.

Figure 2.9 Austin's Type 8 Building—a multistoried structure. (*Courtesy: The Austin Company, Cleveland, Ohio.*)

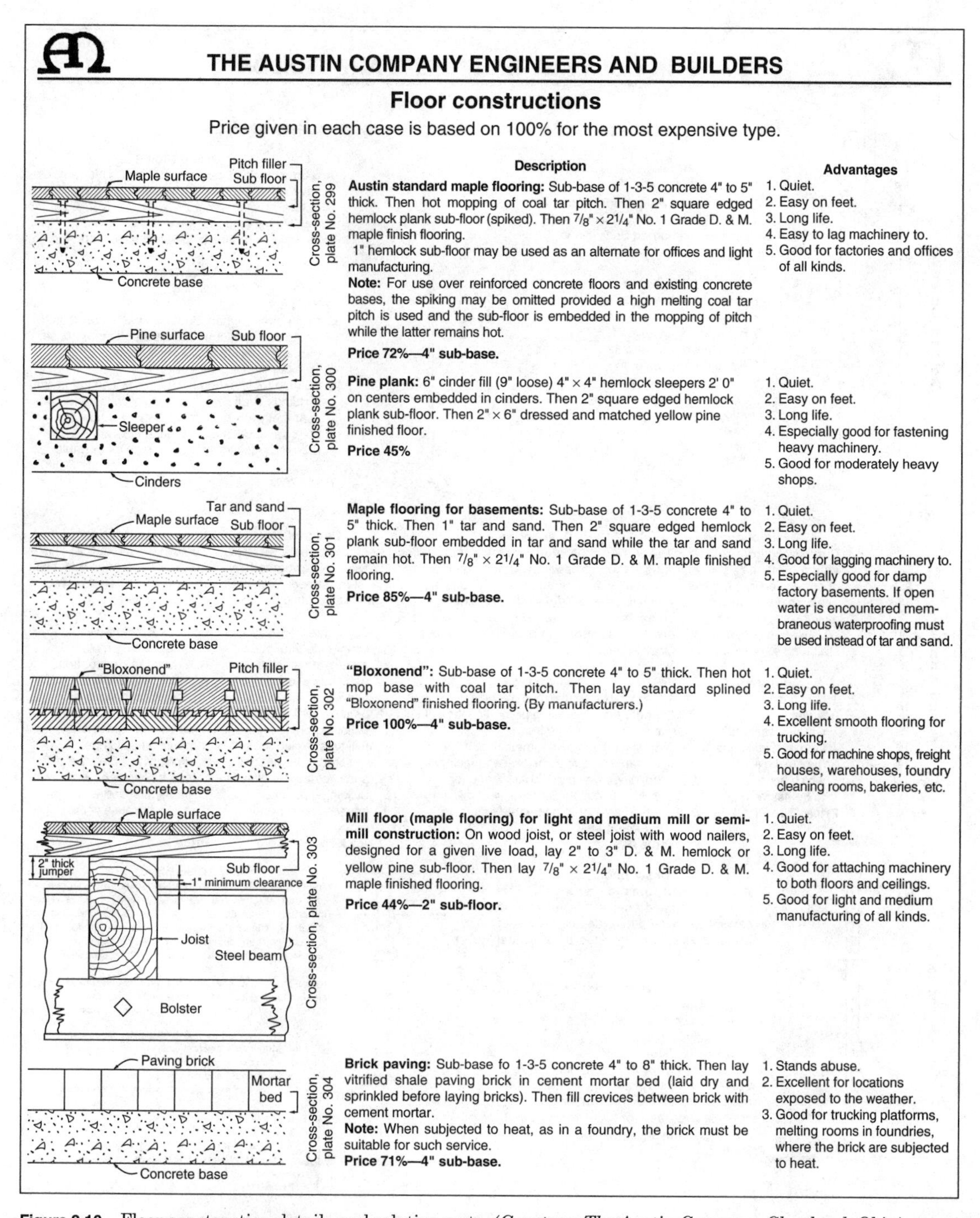

THE AUSTIN COMPANY ENGINEERS AND BUILDERS

Floor constructions

Price given in each case is based on 100% for the most expensive type.

Description	Advantages
Austin standard maple flooring: Sub-base of 1-3-5 concrete 4" to 5" thick. Then hot mopping of coal tar pitch. Then 2" square edged hemlock plank sub-floor (spiked). Then 7/8" × 21/4" No. 1 Grade D. & M. maple finish flooring. 1" hemlock sub-floor may be used as an alternate for offices and light manufacturing. **Note:** For use over reinforced concrete floors and existing concrete bases, the spiking may be omitted provided a high melting coal tar pitch is used and the sub-floor is embedded in the mopping of pitch while the latter remains hot. **Price 72%—4" sub-base.**	1. Quiet. 2. Easy on feet. 3. Long life. 4. Easy to lag machinery to. 5. Good for factories and offices of all kinds.
Pine plank: 6" cinder fill (9" loose) 4" × 4" hemlock sleepers 2' 0" on centers embedded in cinders. Then 2" square edged hemlock plank sub-floor. Then 2" × 6" dressed and matched yellow pine finished floor. **Price 45%**	1. Quiet. 2. Easy on feet. 3. Long life. 4. Especially good for fastening heavy machinery. 5. Good for moderately heavy shops.
Maple flooring for basements: Sub-base of 1-3-5 concrete 4" to 5" thick. Then 1" tar and sand. Then 2" square edged hemlock plank sub-floor embedded in tar and sand while the tar and sand remain hot. Then 7/8" × 21/4" No. 1 Grade D. & M. maple finished flooring. **Price 85%—4" sub-base.**	1. Quiet. 2. Easy on feet. 3. Long life. 4. Good for lagging machinery to. 5. Especially good for damp factory basements. If open water is encountered membraneous waterproofing must be used instead of tar and sand.
"Bloxonend": Sub-base of 1-3-5 concrete 4" to 5" thick. Then hot mop base with coal tar pitch. Then lay standard splined "Bloxonend" finished flooring. (By manufacturers.) **Price 100%—4" sub-base.**	1. Quiet. 2. Easy on feet. 3. Long life. 4. Excellent smooth flooring for trucking. 5. Good for machine shops, freight houses, warehouses, foundry cleaning rooms, bakeries, etc.
Mill floor (maple flooring) for light and medium mill or semi-mill construction: On wood joist, or steel joist with wood nailers, designed for a given live load, lay 2" to 3" D. & M. hemlock or yellow pine sub-floor. Then lay 7/8" × 21/4" No. 1 Grade D. & M. maple finished flooring. **Price 44%—2" sub-floor.**	1. Quiet. 2. Easy on feet. 3. Long life. 4. Good for attaching machinery to both floors and ceilings. 5. Good for light and medium manufacturing of all kinds.
Brick paving: Sub-base fo 1-3-5 concrete 4" to 8" thick. Then lay vitrified shale paving brick in cement mortar bed (laid dry and sprinkled before laying bricks). Then fill crevices between brick with cement mortar. **Note:** When subjected to heat, as in a foundry, the brick must be suitable for such service. **Price 71%—4" sub-base.**	1. Stands abuse. 2. Excellent for locations exposed to the weather. 3. Good for trucking platforms, melting rooms in foundries, where the brick are subjected to heat.

Figure 2.10 Floor construction details and relative costs. (*Courtesy: The Austin Company, Cleveland, Ohio.*)

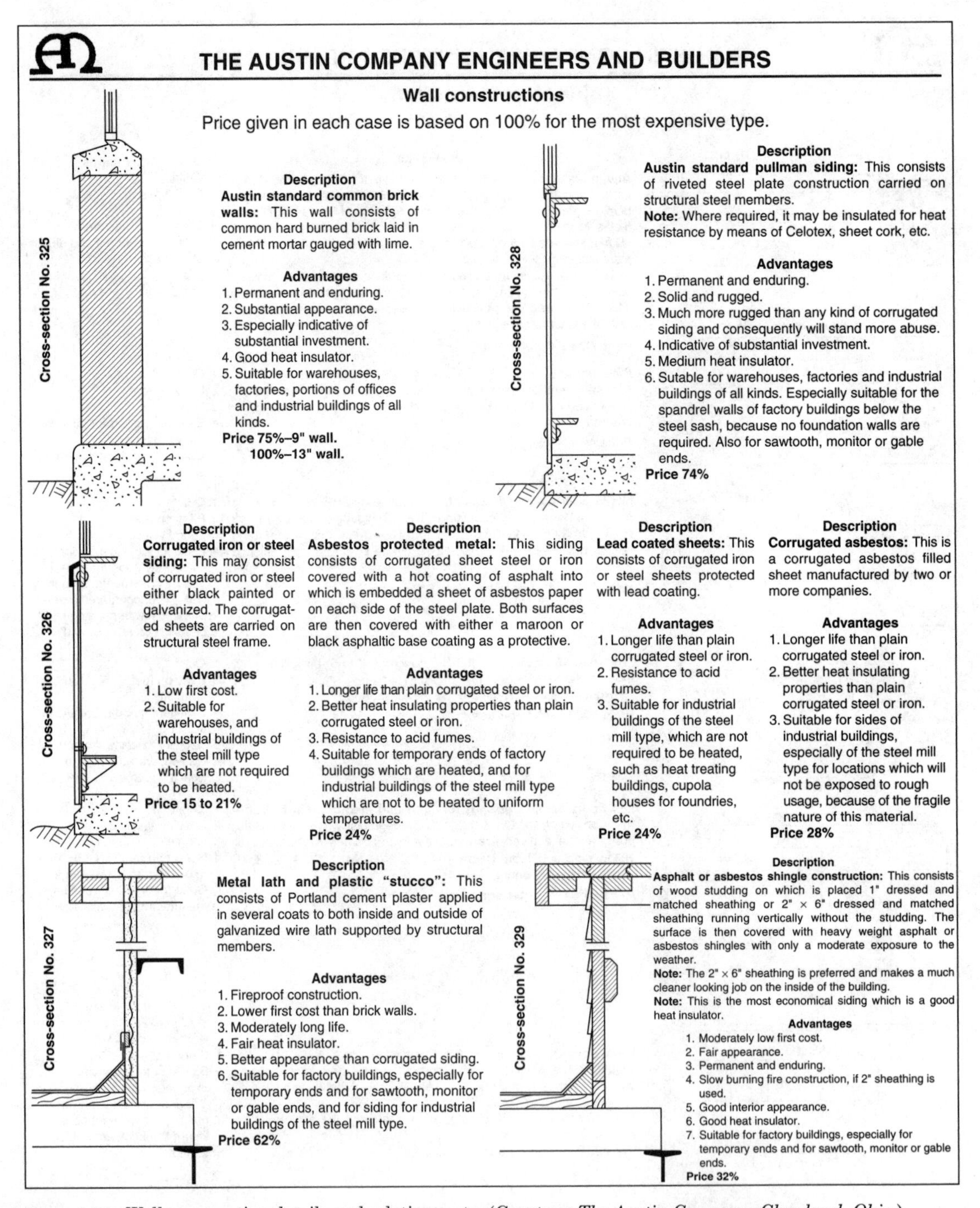

THE AUSTIN COMPANY ENGINEERS AND BUILDERS

Wall constructions

Price given in each case is based on 100% for the most expensive type.

Cross-section No. 325

Description

Austin standard common brick walls: This wall consists of common hard burned brick laid in cement mortar gauged with lime.

Advantages

1. Permanent and enduring.
2. Substantial appearance.
3. Especially indicative of substantial investment.
4. Good heat insulator.
5. Suitable for warehouses, factories, portions of offices and industrial buildings of all kinds.

Price 75%–9" wall.
100%–13" wall.

Cross-section No. 328

Description

Austin standard pullman siding: This consists of riveted steel plate construction carried on structural steel members.

Note: Where required, it may be insulated for heat resistance by means of Celotex, sheet cork, etc.

Advantages

1. Permanent and enduring.
2. Solid and rugged.
3. Much more rugged than any kind of corrugated siding and consequently will stand more abuse.
4. Indicative of substantial investment.
5. Medium heat insulator.
6. Sutable for warehouses, factories and industrial buildings of all kinds. Especially suitable for the spandrel walls of factory buildings below the steel sash, because no foundation walls are required. Also for sawtooth, monitor or gable ends.

Price 74%

Cross-section No. 326

Description

Corrugated iron or steel siding: This may consist of corrugated iron or steel either black painted or galvanized. The corrugated sheets are carried on structural steel frame.

Advantages

1. Low first cost.
2. Suitable for warehouses, and industrial buildings of the steel mill type which are not required to be heated.

Price 15 to 21%

Description

Asbestos protected metal: This siding consists of corrugated sheet steel or iron covered with a hot coating of asphalt into which is embedded a sheet of asbestos paper on each side of the steel plate. Both surfaces are then covered with either a maroon or black asphaltic base coating as a protective.

Advantages

1. Longer life than plain corrugated steel or iron.
2. Better heat insulating properties than plain corrugated steel or iron.
3. Resistance to acid fumes.
4. Suitable for temporary ends of factory buildings which are heated, and for industrial buildings of the steel mill type which are not to be heated to uniform temperatures.

Price 24%

Description

Lead coated sheets: This consists of corrugated iron or steel sheets protected with lead coating.

Advantages

1. Longer life than plain corrugated steel or iron.
2. Resistance to acid fumes.
3. Suitable for industrial buildings of the steel mill type, which are not required to be heated, such as heat treating buildings, cupola houses for foundries, etc.

Price 24%

Description

Corrugated asbestos: This is a corrugated asbestos filled sheet manufactured by two or more companies.

Advantages

1. Longer life than plain corrugated steel or iron.
2. Better heat insulating properties than plain corrugated steel or iron.
3. Suitable for sides of industrial buildings, especially of the steel mill type for locations which will not be exposed to rough usage, because of the fragile nature of this material.

Price 28%

Cross-section No. 327

Description

Metal lath and plastic "stucco": This consists of Portland cement plaster applied in several coats to both inside and outside of galvanized wire lath supported by structural members.

Advantages

1. Fireproof construction.
2. Lower first cost than brick walls.
3. Moderately long life.
4. Fair heat insulator.
5. Better appearance than corrugated siding.
6. Suitable for factory buildings, especially for temporary ends and for sawtooth, monitor or gable ends, and for siding for industrial buildings of the steel mill type.

Price 62%

Cross-section No. 329

Description

Asphalt or asbestos shingle construction: This consists of wood studding on which is placed 1" dressed and matched sheathing or 2" × 6" dressed and matched sheathing running vertically without the studding. The surface is then covered with heavy weight asphalt or asbestos shingles with only a moderate exposure to the weather.

Note: The 2" × 6" sheathing is preferred and makes a much cleaner looking job on the inside of the building.

Note: This is the most economical siding which is a good heat insulator.

Advantages

1. Moderately low first cost.
2. Fair appearance.
3. Permanent and enduring.
4. Slow burning fire construction, if 2" sheathing is used.
5. Good interior appearance.
6. Good heat insulator.
7. Suitable for factory buildings, especially for temporary ends and for sawtooth, monitor or gable ends.

Price 32%

Figure 2.11 Wall construction details and relative costs. (*Courtesy: The Austin Company, Cleveland, Ohio.*)

Insurance Facts
in Relation to
Various Types of Industrial Buildings

One of the many problems to be considered carefully in connection with industrial building, is that of the cost of insurance. Insurance premiums constitute an overhead expense item which frequently can be appreciably lowered by consideration of problems involved before the building is designed and built.

The following table, giving approximate figures only, shows in a general way the effect on rates of various types of construction. The figures are based upon tables in use in Ohio in localities having what is known as "Fourth Class" fire protection, and these figures can only be used as giving the approximate ratio of insurance cost on the various classes of construction. The actual rate for each and every risk is based upon an inspection of that particular risk after it is built and cannot be approximated until full plans and specifications, together with details of occupancy, are ready for consideration.

In this table non-combustible contents have been assumed in each and every building, on the order of the average metal working risk, and the Austin No. 1 Type Building has been used as a base, or 100% rate.

No.	Building	Gross Rate	80% Co-Insurance Rate	90% Co-Insurance Rate	Comparison in % of Insurance Cost
A	One-story, fireproof, all steelwork protected, steel sash, incombustible content.	.18	.072	.065	100%
B	Multi-story, reinforced concrete construction, incombustible content.	.19	.076	.068	105%
C	One-story, steel frame, brick apron wall, steel sash, fireproof roof, incombustible content.	.25	.125	.113	139%
D	One-story, wall bearing, brick apron walls and brick pilasters, metal sash, steel roof supports, wood roof, incombustible content.	.33	.248	.231	183%
E	One-story, steel frame, brick apron walls, metal sash, wood roof, incombustible content.	.41	.369	.349	227%
F	Multi-story, semi-mill construction, steel frame or wood posts, with wood floors.	.59	.443	.413	327%
G	One-story, steel frame, wood siding, wood roof, steel sash, incombustible content.	.86	.774	.731	477%

The installation of a sprinkler system in any of the above buildings reduces the insurance cost from 50 to 90%, dependent upon the class of building; on the fireproof building the least reduction, and on the building entirely unprotected from a fireproofing standpoint, the largest reduction. The contents of the building will seriously affect the reduction made by the sprinkler installation. Generally speaking, the contents of the building rather than type of construction determine if sprinklers are required.

Figure 2.12 Insurance criteria as it relates to construction components. (*Courtesy: The Austin Company, Cleveland, Ohio.*)

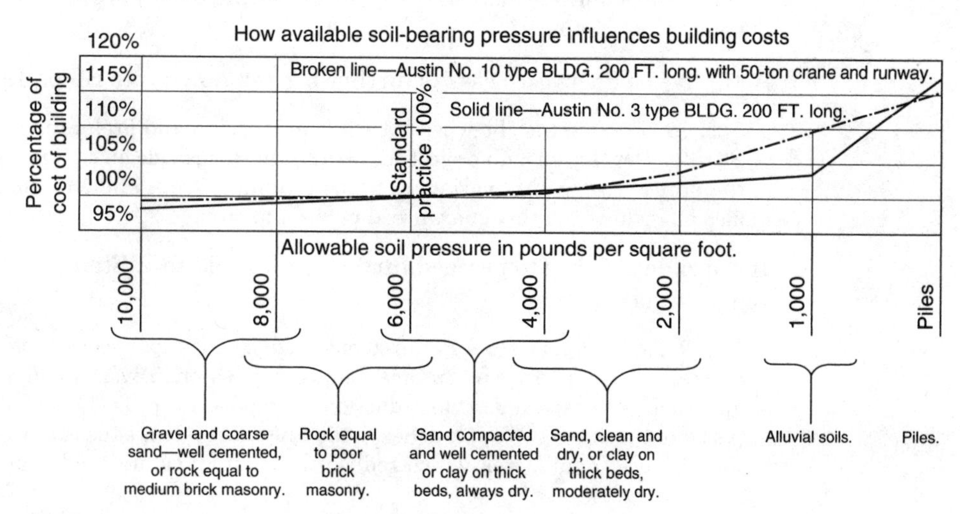

Figure 2.13 How soil bearing capacities affect cost. (*Courtesy: The Austin Company, Cleveland, Ohio.*)

> It has made new customers—three-fourths of last year's work was on repeat orders. It has made new customers, creating a business that now requires the services of more than a thousand men. And it has created a complete organization capable of handling big operations—twelve buildings for one customer erected last year under the Austin Method, and twelve more for the same customer now under way.

The Austin Company in 2005

The *Austin Method* has remained a bedrock of the Austin Company and 84 years later these basic tenets and ethics are as valid as they were when Samuel and Wilbur ran their company back at the turn of the last century.

Today, the Austin Company's design-build approach is much the same. Their sales and marketing efforts mirror this customer-oriented approach, with their two-step design-build program.

Mike Pierce, vice president, Sales and Marketing, said that a prospective client upon reviewing Austin's approach to design-build and deciding to work with them will be presented with a document called *A Standard Engineering Service Agreement*. Mike said that for a negotiated sum, the Austin Company will proceed with their Step 1 process:

> *Step 1.* Austin works in partnership with the client to develop a facility concept and performs sufficient preliminary engineering to establish a definitive cost and schedule for the project. At the completion of Step 1, a very small percentage of the total project cost has been expended, yet the total cost and time at risk have been clearly quantified based upon a mutually agreeable scope of work. Under the design-build approach, cost and schedule are not established until construction documents and bidding are complete. Similarly, under fast-track construction management, the cost is not fixed until after the construction documents are complete and some construction commitments have already been made. Considerably more of the client's time and money are at risk reaching this point.

Actually the Step 1 process is further broken down into Steps 1a and 1b.

> *Step 1a.* Provide the client with a schematic design and preliminary estimate.
> *Step 1b.* Develop enough project scope to be able to provide an estimate guaranteeing the project cost with a variance of plus or minus 5%. A builder's contingency will also be included in this guaranteed maximum sum.

If the client wishes to proceed further, Mike said that Step 2 will carry them through to contract.

> *Step 2.* Austin applies its structured methodology to prepare construction documents and execute competitive procurement and construction in overlapping sequence. This approach results in a sizable reduction in the overall project schedule when compared to the other two approaches. Reducing this cycle means that the client's facility is operational faster, which reduces interest on capital and enables the client

to generate a return on investment. Additionally, Austin's competitive procurement of each project component assures that the client benefits from the best pricing available in the marketplace.

Using the Guaranteed Maximum Price contract format, Mike Pierce said that Austin would complete their buyouts, and once the last major subcontractor group, generally the MEP subs, have been awarded subcontract agreements, the owner has a realistic estimate of what the project's costs will be, barring any major unforeseen events. Each month the Austin-project management team meet for what they call their Project Recapitulation Meeting to discuss cost to date and projected costs to complete the project. This information is then passed on to the owner.

In the lobby of the of their old office building on Euclid Avenue and Nobel Road in Cleveland, there was a stately grandfather's clock with a hand carved case of the finest English walnut directly beneath the office of its president. This clock was Samuel Austin's clock and is one material embodiment of the man that survives to this day—rock solid, functional, displaying a feeling of permanence, much like the company that he created more than 100 years ago.

A Midsize Contractor's View of Design-Build

Mr. Victor Bonardi is the design-build manager at Forrester Construction Company with annual sales in the $150 million range, located just outside the Capital Beltway in Rockville, Maryland.

The company was founded in 1988, and it offers prospective clients a full range of services, from preconstruction consultations to general contracting to construction management to design-build. They have focused on a segment of the market encompassing retail, institutional, laboratory, and biotechnology clients and have specialized expertise in MEP systems implementation and advanced information systems. Forrester has a special projects division and senior estimator Phil Whittaker says this division handles fast track renovation and additional work along with high-end restaurant construction.

Forrester Construction ventured into design-build work in the early 1990s. They typically receive 16 to 20 requests for design-build proposals from prospective clients each year and capture between six and eight projects. The average size of each project is about $6 million. Vic Bonardi says he also conducts about one design-build presentation a month to potential clients.

Forrester's experience in the design-build process, according to Vic, is similar to other studies and surveys which indicate that this type of project-delivery system allows for more rapid completion and usually at little or no additional cost over the original budget unless the owner adds betterments or enhancements to the original project scope. Their record of maintaining initial budgets is pretty good. He discussed a project for a public agency located in Annapolis, Maryland's state capital, which they bid in competition with several other companies. The agency, in their RFP, stated that the cost of the project must not exceed $12.4 million and an award would be made based upon design and compliance with the owner's program. Forrester was awarded the design-build contract, and except for two items of additional work, would have completed the project for $12.4 million or even slightly less. However, additional asbestos abatement was required when more was discovered and a heavy snowfall caused the collapse of an existing roof structure which was outside the scope of the project. Forrester added this work which resulted in increasing the project's total cost to about $13 million. The owner was delighted, having had some previous experience on a design-bid-build project that was not so fortunate. They experienced significant cost overruns from a contractor who took advantage of some perceived design deficiencies to submit a low bid but, on being awarded a contract, proceeded to prepare a number of change orders to increase the project costs by a significant amount.

Bonardi said that design-build projects require considerably more project-management effort to extract, define, and monitor the owner's program, not too dissimilar from other types of negotiated projects. To produce a design-build project, the interaction among project managers, estimators, design consultants, and the owner takes a great deal of time, justifying somewhat higher markups that these types of projects demand.

Forrester does not have in-house design staff but relies on working with outside consultants using a standard for services-type contract that makes the design consultant a subcontractor. They look to the architect to assemble the other consultants—structural, MEP, civil, landscape, interior designers, as each project requires.

Forrester, in the main, uses lump sum or stipulated sum contracts. CM type contracts are frequently used when the client does not have sufficient or qualified professionals on staff to interact with the Forrester design and construction team. Vic Bonardi said that, in his experience, government agencies prefer the CM approach to design-build because they can manage more such projects with less staff. He said a local branch office of the U.S. Corps of Engineers was able to significantly downsize their staff of project managers, primarily, by way of hiring construction management to handle increased work loads. This allowed the Corps to assign one project manager to supervise multiple projects.

Forrester's design-build approach is to prepare a series of phased-in proposals for each client that includes both project scope and price and allows the client to stop the process at each step or continue on to full contract. The Quality Definition Package (QDP) is the first step in a four-step process that commences once the client signs an agreement committing them to the program. Forrester begins to select a design team at this point and starts the process of extracting the client's building program. Forrester charges a fee for this preliminary work which becomes quite intensive as it progresses. The fee will recoup costs if the project is aborted and will also ensure that the client is serious about their interest in working with them on the new project. According to Vic, if a client balks at the initial fee, that's a good sign that they were just on a *fishing expedition* and not really serious about developing the project. This fee includes not only Forrester's cost for project management work but also for their estimating and management-information systems costs and the architectural/structural/MEP fees they incur to prepare the QDP.

Vic said the architectural fees for the entire project will be about 8% of the projected total of the project cost and they negotiate the cost of the design development. Their fee to a client proceeding through Step 1 will be in the range of $25,000 to$60,000, depending upon the nature and complexity of the project.

A closer look at Forrester's QDP approach

Step 1 begins with the preparation of their comprehensive QDP. The QDP will include a floor plan or plans (if multistoried), wall sections, elevations, definition of the structural system, finish schedules, door schedules, one-line electrical drawings, riser diagrams for HVAC, plumbing, and fire protection systems. It will include specifications, a narrative describing the design, and sufficient information to allow the client to clearly see what they are getting.

Figure 2.14 is a cover sheet from one of Forrester's QDPs. As shown this represents a rather concise presentation. Only the price is missing and that would be in Tab 11.

Step 1. Forrester will have developed a total cost for the project, and if given the nod by the owner, will proceed to completely develop the plans and specifications and prepare a contract for construction. If the client would like more project definition prior to signing a firm commitment, they would authorize Forrester to proceed to the next phase.

Step 2. For an additional sum, Forrester will complete the design to 50%, still retaining the same contract sum, but allowing the owner more specific design and systems information.

Step 3. For an additional fee they will produce 100% plans and specifications.

Step 4. This is the construction phase.

At any time during this process, the client can abort the project and upon payment of Forrester's fee take ownership of all of the documents produced to date. Once Forrester receives a fully executed agreement, they will honor their commitment, no matter what forces the market brings to bear.

Vic talked about a client who did abort a project because of funding problems. The client had inherited several rental properties from their father and was considering upgrading these properties so that leasing rates could be increased. The client had $2 million to spend and after Forrester completed the survey they presented their proposal. The client vacillated for months, but in the meantime, the price of structural steel, a prime component in the upgrade work began its

QDP Design Package
NAFBA1-99-D-0011

QUALITY DEFINITION PACKAGE

TABLE OF CONTENTS

Figure 2.14 Table of contents of Forrester's quality definition package. (*Courtesy: Forrester Construction Company, Rockville, Maryland.*)

climb—adding $30, $60, and in some cases $120 per ton to the base per ton price. When the client finally decided to accept Forrester's proposal, they had to decline the job because of the substantial increase in steel. Vic said that if they had had a firm commitment to proceed with the work early on and these steel increases had taken place, they would have proceeded with the work and absorbed all increases, and losses, but now they told the client, "You did not accept our proposal which was based upon acceptance within a reasonable period of time—your month-long delays forces us to withdraw our offer due to those dramatic increases in steel—a major component in the project. When steel prices go back to the level included in our original proposal, we'll do the work for $2 million."

Forrester continues to add more work in their design-build division today due, in large part, to the thoroughness and fairness of this QDP phased approach.

Chapter

3

The Design-Build Team

The integration of design and construction into a single entity needs to focus on the ability of that entity to perform certain tasks effectively. Is a contractor-led team or an architect/engineer-led team best equipped to:

- Market the services of the entity
- Prepare and present a proposal to a perspective client once a sale lead has been developed
- Deliver a product that will include both design and construction
- Follow up promptly on postconstruction matters such as warranty and commissioning to further enhance their marketing or sales development program

The leadership of the design-build team may vary depending upon the nature of the project, the previous working relationship with a client, or the unique qualities of one or both of the members of the team.

Different Approaches to Assembling a Design-Build Team

The holistic approach

An architect can hire a construction professional to head a new construction department, which would also require hiring an estimator, project manager (PM), and field supervisors, at a bare minimum.

A contractor conversely can employ an architect who will, acting as the design team captain, engage other disciplines as subcontractors—a structural design firm, an MEP engineering company, and civil and geotechnical engineers, depending upon the nature of the project being undertaken.

Each of these professionals comes at a cost and often seeks equity incentives before considering leaving a long-term employer for one with little or no track record in a new business venture.

There are ancillary costs associated with either of these two approaches and they can be more than inconsequential—increased office space, additional office equipment including computers and related software, increased payroll for support staff, and other start-up costs. And, of course, if no new business is quickly generated with this new investment, the added overhead costs can have a serious impact on the company's core business.

Collaboration

For the average small to midsized firm, collaboration would seem to be a logical approach to create a design-build entity. Forging an alliance with an architectural firm or construction company, at best, allows the new firm to be up and running with the least amount of overhead exposure and, conversely, would allow for the dissolution of what seemed like a good idea to take place, if need be, at the least cost to overhead.

The New Business Entity—Joint Venture, Teaming Agreement, Limited Liability Corporation, or S Corporation?

There are numerous ways a collaborative approach to creating a design-build team can be accomplished; each one has legal and accounting pluses and minuses that can only be properly addressed by the appropriate professionals. But the basics of a new venture are rather easy to comprehend.

The joint venture

A joint venture (JV), where a one-time project entity can be created to work on a specific project, is one way to provide a vehicle to combine design and construction. But there are a number of issues that must be considered before a JV entity is to be formed. First of all there is the question of licensing for both contractor and design consultants, which varies from state to state. In the absence of specific legislation allowing a JV entity to practice architecture or engage in contracting, any such venture may be in jeopardy. So the first step to consider is how licensing law will affect the JV. The JV must also specify the obligations, rights, and responsibilities of each member of the entity. A few such considerations are rather basic but also reveal how this area of responsibility and obligations is not a simple matter.

- Who will assume the lead in developing the owner's program, and how are design issues and restraints between design and budget resolved?
- Is it the designer's obligation to design or redesign a budget without increasing the cost of their services to the JV?
- Who is responsible for design errors and how is this responsibility covered by insurance? If design errors are made, in what amount(s) and to whom are the proceeds paid?

- If the contractor provides value engineering in order to meet budget, what responsibility and authority does the design consultant have in reviewing, approving, or rejecting any value engineering proposals?
- How is compensation divided and is any upfront money to be provided for conceptual design or design development? If any upfront money is available, then who is required to pay and in what amount?
- During construction does the builder provide all supervisory personnel? What authority does the architect/engineer (A/E) have in the inspection process and how will the A/E reject nonconforming work or poor quality work?
- Postconstruction issues—correction of defective work or design errors, warranty issues and statute of limitation, responsibilities relating to design, and construction work. How are these matters covered in the JV agreement?

This division of rights, duties, obligations, and responsibilities is best thought out and incorporated into the JV agreement. The assigning of obligations and responsibilities is often referred to as a teaming agreement. This is a document that is not unique in the preparation of a JV agreement but is probably a necessity whenever a builder and a design consultant jointly embark on a design-build project.

The teaming agreement

A teaming agreement is generally prepared when the team is initially being assembled for the purpose of developing and presenting a design-build proposal to an owner. A secondary teaming agreement is often used upon notification by the owner that the team's proposal has been accepted, and this agreement will form the basis for contractual relations between builder and design consultants.

The teaming agreement—Part A. When either architect or contractor is considering forming a design-build team for the purpose of responding to an owner's request for proposal (RFP), they must consider some very basic elements, and we might call this *teaming agreement—Part A:*

- Does the architect (or contractor) have the necessary experience required for this project?
- Do we think we have like goals and compatible personnel that can work together? Have we had working experience with each other before on other types of projects?
- Do we think that if this team is assembled, it will have a good chance of winning the competition?
- Does the other member of the team have the financial wherewithal to provide the necessary services required before contract award, and if they don't win the competition will they be able to absorb all associated costs?
- Does either party have a positive past relationship with the owner?

- Is the other party amenable to the type of business structure under consideration, i.e., JV, limited liability corporation (LLC), other?
- How will costs be allocated during the bidding process, and if no award is made, when will costs be apportioned?
- If an award is made, are both parties committing to a continuing relationship?

The teaming agreement—Part B. If the design-build proposal has been accepted by the owner, then the architect and builder must now proceed to contract with each other, in some way, to form a design-build alliance, and a second agreement, let's call it the *teaming agreement—Part B*, must be prepared. This agreement will spell out the parties' obligations, rights, and responsibilities during their entire working relationship on this project. Among issues that need to be considered are:

- How will preconstruction costs be distributed and how will payment be made when the project goes to construction or at various stages in the process?
- If the contract with the owner is a guaranteed maximum price (GMP) contract with a savings clause, how will savings be split among team members?
- How will insurance requirements be allocated, and if there are any uncollected claims or partially collected claims, who will be responsible for the uncollected portion?
- What will be the impact of any escalation of costs during construction; which party(s) will be responsible for these added costs?
- If liquidated damages are included in the owner contract, how will the design-build team deal with them?

These are obviously not all of the topics to be included in a teaming agreement, but are meant to present the complexity of such an agreement, where considerable thought in its preparation may prevent serious disagreements once the project is under way.

The Associated General Contractors of America (AGC) has created a teaming agreement, AGC Document No. 499 (Fig. 3.1). It includes the following provisions:

- Team relationships and responsibilities
- A noncompete clause that prohibits any team member from acting in an independent capacity with the owner
- The need for all team members to prepare a statement of qualifications when requested by the owner
- A confidential agreement between team members preventing confidential matter from being disclosed to third parties
- The right of ownership of design-build documents and passage of title to those documents
- Contractually forming the team upon award by owner

THE ASSOCIATED GENERAL CONTRACTORS OF AMERICA

AGC DOCUMENT NO. 499
STANDARD FORM OF TEAMING AGREEMENT FOR DESIGN-BUILD PROJECT

This Agreement is made this ________________ day of ____________________ in the year __________, ◆

by and between

TEAM LEADER ____ ◆
(Name and Address)

and **TEAM MEMBER** ____ ◆
(Name and Address)

and **TEAM MEMBER** (if applicable) ______________________________________ ◆
(Name and Address)

and **TEAM MEMBER** (if applicable) ______________________________________ ◆
(Name and Address)

the parties collectively referred to as the **TEAM** for services in connection with the following **PROJECT**

____ ◆
(Name, Location and Brief Description)

for **OWNER** ____ ◆
(Name and Address)

1

AGC DOCUMENT NO. 499 • STANDARD FORM OF TEAMING AGREEMENT FOR DESIGN-BUILD PROJECT

Figure 3.1 AGC Document No. 499—Standard form of teaming agreement for design-build projects. (*Source: All materials are displayed or reproduced with the express written permission of the Associated General Contractors of America under License No. 0105.*)

ARTICLE 1

TEAM RELATIONSHIP AND RESPONSIBILITIES

1.1 This Agreement shall define the respective responsibilities of the Team Members for the preparation of responses to the Owner's request for qualifications and request for proposals for the Project. Each Team Member agrees to proceed with this Agreement on the basis of mutual trust, good faith and fair dealing and to use its best efforts in the preparation of the statement of qualifications and proposal for the Project, as required by the Owner, and any contract arising from the proposal.

1.2 The Team Leader, ______________________ ♦
______________________,
shall provide overall direction and leadership for the Team and be the conduit for all communication with the Owner. In addition the Team Leader shall provide expertise in the areas of (a) construction management and construction; (b) the procurement of equipment, materials and supplies; (c) the coordination and tracking of equipment and materials shipping and receiving; (d) construction scheduling, budgeting and materials tracking; and (e) administrative support. The Team Leader's representative shall be: ____________ ♦
______________________.

1.3 The principal design professional is Team Member,
______________________ ♦
who shall perform the following design and engineering services required for the Project: ____________ ♦
____
____
____
____
____
____

In addition this Team Member shall coordinate the design activities of the remaining design professionals, if any. This Team Member's representative shall be: ____________ ♦
____
____
______________________.

1.4 Team Member, ______________________ ♦
____
shall provide expertise in the following areas:
______________________ ♦
____
____
____
____
____
____
______________________.

This Team Member's representative shall be: ____________ ♦
____
______________________.

1.5 Team Member, ______________________ ♦
______________________,
shall provide expertise in the following areas:
______________________ ♦
____
____
____
____
____
____
______________________.

This Team Member's representative shall be: ____________ ♦
____
______________________.

1.6 Each Team Member shall be responsible for its own costs and expenses incurred in the preparation of materials for the statement of qualifications and the proposal and in the negotiation of any contracts arising from the proposal, except as specifically described herein:
______________________ ♦
____
____
____
____
____
____
____
____
____
____
____
______________________.

Any stipends provided by the Owner to the Team shall be shared on the following basis:
______________________ ♦
____
____
____
____
____
____
____
____
______________________.

1.7 **EXCLUSIVITY** No Team Member shall participate in Owner's selection process except as a member of the Team, or participate in the submission of a competing statement of qualifications or proposal, except as otherwise mutually agreed by all Team Members.

Figure 3.1 *(Continued)*

ARTICLE 2

STATEMENT OF QUALIFICATIONS AND PROPOSAL

2.1 The Team Members shall use their best efforts to prepare a statement of qualifications in response to the request of the Owner. Each Team Member shall submit to the Team Leader appropriate data and information concerning its area or areas of professional expertise. Each Team Member shall make available appropriate and qualified personnel to work on its portion of the statement of qualifications in the time frame proscribed, and shall provide reasonable assistance to the Team Leader in preparation of the statement of qualifications.

2.2 The Team Leader shall integrate the information provided by the Team Members, prepare the statement of qualifications and submit it to the Owner. The Team Leader has responsibility for the form and content of the statement of qualifications and agrees to consult with each Team Member, before submission to the Owner, on all matters concerning such Team Member's area of professional expertise. The Team Leader shall represent accurately the qualifications and professional expertise of each Team Member as stated in the submitted materials.

2.3 If requested by the Owner, the Team Members shall prepare and submit a proposal for the Project to the Owner. Each Team Member shall support the Team Leader with a level of effort and personnel, licensed as required by law, sufficient to complete and submit the proposal in the time frame allowed by the Owner. A clear and concise statement of the division of responsibilities between the Team Members will be prepared by the Team Leader. The Team Leader shall make all final determinations as to the form and content of the proposal. The Team Leader shall use its best efforts, after the Team has qualified for the Project, to obtain the contract award, and each Team Member shall assist in such efforts as the Team Leader may reasonably request.

ARTICLE 3

CONFIDENTIAL INFORMATION

3.1 The Team Members may receive from one another Confidential Information, including proprietary information, as is necessary to prepare the statement of qualifications and the proposal. Confidential Information shall be designated as such in writing by the Team Member supplying such information. If required by the Team Member supplying the Confidential Information, a Team Member receiving such information shall execute an appropriate confidentiality agreement. A Team Member receiving Confidential Information shall not use such information or disclose it to third parties except as is consistent with the terms of any executed confidentiality agreement and for the purposes of preparing the statement of qualifications, the proposal and in performing any contract awarded to the Team as a result of the proposal, or as required by law. Unless otherwise provided by the terms of an executed confidentiality agreement, if a contract is not awarded to the Team or upon the termination or completion of any contract awarded to the Team, each Team Member will return any Confidential Information supplied to it.

ARTICLE 4

OWNERSHIP OF DOCUMENTS

4.1 Each Team Member shall retain ownership of property rights, including copyrights, to all documents, drawings, specifications, electronic data and information prepared, provided or procured by it in furtherance of this Agreement or any contract awarded as a result of a successful proposal. In the event the Owner chooses to award a contract to the Team Leader on the condition that a Team Member not be involved in the Project, that Team Member shall transfer in writing to the Team Leader, upon the payment of an amount to be negotiated by the parties in good faith, ownership of the property rights, except copyright, of all documents, drawings, specifications, electronic data and information prepared, provided or procured by the Team Member pursuant to this Agreement and shall grant to the Team Leader a license for this Project alone, in accordance with Paragraph 4.2.

4.2 The Team Leader may use, reproduce and make derivative works from such documents in the performance of any contract. The Team Leader's use of such documents shall be at the Team Leader's sole risk, except that the Team Member shall be obligated to indemnify the Team Leader for any claims of royalty, patent or copyright infringement arising out of the selection of any patented or copyrighted materials, methods or systems by the Team Member.

ARTICLE 5

POST AWARD CONSIDERATIONS

5.1 Following notice from the Owner that the Team has been awarded a contract, the Team Leader shall prepare and submit to the Team Members a proposal for a Project-specific agreement of association among them. (Such agreement may take the form of a design-builder/subcontractor agreement, a joint venture agreement, a limited partnership agreement or an operating agreement for a limited liability company.) The Team Members shall negotiate in good faith such Project-specific agreement of association so

3

AGC DOCUMENT NO. 499 • STANDARD FORM OF TEAMING AGREEMENT FOR DESIGN-BUILD PROJECT

Figure 3.1 *(Continued)*

that a written agreement may be executed by the Team Members on a schedule as determined by the Team Leader or by the Owner, if required by the request for proposal. The Team Leader shall use its best efforts, with the cooperation of all Team Members, to negotiate and achieve a written contract with the Owner for the Project.

ARTICLE 6

OTHER PROVISIONS ♦

This Agreement is entered into as of the date set forth above.

WITNESS: TEAM LEADER: ____________ ♦

____________ ♦ BY: ____________ ♦

PRINT NAME: ____________ ♦

PRINT TITLE: ____________ ♦

WITNESS: TEAM MEMBER: ____________ ♦

____________ ♦ BY: ____________ ♦

PRINT NAME: ____________ ♦

PRINT TITLE: ____________ ♦

WITNESS: TEAM MEMBER: ____________ ♦

____________ ♦ BY: ____________ ♦

PRINT NAME: ____________ ♦

PRINT TITLE: ____________ ♦

WITNESS: TEAM MEMBER: ____________ ♦

____________ ♦ BY: ____________ ♦

PRINT NAME: ____________ ♦

PRINT TITLE: ____________ ♦

12/01
4

AGC DOCUMENT NO. 499 • STANDARD FORM OF TEAMING AGREEMENT FOR DESIGN-BUILD PROJECT

Figure 3.1 *(Continued)*

The limited liability corporation

The limited liability corporation (LLC), is another legal entity that can be formed to create an entity between design consultants and a contractor looking to do design-build work. This LLC offers the liability protection of a corporation and exists as a separate and distinct entity, usually created for one purpose only, such as a specific design-build project. LLCs are usually one-off deals and a different LLC will be used for any further design-build projects involving either a different owner, different builder, or different design consultants. When establishing an LLC, an application is usually required to be filed with the secretary of state in the state where the LLC will be operating. Articles of organization are required to be submitted with the application, and there is a fee associated with the filing. Some states require an operating agreement, similar to corporate bylaws or partnership agreements to be filed along with an application. Some states require public notification that an LLC is being formed.

The advantages of an LLC, apart from its limited liability feature, are

- Unlike a regular corporation, no formal meetings are required, and therefore no minutes of meetings are necessary.
- No corporate resolutions are needed.
- The distribution of profits can be tailored as required.
- All business profits, losses, and expenses flow through the corporation to the individual members of the corporation avoiding the double taxation of paying corporate and individual taxes on money earned.

The disadvantages of the LLC are

- The LLC is dissolved when a member dies or undergoes bankruptcy whereas a conventional corporation can live forever.
- Because of the nature of an LLC, lending institutions are reluctant to provide funds without personal guarantees from its officers.
- Owners of projects may be reluctant to do business with an LLC because they recognize its single-subject nature.

The S corporation

The Internal Revenue Service must first rule on the acceptability of a corporation to meet the S corporation requirements. This "S" status allows the taxation of a company to be similar to that of a partnership or a sole proprietor as opposed to a corporation. The profits and losses of an S corporation pass through the corporation onto the owner(s) personal income tax, thereby avoiding double taxation and allowing any losses to be deducted from other income streams of the owner(s). In the early years of a business, when start-up expenses can be considerable, with the S corporation, these expenses are directly deductible on the owner(s) personal income tax. These are some of this business entity's advantages.

There are several disadvantages of an S corporation:

- S corporation officials can be held personally liable for some of their actions.
- Only one class of stock can be issued, so there is less control over the business.
- It is less attractive to outside investors who may not like the pass-through tax setups afforded by this type of business entity.
- The corporation can have no more than 75 shareholders (this is probably not a problem for many design-build ventures).
- This entity is a corporation and, as such, must conduct regular meetings and maintain company minutes of those meetings.
- Shareholders must be U.S. residents.

The partnership

A partnership can consist of two or more members who prepare a document listing the rights, responsibilities, and obligations of each partner. There is no legal protection against claims and each partner becomes liable for any claims or legal action placed against the partnership. Profits and losses from a partnership accrue directly to each partner in percentages as specified in the partnership agreement. Partnership insurance to cover any claims against it is available but somewhat costly.

The corporation

This business entity is familiar to most business people and is an individual in the eyes of the law. The legal makeup of a corporation and the various advantages and disadvantages of all of these forms of business entities are best discussed with legal and accounting professionals.

Architect- or Contractor-Led Team?

When creating a new design-build team, the question arises who is best suited to be the leader—architect, engineer, or contractor? There are many answers to this question. Possibly the first consideration would be which firm or company developed the lead through their marketing efforts or had been approached by a previous client to present a design-build proposal for that new project under consideration. There is also the issue of whether the formulation of the design-build entity will be created in-house, will rely on collaboration between designer and contractor, or will rely on subcontractors. Obviously the former may require significant changes in the A/E or the general contractor's present structure. Leaving that aside, let's explore the role of contractor as leader in this endeavor.

Contractor as team leader

A contractor may decide to hire full-time design professionals thereby integrating their company into a full-service design-build firm. This entails adding substantial overhead, and the contractor will need to continue to pursue design-build work as part of their overall sales development program, probably by creating a new position—Director of Design-Build Operations.

More small to midsized builders, such the Forrester Construction Corporation mentioned in the previous chapter, will seek an alternative route, hiring design consultants as subcontractors.

The predominance of contractors as team leaders probably derives from the nature of the building business—having significant financial resources and substantial lines of credit is an essential element of the contracting business. Contractors historically have reliable sources for insurance and bonds; the former being essential in design-bid-build projects; the latter essential in public works projects, but often not required for private sector work.

There are other reasons that validate the contractor as leader—experience in dealing with the complexities of the construction process and the relationship with specialty contractors and suppliers of materials and equipment.

Control over scheduling and costs are everyday occurrences for contractors and their people are acutely aware of the results of poor control over scheduling and costs.

Contractors rely on their estimating department for hard bid information, and they also rely on their long-term relationships with specialty contractors and vendors to provide budget estimates. With the abundance of estimating software available in the marketplace providing the contractor with more rapid takeoffs and a computerized database of costs, they seem well suited to provide one of the more essential ingredients for the design-build team—conceptual and final project costs.

As the name implies, a contractor deals with contracts—contracts with owners, contracts with subcontractors, contracts with materials and equipment suppliers, and over the years they have distilled the salient points of each type of agreement into what works and what doesn't.

The contractor maintains a strong nucleus of field supervisors who not only are technically proficient, but who have developed the management skills necessary to orchestrate the complexities of the construction process.

The contractor, with a history of established business relationships with a varied cadre of subcontractors and vendors finds these specialty contractors and vendors more readily available to provide valuable design and cost information during the design-development phase of a project. During construction those favorable relationships with vendors and subcontractors will pay off by affording the design-build team with very competitive pricing.

Contractors considering design-build must have the ability to conceptualize, a process that design consultants may find easier to do. Contractors with a history of negotiated work will find the transition to design-build easier than those coming from a hard bid background.

Contractor as prime contractor, architect as subcontractor

A contractor will have developed considerable experience dealing with subcontractors in the course of their normal working environment, and they will have had considerable experience working with architects and engineers on design-bid-build projects as well, so this subcontracting of design consultants in a design-build program will strike many familiar chords.

This process eliminates many of the liability issues and dangers faced by architects in design-led teams, and provides both designer and builder the experience of working in a familiar environment while entering into a new field. However, the potential for conflicts may occur between architect and builder when design considerations run into the stone wall of the budget. These new roles can only partially be covered by contract language. The contractor must be able to discern the designer's concern over architectural integrity of the project, and the designer must consider the cost structure and be willing to explore more cost-effective ways to achieve their design. The concern of the owner's best interests must be addressed by all parties if this design-build venture is to become a stepping stone to more projects down the road.

If the architect has a subcontract agreement with the contractor, many of the concerns about explicit-design criteria, redesign responsibilities, inspections and other roles during preconstruction, construction, and postconstruction along with fees and reimbursable costs can be negotiated into that subcontract agreement.

Architect as team leader

Owners often have close and trusting relationships with architects, who on previous projects have demonstrated their ability to control costs, scrutinize contractor requests for change orders and generally protect owner's interests. The institutional concept of the architect's role as defender of the owner and the contractor's perceived primary goal of making a profit are sometimes hard to dismiss, and owners may feel more comfortable having the A/E firm in charge.

An architect's proven track record of expertise in designing specific types of projects that provide both outstanding design and functionality can be the rationale behind selecting an A/E firm as the team leader. Architects who have had extensive experience in providing construction services can possibly use as field supervisors the personnel who designed the project and hence best know the plans and specifications. An architect-led design-build team can be formed in one of several ways.

The integrated firm. In 1978, the American Institute of Architects (AIA) lifted its ban on design-build, but it was not until 1985 that the AIA published the first edition of its design-build documents. In fact in both the 1985 and later in the 1996 version of these design-build documents, the AIA wrestled with three concerns relating to design-build in general:

1. *To whom does an architect owe their allegiance?* When dealing with the interests of an owner or the design-build firm, the documents committee decided to include a specific disclosure statement to the effect that the architect's services were being performed in the interest of the design-builder. In reality, if an owner's interests are blatantly disregarded, the A/E and/or design-build firm will have difficulty in obtaining future work.

2. *How can the AIA documents facilitate design-build when the architect is the leader?* AIA created a design-builder/contractor agreement that would cover the issues of leadership and the role the leader will play.
3. *How to assure that these design-build documents will ensure that the owner receives the proper design and a fair price for the work?* The two-part agreement was devised whereby an owner could cease any further relationship with the design-build team if the preliminary design phase in Part 1 was not to their satisfaction.

Design consultants wishing to embark on this route will need considerable resources in capital, manpower, equipment, hardware, and software power. Assuming that the integrated firm will not subcontract construction services, investments will need to be made in personnel—PMs, estimating staff, and field supervisors. While offering these construction experts a "ground floor" opportunity, experienced qualified PMs and project superintendents command salaries in the $75,000 to $125,000 range plus a surfeit of fringe benefits–pension plans, 401Ks, yearly bonuses, and the like. Estimators will also require salaries in close parity to the PMs and superintendent supervisors.

Engineering News Record in its January 17, 2005 edition listed some median construction executive salaries:

Vice president—operations	$123,725
Vice president—business development	$122,500
Vice president—estimating	$113,200
Operations manager	$110,500
IT/MIS manager	$96,250
General superintendent	$95,000

It is of critical importance that the database of construction costs be based on real-time experience, and not on generic databases offered by a wide variety of companies that require geographic and inflation factor upgrading to suit local conditions. One major strength of the design-build concept is its ability to track costs during conceptual or design-development stages, so the acquisition of local cost data will be a daunting task.

The architect as team leader must consider the substantially increased financial responsibilities that will accompany this role.

Most contractor-subcontractor agreements contain a "Paid when Pay clause," which limits the responsibility of the contractor to pay the subcontractor within a short period of time after receiving payment from the owner. Several states have banned this practice in the private sector as well as in the public works projects. Some owners are notoriously late with their payments for a number of reasons, and some subcontractors and vendors may demand payment even though the owner is late in their payment to the general contractor. Placing orders for special or custom materials or equipment frequently requires a down payment and a significant line of credit comes in handy when these events occur.

Not only all costs associated with the design, but also upfront costs such as insurance and bonding, and even building permits that can run into six figures must be considered. This integration concept may afford the design consultant utmost control over the project, both during design and during construction, but it also carries with it also a great deal of new responsibilities, many of which are financial in nature.

When considering whether or not to form an integrated design-build company, an architect will have other business and legal matters to consider:

1. Added risks include insurance coverage for faulty or defective work in the architect's new role as builder. Some states have laws on their books relating to latent defects and structural failures with a prescribed statute of limi-tations.
2. The architect will be liable for any accidents or Occupational Safety and Health Administration (OSHA) violations/fines during the construction process and will require additional insurance coverage in that respect.
3. In case of cost overruns not attributable to justifiable increases in contract cost, the architect may experience a diminution of fee.
4. If a subcontractor or vendor defaults on their contract or declares bankruptcy, the architect may have to engage another subcontractor, often at significantly higher costs to the project.

Architect as prime contractor, builder as subcontractor

A straightforward approach to an architect-led design-build team can be achieved by engaging a builder to perform specific tasks in return for a fee. This is a very common approach in custom home or high-end residential construction projects where an owner hires the design firm based upon design considerations and trusts the architect to contract with a home builder having the credentials to build the residence using the best of materials and workmanship.

The AIA recognized the need for such a contract between architect and builder, and their Document A491—Standard Form of Agreements between Design/Builder and Contractor, provides a good base for such an arrangement. This contract format is in two parts—Part 1, Agreement, covers the contractor's services during the design-development stage of the project, and Part 2 covers the contractor's services during construction.

Although this form of a two-part contract has been superseded by the new AIA A141 contract form issued by AIA in 2004, it is still a valid, some say preferred method of entering into an owner/design-builder contract.

Part 1 requires the contractor to provide a preliminary evaluation of the owner's program, advise on the selection of materials and constructability issues, prepare the schedule, and provide preliminary estimates and detailed estimates. The contractor is responsible for preparing a fixed price or GMP proposal for the cost of construction that could become the basis for a contract for construction, if requested by the design-builder.

Part 2 of AIA A491 is basically a standard contract for construction requiring the contractor to provide all labor, materials, and equipment to complete the work as outlined in the contract documents—the plans and specifications. The contractor is also obligated to continue to provide the services included in Part 1, Agreement, such as updating and refining the detailed cost estimate.

Architect as participant in a JV or an LLC with a contractor on a project-specific basis

Architects may wish to join forces with a contractor of choice on a specific private or public works project and participate as a design-builder while still retaining their core business of architecture. A contact from a valued client wanting to embark on a new construction project might present an opportunity to test the waters of design-build to determine whether this project delivery system has potential for increased volume and profit.

The JV or LLC are two options available to the design firm in this respect.

One basic constituent in a JV or LLC must be the division of responsibilities and this can be best expressed in a teaming agreement similar to the one shown in Fig. 3.1 or the one included at the end of this chapter.

And even if a teaming agreement includes provisions for cash flow, the architect will undoubtedly require more professional accounting services to handle this increase and intensity of cash flow activity.

In the lead role, the architect will have the most direct contact with the owner and must be able to deal with contractual issues and financial concerns that may be far removed from their regular field of architecture, whereas the contractor may be more familiar with these events which are a day-to-day occurrence in the construction industry. On the other hand, an architect's share of profit may be considerably higher if they assume the role of prime contractor rather than the role of subcontractor to a builder.

Initially setting aside issues of contractor licensing, bonding, and insurance, which can be overcome by the creation of the legal design-build entity (joint venturing with a builder), developing a design-build capability within the architect's firm can be explored with a contractor(s) with whom the firm has had prior positive dealings.

The Collaborative Approach

One of the strengths of the design-build process is the collaborative nature of the endeavor—teaming the design concept to a database of costs to meet the goal of the owner's program from an aesthetic, functional, operational, and cost basis. Pairing a designer with a contractor's real-world database of costs would appear to fulfill two key concerns of an owner: tracking design development with realistic costs, and avoiding the need to redesign with concurrent costs and delay implications.

By bringing selected specialty contractors into the design-budgeting process, an even further refinement of the project budget can be effected. Constructability issues, always lurking at critical stages of construction, can also be minimized or eliminated entirely with the input from experienced vendors and subcontractors.

An experienced team of designers and construction people contributing "what works and what doesn't work" will prove invaluable in the successful completion of the project with either architect or contractor in the lead.

Other Essential Elements of the Design-Build Team

Whether a contractor- or architect-led entity, JV, or prime/subcontractor arrangement, this collaborative team effort will require two additional components in the program— a comprehensive safety program and a quality-control program.

The design-build safety program

A formalized safety program is necessary for many reasons:

- To provide for a safe working environment to protect against injuries, fatalities, and damage to property at the construction site.
- To provide guidelines to disseminate safety information to comply with various state and federal agencies including OSHA and the Environmental Protection Agency (EPA), and to develop inspection procedures to ensure compliance with those agency's requirements.
- To avoid monetary penalties imposed due to poor safety performance, fines by enforcement agencies, lawsuits filed by families of injured or deceased workers, and increases in insurance premiums including worker compensation insurance.

Construction sites are dangerous places in which to work. Although construction workers make up 6.6% of the entire U.S. workforce, they account for about 19½ % of all workplace deaths, a figure that has exceeded 1000 annually since 1994.

OSHA inspections, on a federal level or state cloned, are made on an ad hoc basis or upon receipt of information about safety violations at the jobsite, and always after a fatality occurs. According to OSHA the four leading causes of fatal injuries are:

1. Falls from elevated areas
2. Impact by an object or machine
3. Entrapment between objects (such as by machine moving through tight quarters)
4. Electrical hazards (20% of all reported violations!)

Developing a safety program encompasses the following components:

- *A statement of company policy:*

  The ——— Company recognizes that accident prevention is a problem of organization and education which can and must be administered to avoid pain and suffering to our employees and also reduce lost time and operating costs incurred by our company. According I state and pledge my full support to the commitment of a Safety Program. Signed: ________________ (Principal or CEO)

- *The objective of an accident prevention program:*
  1. Planning all work to minimize losses due to detection and correction of unsafe practices and conditions
  2. Maintaining a system for prompt detection and correction of unsafe practices and conditions
  3. Making available and enforcing the use of personal protection equipment, physical and mechanical guards

4. Maintaining an effective system of tool and equipment inspections and maintenance
5. Establishing an educational program to instruct all participants in the basics of accident control and prevention by instituting:
   a. New employee orientation training
   b. Periodic safety meetings
   c. Use and distribution of safety bulletins and related materials
   d. Instruction in the proper and prompt reporting of all accidents and a system for immediate investigation to determine the cause of the accident and taking steps to prevent occurrence

- *The appointment, duties, and responsibilities of a safety director or safety coordinator.* The duties of a safety director/safety coordinator would include:
  1. Responsibility for coordinating and monitoring the accident prevention program
     a. Overseeing accident investigations. All accident investigations involving serious injuries or those that could have resulted in serious accidents will be investigated by the safety director
     b. Overseeing the proper use of safety equipment
     c. Performing frequent and unannounced jobsite safety inspections
     d. Attending and participating in regular safety meetings
  2. Continual review of job safety reports and preparation and dissemination of monthly summaries of safety violations, field inspections, and general program administration items
  3. Immediate documentation of critical conditions and steps to be taken, and by whom, to correct these conditions
  4. Maintaining liaison with insurance carriers regarding accident prevention problems
  5. Reviewing and taking action, as required, on all safety program violators
- *Responsibilities of field supervisors in administering the program and their relationship with the safety director / safety coordinator*
  1. Field supervisors are the first line of defense in accident prevention and they must develop, as part of their daily routine, a method of communicating the accident and safety program to everyone on the site—both their own employees and those of their subcontractors. During their daily tours of the site, safety conditions must be observed and any unsafe conditions corrected immediately.
  2. *Tool box talks.* Those short, 15- to 20-minutes weekly meetings in which one or two safety-related topics are discussed will be the responsibility of the field supervisor or their designated appointee.
- *Procedures for reporting job-related injuries and illnesses.* All job-related accidents and injuries are to be reported, with emphasis on ALL. Two types of accident reporting forms are required—one required for the company's safety director and one required to comply with OSHA reporting requirements. Copies of both forms are to be in the field office and completed not later than the end of the day of the injury or illness.
- *Working rules and regulations of the safety program.* This is the "nuts and bolts" of the safety program outlining the specific items of personal protection

equipment for all general and specialized operations, the training and use of power actuated tools, safe electrical extension cords, temporary electrical connections requirements, compliance with OSHA scaffolding and trenching requirements, the need for a competent person, and so forth.

- *A hazard communication (HazCOM) program as required by OSHA and most local or state government agencies.* HazCom, another OSHA requirement deals with hazardous materials, either en route to the site, stored on site, or incorporated in a construction material or component. Manufacturers of materials deemed to contain hazardous materials are required, by law, to prepare a material safety data sheet (MSDS) that describes its hazardous nature, proper handling and storage instructions, and, in case of contact or ingestion by a worker, the necessary first aid and/or medical procedures to follow. These MSDS sheets must be on site prior to the arrival of the product which they cover.
- *Procedures for dealing with safety violations and violators.* A model policy needs to be established for dealing with violations. A first warning is usually an oral one accompanied by a written safety violation notation to be placed in the offender's personnel file. A second warning will result in the offender being given another verbal warning and another written notification to be placed in their personnel file. They will be required to meet with the safety director for counseling. The third warning to someone committing another violation will consist of a written warning, and a requirement to meet with the safety director and a member of top management to determine why the employee continues to violate safety rules. Depending upon company policy that might state "Employees who accumulate three warnings in a 12-month period may be suspended from work, without pay, for up to one week" or the program may include other disciplinary action. A fourth violation will also be a cause for a written notification to be placed in the worker's file and may, at the company's discretion, be a cause for dismissal.

OSHA has several reporting requirements that are often given less than full attention by contractors and these violations can result in the levying of fines. The five most frequently cited reporting and paperwork violations are:

1. Failure to provide the log and summary of occupational injuries and illnesses
2. Failure to adhere to the general duty clause of the OSHA act (a citation based on no specific violation, or a citation issued after a previous one has been ignored)
3. Failure to report a fatality or multiple hospitalization incidents
4. Failure to record occupational injuries and illnesses on the supplementary record form
5. Failure to record and report occupational injuries and illnesses on the required OSHA log form

There are a number of safety consultants who can not only develop a safety program but also can monitor it. Some companies prefer to have these safety firms prepare the safety program, conduct an initial-training session, monitor the project intermittently, and be available for consultation, when required. The ongoing monitoring of the program can be achieved by appointing a safety director.

Whichever method is used by the design-build entity, close attention to safety on the jobsite is both an economic and a moral imperative.

The Quest for Quality

Quality control, the quality standards imposed by the project plans and specifications, and quality assurance, those inspections to ensure or "assure" that these quality standards are being met, play an important role in the design-build process. One of an owner's greatest concerns in the design-build process is the apparent lack of a "gatekeeper" to ensure that they are receiving a quality project. In the more conventional design-bid-build-type project, the architect, as the owner's agent, is their assurance. When a construction manager or owner's representative is on board, quality issues will be part of their responsibilities. But in either, rather, in any case, the design-build team needs to pay more than lip service to quality. The collective knowledge of the designers and the contractors, working together during conceptual development, possibly enhanced by specialty contractor input, should by its very nature reduce or eliminate "constructability issues."

The design-build design development process generally includes a guarantee of performance, particularly in the electrical, mechanical, and plumbing design thereby motivating the design-build team to meet or exceed the standards required by the owner of a project.

The level of quality is often a function of the complexity of the project and procedures for including quality in the design can be outlined in the very beginning of that design. These quality expectations must then be communicated during design-development meetings attended by architect, engineer, builder, subcontractors, and material and equipment vendors. At these meetings commitments to quality levels can be obtained, and any suggestions, improvements, or recommendations viewed, considered, and acted upon. By its very nature, these quality issues can be incorporated into the design once they have been reviewed and accepted by the appropriate parties. The quality-assurance aspect, carried out during applicable shop drawing reviews, and inspections by the design consultants during construction should provide an owner with a high comfort level.

The concept of total quality management (TQM) is one that can also be rather easily incorporated into the design-build process. There is quality of design and quality of construction, but there are other quality issues that relate to TQM:

- An orderly review, resolution, and execution of the design-build contract
- Well-organized and disciplined meetings between the design-build team and the owner during the project's genesis to complete the required work in an orderly fashion quickly recognizing the fact that everyone's time is valuable

- Conduct organized meetings during construction where questions are received, reviewed, and responded to quickly
- Prepare monthly requisitions to the owner, on time, and with sufficient documentation that will allow their prompt review
- When the requirement for change orders arises, furnish the owner with sufficient detailed information to allow for a complete understanding of the nature of the change and its related cost
- Approach all close-out procedures promptly, professionally to complete the project in a timely manner, including commissioning

(Quite often the fine performance during construction is forgotten entirely by an owner when close-out procedures drag on and appear to be haphazard and incomplete.)

Is It All Worth It?

Zweig White, the well-respected management consulting firm headquartered in Chicago, provides business insight and expertise for the architecture, engineering, and construction industries, which they say encompass 700,000 firms employing 7 million people and generating $850 billion in revenue.

Laura Rothman, an executive at Zweig White Research in Natick, Massachusetts, pointed to their *2005 Design/Build Survey of Design and Construction Firms*, which asked the question, "Are design-build projects more or less profitable than those projects completed using traditional project delivery methods?"

The results of their survey were:

> The vast majority of firm leaders (84%) believe that design-build projects are more profitable than traditional projects. One of the most common reasons firms cited for this belief was that design-build projects allow the builder more control over the entire process.
>
> Firms that consider design-build projects to be less profitable than traditional ones did so because of higher costs and increased risk.
>
> It would appear from looking at these results that the 16% that incurred higher costs and possibly higher risks were also firms where the necessary controls to reign in costs and risks were not what they should be.

The Zweig White executive summary for this study includes a breakdown of the survey sample (Fig. 3.2), mainly integrated design-build (43% of respondents), growth projections (Fig. 3.3) where the vast majority predicted growth in the next five years, and design-build profitability (Fig. 3.4) where the 84% majority indicated more profitability.

Breakdown of the survey sample

Sample size

98 design and construction firms completed and returned a valid questionnaire.

Firm type

Integrated design/build	43%
Design services/consulting	27%
Construction	31%

Year founded

Prior to 1945	29%
1945 - 1959	16%
1960 - 1969	11%
1970 - 1979	9%
1980 - 1989	15%
1990 - 1999	14%
2000 to present	3%
Unspecified	2%

Region of headquarters

New England	6%
Middle Atlantic	10%
South Atlantic	18%
North Central	28%
South Central	6%
Mountain	12%
Pacific	17%
Unspecified	2%

Staff size

Minimum	2
Lower Quartile	30
Median	150
Mean	1,392
Upper Quartile	650
Maximum	35,000

Staff size *(breakdown)*

1 - 49	33%
50 - 99	11%
100 - 249	14%
250 - 499	13%
500 - 999	5%
1,000 +	20%
Unspecified	3%

2004 gross revenue

Minimum	$720,000
Lower Quartile	$10,000,000
Median	$50,000,000
Mean	$321,818,352
Upper Quartile	$220,000,000
Maximum	$5,000,000,000

Note: due to rounding, percentages for some questions do not total 100.

Breakdown of the survey sample *(continued)*

Legal form of ownership

Private Corp. (S-Corp.)	42%
Private Corp. (C-Corp.)	31%
LLC or LLP	8%
Professional Corp. (PC)	5%
Public Corp.	4%
Partnership	3%
Proprietorship	1%
Unspecified	6%

Client base

More than 50% private	58%
50% or less private	40%
Unspecified	2%

Growth rate

Fast growth	13%
Slow growth	55%
Stable	22%
Decline	8%
Unspecified	1%

Are 10% or more of your firm's projects completed using design/build as the project delivery method?

Yes	83%
No	17%

Note: due to rounding, percentages for some questions do not total 100.

Figure 3.2 Zweig White breakdown of survey sample. (*Source: 2005 Design/Build Survey of Design and Construction Firms. Copyright 2005, Zweig White Information Services, LLC.*)

Growth projections

Issues

Do firms think there will be an overall increase in the use of design/build in the next five years?

Do firms think there will be an increase in the use of design/build in the public sector within the next five years?

Background

Whether or not they project increased amounts of design/build work for their own firms, we wanted to know if firm leaders thought there would be an overall increase in the use of design/build in the industry.

As discussed in Chapter 2, procurement laws in some states can effectively shut firms out of acquiring public-sector design/build work. Knowing this, we wanted to find out whether firms believed there would be an increase in the use of design/build in the public sector within the next five years.

Survey Findings

- The vast majority of firms predict an overall industry increase in the use of design/build in the next five years.
- Nearly all firms also believe there will be an increase in the use of design/build in the public sector within the next five years.
- Firms that project an increase in public-sector design/build work cite cost control or savings, single-source responsibility, and time efficiency among the reasons for their predictions. Those that disagree cite procurement laws and decreasing quality among the reasons for no increase.

Growth projections

	Integrated design/build	Design services/ consulting	Construction
In the next five years, do you think the use of design/build as a project delivery method will... ?			
Increase	98%	65%	83%
Remain the same	2%	31%	17%
Decrease	0%	4%	0%
Do you think that there will be an increase in the use of design/build in the public sector within the next five years?			
Yes	95%	73%	93%
No	2%	23%	7%
Unspecified	2%	4%	0%

Note: due to rounding, percentages for some questions do not total 100.

Figure 3.3 Zweig White design-build growth projections. (*Source: 2005 Design/Build Survey of Design and Construction Firms. Copyright 2005, Zweig White Information Services, LLC.*)

Design/build profitability

In your opinion, are design/build projects in general more or less profitable than projects completed using "traditional" project delivery methods?

More profitable 84%

Less profitable 12%

Unspecified 4%

Design/build profitability

The vast majority of firm leaders believe design/build projects are more profitable than traditional projects.

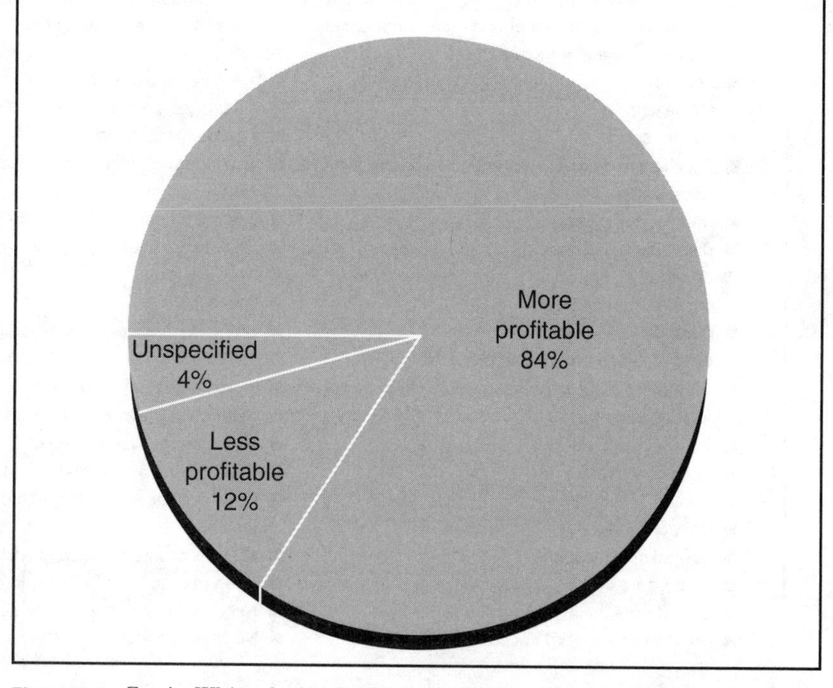

Figure 3.4 Zweig White design-build profitability. (*Source: 2005 Design/Build Survey of Design and Construction Firms. Copyright 2005, Zweig White Information Services, LLC.*)

Design/build profitability *(continued)*

Why are design/build projects in general *more* profitable than projects completed using "traditional" project delivery methods?

- Ability to directly manage all costs and to plan and design to cost
- Able to control costs more
- Assuming owner/regulators don't hamstring design/build on project submittal requirements
- Assumption of greater risk if managed properly should produce greater reward
- Because of "team effort," fewer documents are used
- Best value selection
- Better control (2)
- Better coordination between A/E and contractor makes project more efficient and therefore more profitable
- Better defined scope and more efficiencies provide more value to owner and more room under budget for higher contractor fees
- Better risk control when true "team" relationship exists
- Budget control
- But not enough yet— market still learning the value
- Can control schedule
- Client is buying value
- Constant communication with owner prevents out-of-budget changes
- Control
- Control added early in the process
- Control is in hands of those who can keep overall project on track
- Control materials
- Control of cost
- Control schedule
- Control scope
- Control specifications
- Cost savings
- Cost-control ability, but more risk
- Costs are controlled continuously
- Design firms paid for ideas, not hours
- Design/build firm has opportunity to meet scope more efficiently with right design partners
- Faster delivery speed
- Flexibility in design solutions with cost savings and still maintain quality
- For both design/build team and owner provides increased time savings, reduced claims, improved community relations
- For integrated firms— control; client satisfaction = repeat business = profit
- For our firm, we are entering the construction side of the equation, which we didn't do before
- Full control of the project
- Higher design and construction fee
- Higher fee
- Higher profits for partners by adjusting project costs to accommodate profit
- Higher risk
- If estimated properly
- If managed properly, more profit can be made
- If properly managed
- If you learn to identify, quantify, and manage risk
- It is not (usually) purely fee driven
- Less competition (3)
- Less litigation
- Less time
- Longer-term customer relationship
- More control (2)
- More control during project cycle
- More control equals more cost/profit control
- More control over schedule
- More control over scope
- More cushion/contingency to be spread out on project
- More input into the means and methods
- Need to reduce risk by defining the project
- Negotiated
- No "bad" projects

Figure 3.4 *(Continued)*

A Sample Teaming Agreement for Architect and Contractor

Any such agreement has significant legal consequences and should be prepared only after consultation with legal and accounting professionals:

The Agreement

This Agreement between (architect) and (contractor), referred to thereafter as the Parties, represents the teaming arrangement for the purpose of creating a design-build entity to market, design and construct the project known as (project name). The intent of this Agreement is to establish a working arrangement between (the architect) and the (contractor) resulting in a strong, integrated, collaborative effort by both designer and contractor. All Parties agree to set forth their best effort in pursuing and executing a contract with design and construction of (project), and to work exclusively with each other to achieve that goal.

It is understood that the (architect or contractor) will act as the prime contractor and will exercise full control over the entire program during the length of this Agreement. The (architect) is designated as designer of record in this Agreement.

In the event that the Parties are successful in their attempt to gain a contract for design-build and are awarded a contract with the client, the Parties agree to enter into a new contract agreement based upon the terms and conditions contained in the proposal submitted by this team.

The new agreement's architectural fee will be $__________.

(Architect) maintains a $__________ errors and omissions professional liability insurance policy, the proportional cost of which is included in the architectural fee stated above. If additional coverage is required, the incremental costs will be apportioned to the fee stated above.

Terms and conditions will be negotiated between the Parties during the negotiation of the prime contract with the (client) and will be in agreement with this Agreement and the terms and conditions of the contract with (client).

This Agreement is not intended to constitute or create, or otherwise recognize any formal business entity such as a joint venture, limited liability corporation, partnership, and the rights and responsibilities of the Parties shall be limited to those implied in this Agreement. Neither party shall be liable to the other for any costs, expenses, exposure to risk, or liabilities arising out of or from the other party's efforts in connection with any preliminary proposal or prebid effort except as provided herein.

In the event that the Parties are not successful in obtaining a design-build contract with (client), (architect) will be reimbursed $____________ in full for their design work and (contractor) will be reimbursed $______________ for all services rendered to date.

Should the team be unsuccessful because one party decided to discontinue its involvement in the project under the terms and conditions of this Agreement, the party desirous of discontinuing their involvement will pay the other party out-of-pocket costs for all labor and materials up to a maximum of $____________. Payment for such costs and expenses will be made within five working days of the decision to discontinue.

Reimbursable costs will include expenditures directly related to this proposal and Agreement such as

- Actual costs of salaries of design professionals and staff exclusive of any overhead, general, or administrative costs. All such requests to be fully documented with payroll ledgers, receipts, and so forth.
- Actual cost of salaries of contractor and staff exclusive of any overhead, general, or administrative costs, out-of-pocket costs.
- Expenses such as voice/data communication costs, postage, reproductions, supplies, and reasonable, documented travel expenses.

This Agreement shall terminate with the occurrence of either of the following events:

1. Award of a contract for the work to an entity other than (architect or contractor) and payment by the architect or contractor to the architect or contractor of the amounts stated above
2. Decision by the client to abort the project
3. Award of contract to architect or contractor and the issuance of a new agreement acceptable to both Parties
4. Such changes to the program as directed by the client to substantially change or eliminate the scope of work as originally contemplated.

Notwithstanding the above, none of the basis for termination contained in this Agreement shall relieve the Parties of their respective rights and responsibilities under any new agreement entered into or contemplated by this Agreement.

Any new agreement shall be governed exclusively by the terms and conditions set forth in that agreement.

(Architect)		(Contractor)	
By:	________________	By:	________________
Title:	________________	Title:	________________
Date:	________________	Date:	________________

Chapter

4

The Design and Construction Industries

When considering design-build, the contractor-led or the architect/engineer-led team approach, it might be of interest to step back and take a look at both the design and construction industries, through the eyes of the U.S. Department of Labor, a professional trade organization, and two independent research organizations.

Contractors, architects, and engineers make up a considerable portion of the workforce in the United States and, through their efforts, contribute significantly not only to the economy of this country but also to the well-being of its citizenry.

The Industry According to the U.S. Department of Labor

Government statistics place the value of construction for the year 2005 at $1047.3 billion, based on construction spending as of February 2005 and projected to the end of the year. The construction industry employed 12,256,000 workers, and the architectural and engineering professions employment was 2,659,298, according to the U.S. census in the year 2000. That's a total of 14,915,000 people, which is 11.4% of the entire civilian working population.

The government reported 697,747 firms engaged in construction—223,114 were building, developing, and general contractors; 36,647 were engaged in heavy construction; and 437,986 were listed as specialty contractors (subcontractors).

Median earnings in construction were $32,000. The Census Bureau groups architects and engineers in their management, professional, and related occupations category. The Bureau reported a $50,034 median annual earnings for this group.

The Construction Industry

The construction business is a high-risk endeavor—some projects yield an above average profit, others lesser profits, and some are dead losses. According to figures provided by the Construction Financial Management Association (CFMA) of Princeton, New Jersey, the average net profit for general contractors as reported by their members in a 2004 survey was, on average, between 1.5% and 2.3% after taxes; specialty contractors included in the survey reported a slightly higher net profit between 1.6% and 3.1%. These relatively small profit percentages coupled with the competitive nature of the industry are major factors to be considered in assessing overall risk for the industry.

Construction financial management association

The CFMA is a nonprofit organization, acting as a repository and a resource for construction financial professionals. Established in 1981, CFMA currently has 7000 members across the country, which include lawyers, accountants, lending institutions, contractors, architects, engineers, and materials and equipment suppliers.

Each year CFMA publishes its *Construction Industry Annual Financial Survey* that includes financial and benchmarking information for the construction industry. This survey and its accompanying report provide a look at this industry, based on information received from its members. The survey includes a wide variety of topics—from strictly financial reporting to "best practices" to company organization structures and policies. Recently CFMA added another category to its annual report, called *Hot Topics*, which is a result of a survey that asked the question: "What do you consider the key issues for the industry in the year ahead?" The answer was strategic planning.

Strategic planning—the Hot Topic

In the 2004 CFMA report, strategic planning is viewed by CFMA members as an important issue because it reflects on a company's ability to react to the increased pressures most construction firms face in gaining or maintaining a competitive edge. Develop a sound business plan and beat the odds in business survival rates. This is a contractor-oriented survey, but it would seem that design consultants will also be interested in the results of the survey—since parts of it are fairly generic in nature and would apply across a wide spectrum of businesses. The survey will be of special interest to architects and engineers considering teaming with a contractor on a design-build basis. The very low profit percentages that accrue to contractors will place their concern about risk in perspective when viewed by an architect/engineer team member. This process of developing a strategic plan has probably been on the minds of contractors, engineers, and architects considering ways to expand their businesses and increase market share and profits. The CFMA survey results and planning process will only fortify their opinion that they ought to seriously think about formalizing their plans and develop their own program. CFMA's survey indicated that companies with a strategic plan were larger and more successful than those that had no plan. Developing a strategic plan is not that difficult, but it requires a dedicated and concerted effort to formulate and implement a strategy to make the company stronger.

Figure 4.1 illustrates the key financial characteristics of companies with a strategy (light gray) and those without it (black), revealing that firms with a game plan excel those entities that do not.

Developing the plan. Strategic planning is not a one-time event, but must be tweaked now and then as market conditions change. Ninety-one surveyed participants said they met at least once a year to review and upgrade their plan.

The planning process begins with a well-thought-out goal in mind, an enforceable agenda, and a well-defined decision-making process. Top management commitment and encouragement is an essential ingredient. The plan will incorporate short-term goals that require immediate attention and long-term plans with performance milestones to be met along the way.

The implementation process, once defined and memorialized, must be properly communicated to all employees within the context of its goal, which is simply to create a strong, more competitive company, increase employment stability, and award performers.

The basic reason for developing a strategic plan is to evaluate the company's strengths and weaknesses, where its unique strengths lie, and where some activities need to be shored up, or possibly discarded. These unique strengths, referred to as primary strategic differentiators by CFMA, are the qualities that differentiate an individual company from others in the marketplace. A company may have already achieved one or more of these differentiators.

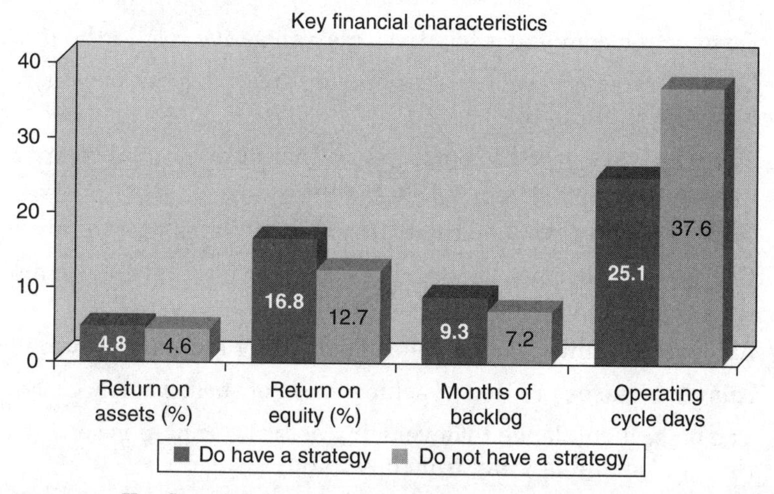

Figure 4.1 Key financial characteristics of firms with and without a strategic plan. (*Copyright of Construction Financial Management Association. All rights reserved. Reprinted with permission of CFMA.*)

- *Dominance.* A company that fills a void caused by the lack of a leader in its market area, a situation that is rather rare in the mature business environment of today
- *Superiority.* Customers of companies that employ this strategy expect the highest quality service, one that is least driven by price
- *Niche Market.* A client-centric, client-driver, client-focused strategy that targets a specific customer or type of customer
- *Technical Specialty.* A company with a high level of quality expertise required to handle complex projects
- *Low-Cost Provider.* A company that excels in providing a service to the most cost-sensitive customers
- *Unique Distribution Method.* A strategy that primarily focuses on current customers and seeks to derive business from the relationship(s) they have developed with one or more particular clients—another name for repeat business
- *Unique Natural or Human Resource.* A company with a unique source to a product or material, or having a known personality with particular expertise in their given field

It is difficult to be all things to all people, and strategic planning should point the way to revealing those activities that the company performs well and should continue doing; those activities that the company needs to improve; those new activities that the company needs to embrace to remain competitive; or just as importantly, those nonperforming activities that should be discarded.

Figure 4.2 contains the results of the *Hot Topic*—Strategic Planning Questionnaire sent to CFMA members. It reveals the following:

1. Owners and company executives play an active role in the process.
2. These strategic planning sessions are held at least once a year and most likely twice annually.
3. The plan is conveyed to employees either at an annual company meeting or during each employee's annual review.
4. Most companies evaluate their current capabilities when analyzing their plan.
5. Changes in labor markets and in economic forecasts and market cycles are the two dominant issues impacting the plan.
6. An overwhelming majority of companies set annual performance goals.
7. Annual bonuses are tied to achievement of these goals.
8. The biggest challenge to developing a strategic plan is setting aside enough time and stepping away from day-to-day issues.

One important item the CFMA strategic plan survey did not include. Management must turn their attention to our aging workforce, not only in-field personnel but also

Hot Topic - Strategic Planning

Do you have a strategic plan?

Yes	62%
No	38%
	100%

The remaining data refers to those participants who answered "Yes" to question 1.

Do you have strategic planning sessions?

Yes	97%
No	3%
	100%

Who attends strategic planning sessions?

Owners	87%
Executives	85%
Managers	63%
Board members	43%
Other staff	17%
External Associates	5%
Of those:	
Strategic Planning Consultant	36%
Lawyer	9%
Marketing / Business Development Consultant	9%
Insurance Agent	9%
Surety Agent	9%
Accountant / CPA	0%
Technology Consultant	0%
Banker	0%

How many individuals participate? 8 *(median number of participants)*

How frequently do you have strategic planning sessions?

More than once per year	50%
Once per year	41%
Once every two years	3%
Once every three years	6%
	100%

Have you ever used an outside consultant/coach to facilitate the planning process?

Yes	67%
No	33%
	100%

How is the strategic plan communicated to employees?

At an annual meeting	74%
In person during annual employee reviews	62%
Mailed electronically	43%
Posted in communal areas	38%
Mailed in hard copy	31%
Other	14%
Not communicated in an organized way	10%

Which strategy best describes how you differentiate your business?

Dominant Local Firm (among the top 3 in the local market)	33%
Product or Service Superiority (the "Rolex")	28%
Niche Market Firm (responding to the needs of specific, named market)	20%
Technical Specialty Firm (known for a particular specialty expertise)	10%
The Low Cost Provider (the "Timex")	4%
Unique Distribution Method (leverages the existing customer)	3%
Unique Natural or Human Resource (access to a scarce good or a "star" in the industry)	1%
Unique Sales Method (drives majority of new business)	1%
	100%

Figure 4.2 CFMA's *HotTopic* Questionnaire. (*Copyright of Construction Financial Management Association. All rights reserved. Reprinted with permission of CFMA.*)

Hot Topic - Strategic Planning

To what extent do you use the following analyses to develop your strategic plan?

	Deep	Some	None
Evaluation of current capabilities of the business	59%	30%	11%
Evaluation of the market and development of a forecast	39%	47%	14%
Traditional SWOT analysis	41%	42%	17%
Evaluation of local competition	19%	61%	20%
Evaluation of key individuals' personal definition of success	26%	50%	25%
Evaluation of national competition	2%	31%	67%
Other	3%	8%	89%

To what extent do the following current issues impact your strategic plan?

	Large	Moderate	Small
Changes in the labor market	34%	53%	13%
Economic forecasts and market cycles	48%	38%	14%
Increasing costs / complexity of insurance	44%	41%	15%
Fundamental changes in sources of future work	36%	46%	18%
Adoption rate of new technology in the industry	9%	64%	27%
Increasing costs / complexity of bonding	24%	41%	35%
Changing regulation	9%	51%	40%
Increasing use of web-based services	5%	49%	46%
Consolidation in the construction industry	8%	42%	51%
Other	3%	6%	92%

Do you set annual company performance goals (e.g., revenue, profitability)?

Yes	89%
No	11%
	100%

How are company goals related to the strategic plan?

Goals driven by strategy	24%
Goals driven by strategy, incorporated into individual goals	31%
Goals developed concurrent with strategy	34%
Goals drive the strategy	6%
Unrelated	6%
	100%

Is some portion of annual bonuses tied to achievement of elements of the strategic plan?

Yes	56%
No	44%
	100%

What are the biggest challenges to developing a strategic plan?

Setting aside time	79%
Stepping away from day-to-day operations / issues	74%
Implementing the strategic plan	73%
Communicating the strategic plan in an understandable way	42%
Large number of strategic options	17%
Lack of buy-in from staff	14%
Lack of business leadership experience	11%
Limited number of strategic options	5%
Other	2%
There is no need for a strategic plan	0%

Figure 4.2 *(Continued)*

in managers. The federal government anticipates that 40% of all U.S. workers will be 55 or older by 2010, and that the size of the workforce in this country will grow at less than 0.5% annually over the next several decades. Coupled with rising employee mobility, early retirements, and fewer replacements, management ought to include short- and long-term replacement personnel policies in their strategic plan.

This part of the plan should cover a five-year period, taking into account past attrition activity, the reasons managers or key personnel left the company, the current succession plan, and current and projected future manager positions and requirements.

Topics to be considered in such a plan:

1. What are our current policies for retaining key people? Are these policies adequate in today's marketplace or do they need to be updated and revised?
2. What programs and recruitment policies do we have to attract younger people?
3. Should we devise an internship program to attract and observe potential managers?
4. Is our succession plan realistic or do we need to rethink the transfer of authority?
5. What are the retirement plans of our current group of managers?
6. How can we convince them to stay on, if needed?
7. How can we supplement our current workforce on a temporary basis, if needed?

Because we are living longer and healthier, not only is the retirement age rising, but many retirees also find that they need to reenter the workforce because they either need some extra money or need to remain active in the field they know best. Several contracting and design firms have rehired these experienced and knowledgeable retirees as consultants, calling on them when needed for job-specific assignments. The aging workforce problem can be attacked from many ends by a well-thought-out strategic plan—make the company more attractive to newcomers, strengthen existing company policies to retain valued employees, and consider ways to retain older managers by creating a flexible working environment.

CFMA financial snapshots. CFMA's annual survey presents a snapshot of the financial side of the industry, based on annual sales volume of their general contractor and specialty contractor respondents, including by geographic region. They also have a best-in-class financial profile based on five indicators of financial health—return on assets, return on equity, fixed asset ratio, debt to equity, and working capital turnover.

Five Indicators of Financial Health

	Best in class	All companies
Return on assets	8.8%	4.7%
Return on equity	27.0%	14.9%
Fixed asset ratio	39.4%	43.5%
Debt to equity	2.1%	2.2%
Working capital turnover	15.4%	13.2%

Other financial data (Fig. 4.3) are grouped by annual sales volume for industrial and nonresidential contractors. Some highlights are listed as follows:

Revenues, Costs, Gross, and Net Profit Figures

	Less than $10 million	$10–$25 million	$25–$50 million
Contract revenue	$6,231,683	$16,081,493	$34,359,148
Other revenue	31,892	36,612	404,126
Total revenue	$6,263,575	$16,117,983	$34,763,274
Total costs	($5,517,439)	($14,492,436)	($32,030,644)
Gross profit	$746,136 (11.9%)	$1,625,548 (10.1%)	$2,732,630 (7.9%)
Net earnings (after taxes)	$101,696 (1.6%)	$236,466 (1.5%)	$479,228 (1.4%)

Other data collected from CFMA's respondents may also be of interest to design-builders.

Figure 4.4 shows that 39% of all companies were looking to design-build to improve profitability and 37% of specialty trade contractors were looking to design-build for greater profitability.

Selected Financial Data by Revenue

		ANNUAL REVENUE				
	All Specialty Trade	$0-$10M	$10-$25M	$25-$50M	$50-$100M	>$100M
Number of Companies	171	25	56	44	28	18
Assets ($)	17,172	2,386	5,818	11,605	27,970	69,847
Liabilities ($)	10,626	1,250	3,403	7,119	15,848	46,572
Net Worth ($)	6,539	1,136	2,415	4,485	12,118	23,216
Net Worth to Assets	38.1 %	47.6 %	41.5 %	38.6 %	43.3 %	33.2 %
Revenues ($)	46,158	5,671	16,431	34,359	70,537	185,789
Gross Profit ($)	6,414	1,084	2,517	5,361	9,101	24,331
Gross Profit Margin	13.9 %	19.1 %	15.3 %	15.6 %	12.9 %	13.1 %
SG&A Expense ($)	5,317	1,067	2,190	4,588	7,928	18,669
SG&A Expense Margin	11.5 %	18.8 %	13.3 %	13.4 %	11.2 %	10.0 %
Net Income ($)	1,079	35	255	688	1,277	5,737
Net Income Margin	2.3 %	0.6 %	1.6 %	2.0 %	1.8 %	3.1 %
Current Ratio	1.7	1.9	1.7	1.6	1.6	1.7
Return on Assets	6.3 %	1.5 %	4.4 %	5.9 %	4.6 %	8.2 %
Return on Equity	16.5 %	3.1 %	10.6 %	15.4 %	10.5 %	24.7 %

Note: All $ amounts are in thousands.

Figure 4.3 Selected financial data by revenue. (*Copyright of Construction Financial Management Association. All rights reserved. Reprinted with permission of CFMA.*)

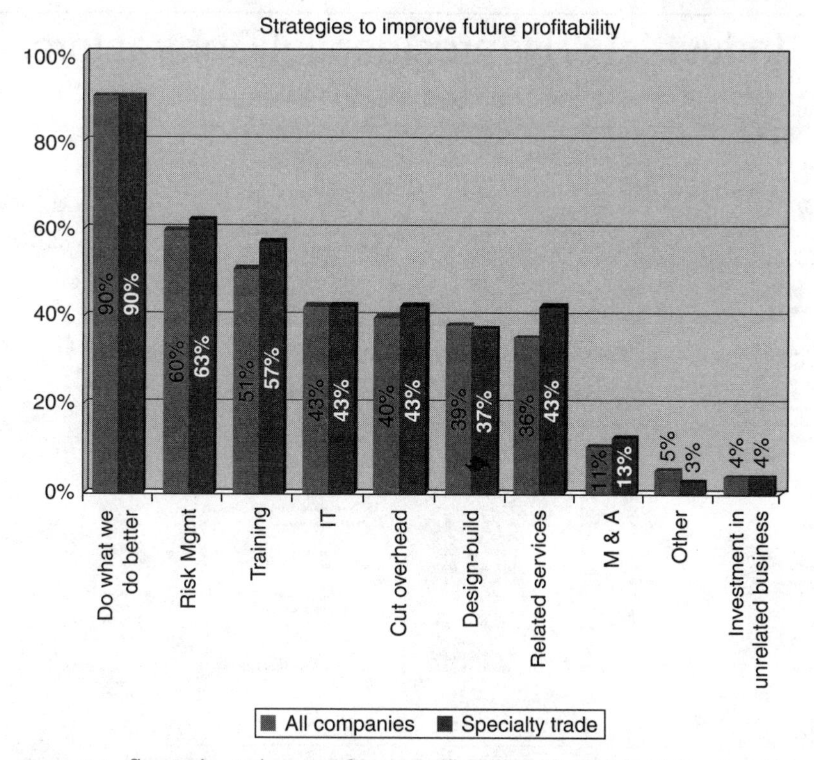

Figure 4.4 Strategies to improve future profitability. (*Copyright of Construction Financial Management Association. All rights reserved. Reprinted with the permission of CFMA.*)

Full CFMA financial profiles included in this chapter are shown in Figs. 4.5 to 4.14.

The geographic areas outlined in Fig. 4.13 encompass the following states:

Northeast. Connecticut, Delaware, District of Columbia, Maine, Maryland, Massachusetts, New Hampshire, New Jersey, New York, Pennsylvania, Rhode Island, Vermont

Southeast. Alabama, Florida, Georgia, Kentucky, Mississippi, North Carolina, South Carolina, Tennessee, Virginia, West Virginia

Southwest. Arkansas, Kansas, Louisiana, Missouri, Oklahoma, Texas

Midwest. Illinois, Indiana, Iowa, Michigan, Minnesota, Nebraska, North Dakota, South Dakota, Ohio, Wisconsin

West. Arizona, Colorado, Idaho, Montana, Nevada, New Mexico, Utah, Wyoming

Far West. Alaska, California, Hawaii, Oregon, Washington

Industrial & Nonresidential Contractors

Less Than $10 Million Revenue

Balance Sheet

	2004 Participants		2003 Participants	
	Amount	Percent	Amount	Percent
Current assets:				
Cash and cash equivalents	$ 423,652	21.4 %	$ 243,532	12.7 %
Marketable securities & short-term investments	51,451	2.6	64,690	3.4
Receivables:				
Contract receivables currently due	842,179	42.5	906,646	47.2
Retainages on contracts	161,671	8.1	94,699	4.9
Unbilled work	669	0.0	10,481	0.5
Other receivables	21,382	1.1	20,803	1.1
Less allowance for doubtful accounts	(250)	(0.0)	(5,683)	(0.3)
Total receivables, net:	1,025,651	51.7	1,026,946	53.5
Inventories	4,007	0.2	10,979	0.6
Costs and recognized earnings in excess of billings on uncompleted contracts	167,667	8.5	120,914	6.3
Investments in and advances to construction joint ventures	0	0.0	0	0.0
Income taxes:				
Current/refundable	7,027	0.4	7,376	0.4
Deferred	2,330	0.1	649	0.0
Other current assets	74,759	3.8	78,349	4.1
Total current assets	1,756,544	88.5	1,560,390	81.3
Property, plant and equipment	522,601	26.3	664,966	34.7
Less accumulated depreciation	(332,732)	(16.8)	(435,532)	(22.7)
Property, plant and equipment, net	189,868	9.6	229,433	12.0
Noncurrent assets:				
Long-term investments	60	0.0	22,382	1.2
Deferred income taxes	0	0.0	2,041	0.1
Other assets	37,371	1.9	104,686	5.5
Total noncurrent assets	37,431	1.9	129,109	6.7
Total assets	$ 1,983,844	100.0 %	$ 1,918,932	100.0 %

	2004 Participants		2003 Participants	
	Amount	Percent	Amount	Percent
Current liabilities:				
Current maturity on long-term debt	$ 21,337	1.1 %	$ 40,685	2.1 %
Notes payable and lines of credit	80,999	4.1	66,002	3.4
Accounts payable:				
Trade, including currently due to subcontractors	665,589	33.6	736,916	38.4
Subcontracts retainages	119,485	6.0	67,396	3.5
Other	9,730	0.5	20,157	1.1
Total accounts payable	794,804	40.1	824,469	43.0
Accrued expenses	59,458	3.0	41,831	2.2
Billings in excess of costs and recognized earnings on uncompleted contracts	200,425	10.1	135,225	7.0
Income taxes:				
Current	4,871	0.2	5,005	0.3
Deferred	2,785	0.1	4,252	0.2
Other current liabilities	2,894	0.1	3,961	0.2
Total current liabilities	1,167,574	58.9	1,121,429	58.4
Noncurrent liabilities				
Long-term debt, excluding current maturities	98,833	5.0	107,466	5.6
Deferred income taxes	10,130	0.5	996	0.1
Other	700	0.0	9,362	0.5
Total liabilities	1,277,237	64.4	1,239,254	64.6
Minority interests	0	0.0	494	0.0
Net worth:				
Common stock, par value	24,236	1.2	63,115	3.3
Preferred stock, stated value	7,341	0.4	2,314	0.1
Additional paid-in capital	76,124	3.8	86,841	4.5
Retained earnings	655,157	33.0	491,908	25.6
Treasury stock	(100,481)	(5.1)	(18,522)	(1.0)
Excess value of marketable securities	0	0.0	(1,636)	(0.1)
Other equity	44,230	2.2	55,163	2.9
Total net worth	706,607	35.6	679,184	35.4
Total liabilities and net worth	$ 1,983,844	100.0 %	$ 1,918,932	100.0 %

Statement of Earnings

	2004 Participants		2003 Participants	
	Amount	Percent	Amount	Percent
Contract revenue	$ 6,231,683	99.5 %	$ 6,819,182	99.9 %
Other revenue	31,892	0.5	8,044	0.1
Total Revenue	6,263,575	100.0	6,827,226	100.0
Contract cost	(5,498,817)	(87.8)	(5,952,327)	(87.2)
Other cost	(18,622)	(0.3)	(5,756)	(0.1)
Total cost	(5,517,439)	(88.1)	(5,958,083)	(87.3)
Gross Profit	746,136	11.9	869,143	12.7
Selling, general & administrative expenses:				
Payroll	(319,406)	(5.1)	(345,698)	(5.1)
Professional fees	(15,459)	(0.2)	(31,367)	(0.5)
Sales & marketing costs	(15,344)	(0.2)	(22,784)	(0.3)
Technology costs	(8,679)	(0.1)	(15,772)	(0.2)
Administrative bonuses	(46,070)	(0.7)	(31,553)	(0.5)
Other	(248,329)	(4.0)	(285,439)	(4.2)
Total SG&A expenses	(653,288)	(10.4)	(732,612)	(10.7)
Income from operations	92,848	1.5	136,531	2.0
Interest income	6,964	0.1	11,609	0.2
Interest expense	(7,739)	(0.1)	(11,716)	(0.2)
Other income / (expense), net	12,372	0.2	(880)	(0.0)
Net earnings / (loss) before income taxes	104,445	1.7	135,544	2.0
Income tax (expense) / benefit	(2,749)	(0.0)	4,019	0.1
Net earnings	$ 101,696	1.6 %	$ 139,564	2.0 %

Number of Participants

	Number
2004	20
2003	24

Financial Ratios

	2004 Participants		2003 Participants	
	Average	Median	Average	Median
Liquidity Ratios				
Current Ratio	1.5	1.5	1.4	1.3
Quick Ratio	1.3	1.2	1.2	1.2
Days of Cash	24.3	17.1	12.8	11.3
Working Capital Turnover	10.6	9.9	15.6	14.8
Profitability Ratios				
Return on Assets	5.1 %	4.2 %	7.3 %	2.3 %
Return on Equity	14.4 %	14.2 %	20.5 %	5.4 %
Times Interest Earned	14.5	13.6	12.6	0.2
Leverage Ratios				
Debt to Equity	1.8	1.6	1.8	1.8
Revenue to Equity	8.9	9.9	10.1	11.7
Asset Turnover	3.2	3.4	3.6	3.7
Fixed Asset Ratio	26.9 %	24.8 %	33.8 %	31.3 %
Equity to SG&A Expense	1.1	1.0	0.9	1.1
Underbillings to Equity	23.8 %	9.1 %	19.3 %	11.7 %
Backlog to Equity	9.2	5.5	5.1	4.4
Efficiency Ratios				
Backlog to Working Capital	10.7	8.1	7.9	5.7
Months in Backlog	10.6	6.8	6.1	5.3
Days in Accounts Receivable	49.6	46.0	48.6	49.8
Days in Inventory	0.3	0.9	0.7	0.0
Days in Accounts Payable	44.1	35.4	45.7	43.8
Operating Cycle	30.2	23.8	16.4	17.2

Figure 4.5 Industrial and nonresidential contractors with less than $10 million revenue.

Industrial & Nonresidential Contractors

$10 - $25 Million Revenue

Balance Sheet

	2004 Participants		2003 Participants	
	Amount	Percent	Amount	Percent
Current assets:				
Cash and cash equivalents	$ 965,994	18.8 %	$ 1,132,732	22.3 %
Marketable securities & short-term investments	285,103	5.6	288,543	5.7
Receivables:				
Contract receivables currently due	2,138,614	41.7	2,014,829	39.7
Retainages on contracts	465,176	9.1	494,314	9.7
Unbilled work	17,421	0.3	5,219	0.1
Other receivables	72,591	1.4	91,831	1.8
Less allowance for doubtful accounts	(2,179)	(0.0)	(4,423)	(0.1)
Total receivables, net:	2,691,622	52.4	2,601,770	51.2
Inventories	44,558	0.9	43,684	0.9
Costs and recognized earnings in excess of billings on uncompleted contracts	221,860	4.3	219,825	4.3
Investments in and advances to construction joint ventures	71,909	1.4	6,550	0.1
Income taxes:				
Current/refundable	3,880	0.1	13,090	0.3
Deferred	3,159	0.1	3,163	0.1
Other current assets	160,725	3.1	128,813	2.5
Total current assets	4,448,810	86.7	4,438,170	87.3
Property, plant and equipment	1,249,483	24.3	1,193,796	23.5
Less accumulated depreciation	(790,513)	(15.4)	(736,353)	(14.5)
Property, plant and equipment, net	458,969	8.9	457,443	9.0
Noncurrent assets:				
Long-term investments	92,952	1.8	38,804	0.8
Deferred income taxes	12,416	0.2	1,791	0.0
Other assets	119,181	2.3	145,125	2.9
Total noncurrent assets	224,549	4.4	185,721	3.7
Total assets	$ 5,132,328	100.0 %	$ 5,081,334	100.0 %

	2004 Participants		2003 Participants	
	Amount	Percent	Amount	Percent
Current liabilities:				
Current maturity on long-term debt	$ 38,189	0.7 %	$ 45,727	0.9 %
Notes payable and lines of credit	85,886	1.7	36,694	0.7
Accounts payable:				
Trade, including currently due to subcontractors	1,725,648	33.6	1,686,224	33.2
Subcontracts retainages	379,037	7.4	397,239	7.8
Other	36,144	0.7	38,142	0.8
Total accounts payable	2,140,829	41.7	2,121,605	41.8
Accrued expenses	171,031	3.3	186,104	3.7
Billings in excess of costs and recognized earnings on uncompleted contracts	557,857	10.9	565,579	11.1
Income taxes:				
Current	14,165	0.3	8,748	0.2
Deferred	3,170	0.1	19,021	0.4
Other current liabilities	57,691	1.1	43,545	0.9
Total current liabilities	3,068,818	59.8	3,027,024	59.6
Noncurrent liabilities				
Long-term debt, excluding current maturities	127,122	2.5	228,912	4.5
Deferred income taxes	11,252	0.2	13,850	0.3
Other	51,431	1.0	17,674	0.3
Total liabilities	3,258,623	63.5	3,287,460	64.7
Minority interests	0	0.0	27,114	0.5
Net worth:				
Common stock, par value	101,248	2.0	94,489	1.9
Preferred stock, stated value	22,633	0.4	2,368	0.0
Additional paid-in capital	129,223	2.5	108,135	2.1
Retained earnings	1,519,872	29.6	1,566,512	30.8
Treasury stock	(54,413)	(1.1)	(34,929)	(0.7)
Excess value of marketable securities	4,595	0.1	(7,660)	(0.2)
Other equity	150,546	2.9	37,846	0.7
Total net worth	1,873,705	36.5	1,766,761	34.8
Total liabilities and net worth	$ 5,132,328	100.0 %	$ 5,081,334	100.0 %

Statement of Earnings

	2004 Participants		2003 Participants	
	Amount	Percent	Amount	Percent
Contract revenue	$ 16,081,493	99.8 %	$ 15,857,836	98.5 %
Other revenue	36,490	0.2	234,612	1.5
Total Revenue	16,117,983	100.0	16,092,448	100.0
Contract cost	(14,480,683)	(89.8)	(14,378,201)	(89.3)
Other cost	(11,753)	(0.1)	(41,592)	(0.3)
Total cost	(14,492,436)	(89.9)	(14,419,792)	(89.6)
Gross Profit	1,625,548	10.1	1,672,656	10.4
Selling, general & administrative expenses:				
Payroll	(732,466)	(4.5)	(676,280)	(4.2)
Professional fees	(61,045)	(0.4)	(43,035)	(0.3)
Sales & marketing costs	(33,847)	(0.2)	(62,015)	(0.4)
Technology costs	(15,273)	(0.1)	(19,790)	(0.1)
Administrative bonuses	(73,275)	(0.5)	(54,047)	(0.3)
Other	(489,588)	(3.0)	(590,209)	(3.7)
Total SG&A expenses	(1,405,494)	(8.7)	(1,445,376)	(9.0)
Income from operations	220,053	1.4	227,280	1.4
Interest income	24,767	0.2	30,072	0.2
Interest expense	(11,722)	(0.1)	(7,129)	(0.0)
Other income / (expense), net	15,176	0.1	(2,116)	(0.0)
Net earnings / (loss) before income taxes	248,274	1.5	248,107	1.5
Income tax (expense) / benefit	(11,808)	(0.1)	(18,127)	(0.1)
Net earnings	$ 236,466	1.5 %	$ 229,981	1.4 %

Number of Participants

	Number
2004	67
2003	57

Financial Ratios

	2004 Participants		2003 Participants	
	Average	Median	Average	Median
Liquidity Ratios				
Current Ratio	1.4	1.3	1.5	1.5
Quick Ratio	1.3	1.2	1.3	1.3
Days of Cash	21.6	16.3	25.3	19.4
Working Capital Turnover	11.7	17.3	11.4	12.0
Profitability Ratios				
Return on Assets	4.6 %	4.1 %	4.5 %	3.0 %
Return on Equity	12.6 %	10.7 %	13.0 %	8.8 %
Times Interest Earned	22.2	23.4	35.8	8.1
Leverage Ratios				
Debt to Equity	1.7	2.1	1.9	1.7
Revenue to Equity	8.6	12.0	9.1	10.2
Asset Turnover	3.1	3.7	3.2	3.6
Fixed Asset Ratio	24.5 %	21.1 %	25.9 %	24.2 %
Equity to SG&A Expense	1.3	1.0	1.2	1.1
Underbillings to Equity	12.8 %	10.7 %	12.7 %	8.5 %
Backlog to Equity	9.1	7.2	4.2	3.3
Efficiency Ratios				
Backlog to Working Capital	18.3	8.1	5.3	4.8
Months in Backlog	8.9	6.6	5.5	3.8
Days in Accounts Receivable	49.3	48.4	47.0	44.1
Days in Inventory	1.1	0.7	1.1	0.0
Days in Accounts Payable	43.8	39.6	43.0	35.2
Operating Cycle	28.3	25.2	30.4	26.9

Figure 4.6 Industrial and nonresidential contractors with $10–$25 million revenue.

Industrial & Nonresidential Contractors

$25 - $50 Million Revenue

Balance Sheet

	2004 Participants		2003 Participants	
	Amount	Percent	Amount	Percent
Current assets:				
Cash and cash equivalents	$ 2,089,823	22.0 %	$ 2,056,574	20.8 %
Marketable securities & short-term investments	260,527	2.7	339,694	3.4
Receivables:				
Contract receivables currently due	4,321,976	45.4	4,272,135	43.2
Retainages on contracts	930,992	9.8	1,038,988	10.5
Unbilled work	14,544	0.2	43,456	0.4
Other receivables	45,001	0.5	253,006	2.6
Less allowance for doubtful accounts	(6,478)	(0.1)	(6,753)	(0.1)
Total receivables, net:	5,306,035	55.8	5,600,833	56.6
Inventories	157,714	1.7	68,244	0.7
Costs and recognized earnings in excess of billings on uncompleted contracts	440,132	4.6	328,375	3.3
Investments in and advances to construction joint ventures	26,132	0.3	37,194	0.4
Income taxes:				
Current/refundable	21,959	0.2	29,321	0.3
Deferred	3,516	0.0	3,860	0.0
Other current assets	313,133	3.3	248,732	2.5
Total current assets	8,618,972	90.6	8,712,826	88.0
Property, plant and equipment	1,663,323	17.5	1,932,129	19.5
Less accumulated depreciation	(1,063,976)	(11.2)	(1,251,000)	(12.6)
Property, plant and equipment, net	599,347	6.3	681,130	6.9
Noncurrent assets:				
Long-term investments	85,094	0.9	206,674	2.1
Deferred income taxes	53,832	0.6	3,189	0.0
Other assets	157,444	1.7	292,261	3.0
Total noncurrent assets	296,370	3.1	502,124	5.1
Total assets	$ 9,514,690	100.0 %	$ 9,896,080	100.0 %

	2004 Participants		2003 Participants	
	Amount	Percent	Amount	Percent
Current liabilities:				
Current maturity on long-term debt	$ 64,156	0.7 %	$ 33,015	0.3 %
Notes payable and lines of credit	247,194	2.6	131,968	1.3
Accounts payable:				
Trade, including currently due to subcontractors	3,760,110	39.5	3,482,907	35.2
Subcontracts retainages	893,161	9.4	1,055,624	10.7
Other	30,389	0.3	25,075	0.3
Total accounts payable	4,683,661	49.2	4,563,607	46.1
Accrued expenses	450,854	4.7	373,364	3.8
Billings in excess of costs and recognized earnings on uncompleted contracts	1,055,941	11.1	1,127,263	11.4
Income taxes:				
Current	20,578	0.2	45,161	0.5
Deferred	4,068	0.0	1,183	0.0
Other current liabilities	42,490	0.4	96,777	1.0
Total current liabilities	6,568,941	69.0	6,372,338	64.4
Noncurrent liabilities				
Long-term debt, excluding current maturities	300,640	3.2	313,222	3.2
Deferred income taxes	13,368	0.1	7,013	0.1
Other	12,846	0.1	45,145	0.5
Total liabilities	6,895,596	72.5	6,737,719	68.1
Minority interests	3,251	0.0	565	0.0
Net worth:				
Common stock, par value	183,233	1.9	143,565	1.5
Preferred stock, stated value	0	0.0	154	0.0
Additional paid-in capital	255,923	2.7	248,077	2.5
Retained earnings	1,939,250	20.4	2,637,060	26.6
Treasury stock	(38,915)	(0.4)	(67,873)	(0.7)
Excess value of marketable securities	36,112	0.4	(389)	(0.0)
Other equity	240,241	2.5	197,214	2.0
Total net worth	2,615,843	27.5	3,157,797	31.9
Total liabilities and net worth	$ 9,514,690	100.0 %	$ 9,896,080	100.0 %

Statement of Earnings

	2004 Participants		2003 Participants	
	Amount	Percent	Amount	Percent
Contract revenue	$ 34,359,148	98.8 %	$ 35,566,375	99.9 %
Other revenue	404,126	1.2	43,016	0.1
Total Revenue	34,763,274	100.0	35,609,391	100.0
Contract cost	(31,739,441)	(91.3)	(32,563,776)	(91.4)
Other cost	(291,203)	(0.8)	(30,891)	(0.1)
Total cost	(32,030,644)	(92.1)	(32,594,667)	(91.5)
Gross Profit	2,732,630	7.9	3,014,724	8.5
Selling, general & administrative expenses:				
Payroll	(1,299,540)	(3.7)	(1,258,685)	(3.5)
Professional fees	(73,261)	(0.2)	(85,104)	(0.2)
Sales & marketing costs	(84,768)	(0.2)	(164,356)	(0.5)
Technology costs	(39,204)	(0.1)	(45,219)	(0.1)
Administrative bonuses	(132,851)	(0.4)	(209,101)	(0.6)
Other	(681,958)	(2.0)	(705,945)	(2.0)
Total SG&A expenses	(2,311,582)	(6.6)	(2,468,410)	(6.9)
Income from operations	421,048	1.2	546,314	1.5
Interest income	32,994	0.1	53,203	0.1
Interest expense	(25,297)	(0.1)	(15,765)	(0.0)
Other income / (expense), net	67,960	0.2	122,324	0.3
Net earnings / (loss) before income taxes	496,705	1.4	706,076	2.0
Income tax (expense) / benefit	(17,476)	(0.1)	(65,099)	(0.2)
Net earnings	$ 479,228	1.4 %	$ 640,977	1.8 %

Number of Participants

	Number
2004	60
2003	64

Financial Ratios

	2004 Participants		2003 Participants	
	Average	Median	Average	Median
Liquidity Ratios				
Current Ratio	1.3	1.3	1.4	1.4
Quick Ratio	1.2	1.2	1.3	1.3
Days of Cash	21.6	20.2	20.8	18.5
Working Capital Turnover	17.0	14.8	15.2	15.2
Profitability Ratios				
Return on Assets	5.0 %	3.7 %	6.5 %	5.4 %
Return on Equity	18.3 %	17.4 %	20.3 %	20.7 %
Times Interest Earned	20.6	8.4	45.8	24.0
Leverage Ratios				
Debt to Equity	2.6	2.7	2.1	2.2
Revenue to Equity	13.3	14.1	11.3	12.0
Asset Turnover	3.7	3.7	3.6	4.0
Fixed Asset Ratio	22.9 %	19.8 %	21.6 %	18.0 %
Equity to SG&A Expense	1.1	1.1	1.3	1.2
Underbillings to Equity	17.4 %	12.6 %	11.8 %	9.4 %
Backlog to Equity	11.8	10.4	5.7	4.5
Efficiency Ratios				
Backlog to Working Capital	15.9	12.1	7.6	6.0
Months in Backlog	8.4	6.9	6.0	4.7
Days in Accounts Receivable	45.2	43.2	45.7	40.0
Days in Inventory	1.8	0.6	0.8	0.0
Days in Accounts Payable	42.6	38.0	38.7	34.7
Operating Cycle	26.0	27.2	28.5	26.4

Figure 4.7 Industrial and nonresidential contractors with $25–$50 million revenue.

Industrial & Nonresidential Contractors

Less Than $50 Million Revenue: Best in Class

Balance Sheet

	All Participants Amount	Percent	Best in Class Amount	Percent
Current assets:				
Cash and cash equivalents	$ 1,350,912	20.8 %	$ 1,584,333	26.2 %
Marketable securities & short-term investments	243,283	3.7	234,493	3.9
Receivables:				
Contract receivables currently due	2,853,396	43.9	2,439,197	40.3
Retainages on contracts	614,012	9.5	593,823	9.8
Unbilled work	13,968	0.2	7,633	0.1
Other receivables	54,362	0.8	38,777	0.6
Less allowance for doubtful accounts	(3,671)	(0.1)	(3,247)	(0.1)
Total receivables, net	3,532,067	54.4	3,076,183	50.8
Inventories	85,227	1.3	14,933	0.2
Costs and recognized earnings in excess of billings on uncompleted contracts	303,578	4.7	338,475	5.6
Investments in and advances to construction joint ventures	43,441	0.7	25,563	0.4
Income taxes:				
Current/refundable	11,688	0.2	7,753	0.1
Deferred	3,192	0.0	1,738	0.0
Other current assets	211,237	3.3	186,188	3.1
Total current assets	5,784,622	89.1	5,469,659	90.3
Property, plant and equipment	1,319,502	20.3	1,212,557	20.0
Less accumulated depreciation	(839,847)	(12.9)	(772,181)	(12.8)
Property, plant and equipment, net	479,654	7.4	440,375	7.3
Noncurrent assets:				
Long-term investments	77,107	1.2	56,315	0.9
Deferred income taxes	27,631	0.4	588	0.0
Other assets	123,668	1.9	88,299	1.5
Total noncurrent assets	228,405	3.5	145,202	2.4
Total assets	$ 6,492,682	100.0 %	$ 6,055,236	100.0 %

	All Participants Amount	Percent	Best in Class Amount	Percent
Current liabilities:				
Current maturity on long-term debt	$ 46,495	0.7 %	$ 31,532	0.5 %
Notes payable and lines of credit	151,061	2.3	54,202	0.9
Accounts payable:				
Trade, including currently due to subcontractors	2,411,815	37.1	2,178,841	36.0
Subcontracts retainages	553,571	8.5	531,585	8.8
Other	30,201	0.5	49,693	0.8
Total accounts payable	2,995,587	46.1	2,760,120	45.6
Accrued expenses	270,064	4.2	278,174	4.6
Billings in excess of costs and recognized earnings on uncompleted contracts	712,526	11.0	711,186	11.7
Income taxes:				
Current	15,518	0.2	17,233	0.3
Deferred	3,484	0.1	1,805	0.0
Other current liabilities	44,031	0.7	28,066	0.5
Total current liabilities	4,238,767	65.3	3,882,317	64.1
Noncurrent liabilities				
Long-term debt, excluding current maturities	194,097	3.0	95,982	1.6
Deferred income taxes	11,963	0.2	12,723	0.2
Other	28,698	0.4	9,232	0.2
Total liabilities	4,473,526	68.9	4,000,254	66.1
Minority interests	1,327	0.0	3,483	0.1
Net worth:				
Common stock, par value	124,233	1.9	150,741	2.5
Preferred stock, stated value	11,314	0.2	8,872	0.1
Additional paid-in capital	173,713	2.7	152,222	2.5
Retained earnings	1,573,398	24.2	1,644,946	27.2
Treasury stock	(54,355)	(0.8)	(26,283)	(0.4)
Excess value of marketable securities	16,834	0.3	1,332	0.0
Other equity	172,691	2.7	119,671	2.0
Total net worth	2,017,830	31.1	2,051,499	33.9
Total liabilities and net worth	$ 6,492,682	100.0 %	$ 6,055,236	100.0 %

Statement of Earnings

	All Participants Amount	Percent	Best in Class Amount	Percent
Contract revenue	$ 22,201,650	99.2 %	$ 24,503,772	99.7 %
Other revenue	185,920	0.8	73,769	0.3
Total Revenue	22,387,570	100.0	24,577,540	100.0
Contract cost	(20,303,052)	(90.7)	(22,322,431)	(90.8)
Other cost	(126,749)	(0.6)	(18,402)	(0.1)
Total cost	(20,429,800)	(91.3)	(22,340,833)	(90.9)
Gross Profit	1,957,770	8.7	2,236,707	9.1
Selling, general & administrative expenses:				
Payroll	(907,726)	(4.1)	(925,839)	(3.8)
Professional fees	(59,829)	(0.3)	(55,264)	(0.2)
Sales & marketing costs	(52,114)	(0.2)	(64,281)	(0.3)
Technology costs	(24,144)	(0.1)	(25,436)	(0.1)
Administrative bonuses	(93,890)	(0.4)	(119,162)	(0.5)
Other	(535,282)	(2.4)	(502,736)	(2.0)
Total SG&A expenses	(1,672,985)	(7.5)	(1,692,719)	(6.9)
Income from operations	284,785	1.3	543,989	2.2
Interest income	25,703	0.1	27,279	0.1
Interest expense	(16,721)	(0.1)	(8,019)	(0.0)
Other income / (expense), net	36,339	0.2	18,903	0.1
Net earnings / (loss) before income taxes	330,106	1.5	582,151	2.4
Income tax (expense) / benefit	(12,889)	(0.1)	(9,974)	(0.0)
Net earnings	$ 317,217	1.4 %	$ 572,178	2.3 %

Number of Participants

All Participants Number	Best in Class Number
147	56

Financial Ratios

	All Participants Average	Median	Best in Class Average	Median
Liquidity Ratios				
Current Ratio	1.4	1.3	1.4	1.4
Quick Ratio	1.2	1.2	1.3	1.3
Days of Cash	21.7	16.4	23.2	21.7
Working Capital Turnover	14.5	14.9	15.5	17.4
Profitability Ratios				
Return on Assets	4.9 %	4.0 %	9.4 %	7.8 %
Return on Equity	15.7 %	13.5 %	27.9 %	25.0 %
Times Interest Earned	20.7	15.9	73.6	48.9
Leverage Ratios				
Debt to Equity	2.2	2.3	1.9	1.9
Revenue to Equity	11.1	12.6	12.0	12.6
Asset Turnover	3.4	3.7	4.1	4.1
Fixed Asset Ratio	23.8 %	21.3 %	21.5 %	19.0 %
Equity to SG&A Expense	1.2	1.0	1.2	1.1
Underbillings to Equity	15.7 %	11.1 %	16.9 %	0.0 %
Backlog to Equity	10.2	7.5	7.2	6.1
Efficiency Ratios				
Backlog to Working Capital	16.4	9.3	9.7	8.2
Months in Backlog	8.9	6.9	6.6	5.9
Days in Accounts Receivable	46.7	45.3	36.2	33.7
Days in Inventory	1.5	0.7	0.2	0.6
Days in Accounts Payable	43.0	38.7	35.9	32.8
Operating Cycle	26.9	25.8	23.8	24.7

Figure 4.8 Industrial and nonresidential contractors with less than $50 million—best in class.

Selected Financial Data

	All Companies	All Specialty Trade
Number of Companies	537	171
Assets ($)	28,165	17,172
Liabilities ($)	19,264	10,626
Net Worth ($)	8,876	6,539
Net Worth to Assets	31.5 %	38.1 %
Revenues ($)	82,160	46,158
Gross Profit ($)	6,689	6,414
Gross Profit Margin	8.1 %	13.9 %
SG&A Expense ($)	5,286	5,317
SG&A Expense Margin	6.4 %	11.5 %
Net Income ($)	1,323	1,079
Net Income Margin	1.6 %	2.3 %
Current Ratio	1.4	1.7
Return on Assets	4.7 %	6.3 %
Return on Equity	14.9 %	16.5 %

Note: All $ amounts are in thousands.

Figure 4.9 Selected financial data for all companies—specialty contractors.

The Architect-Engineering Professions

PSMJ Resources Inc., headquartered in Newton, Massachusetts, has been offering management consulting services to the architecture and engineering profession for more than 30 years. They provide educational programs and in-house training covering a wide array of the topics of key importance to these two professions. Each year they issue their A/E Financial Performance Survey and the 2004 report is the 24th edition of this in-depth look at the design industry.

The full survey contains more than 290 pages covering important information such as income statements, balance sheets, marketing costs, staff ratios, and automation analysis benchmarks.

Between the early 1990s and mid-2001, the engineering and design industries reached the end of their longest period of expansion. Business expansion during that time created lots of opportunities in the design and construction industry and well-managed design firms showed growth in revenue, profit, and backlogs. The recession that piggybacked on the technology bust, the 2000 election, the September 11 disaster, and the war in Iraq created uncertainty in both the design and construction industries. Construction continues to drive the engineering and architectural design industries, and all three depend heavily on government spending and an upbeat economic outlook. It is against this background

Specialty Trade Contractors

Less Than $10 Million Revenue

Balance Sheet

	2004 Participants Amount	2004 Participants Percent	2003 Participants Amount	2003 Participants Percent
Current assets:				
Cash and cash equivalents	$ 329,327	13.8 %	$ 223,522	10.7 %
Marketable securities & short-term investments	76,374	3.2	35,939	1.7
Receivables:				
Contract receivables currently due	982,809	41.2	919,004	44.1
Retainages on contracts	122,254	5.1	121,308	5.8
Unbilled work	9,962	0.4	10,221	0.5
Other receivables	8,231	0.3	18,325	0.9
Less allowance for doubtful accounts	(13,927)	(0.6)	(7,326)	(0.4)
Total receivables, net:	1,109,330	46.5	1,061,533	51.0
Inventories	69,398	2.9	43,985	2.1
Costs and recognized earnings in excess of billings on uncompleted contracts	104,972	4.4	145,821	7.0
Investments in and advances to construction joint ventures	0	0.0	1,286	0.1
Income taxes:				
Current/refundable	11,755	0.5	3,157	0.2
Deferred	2,139	0.1	8,021	0.4
Other current assets	84,889	3.6	116,632	5.6
Total current assets	1,788,185	74.9	1,639,896	78.8
Property, plant and equipment	1,510,477	63.3	1,163,172	55.9
Less accumulated depreciation	(967,409)	(40.5)	(795,326)	(38.2)
Property, plant and equipment, net	543,068	22.8	367,846	17.7
Noncurrent assets:				
Long-term investments	0	0.0	2,387	0.1
Deferred income taxes	6,160	0.3	1,118	0.1
Other assets	48,873	2.0	70,871	3.4
Total noncurrent assets	55,033	2.3	74,376	3.6
Total assets	$ 2,386,285	100.0 %	$ 2,082,118	100.0 %

	2004 Participants Amount	2004 Participants Percent	2003 Participants Amount	2003 Participants Percent
Current liabilities:				
Current maturity on long-term debt	$ 98,300	4.1 %	$ 66,509	3.2 %
Notes payable and lines of credit	86,732	3.6	246,923	11.9
Accounts payable:				
Trade, including currently due to subcontractors	344,031	14.4	308,326	14.8
Subcontracts retainages	3,795	0.2	2,758	0.1
Other	7,055	0.3	26,982	1.3
Total accounts payable	354,881	14.9	338,065	16.2
Accrued expenses	115,983	4.9	133,661	6.4
Billings in excess of costs and recognized earnings on uncompleted contracts	229,594	9.6	128,739	6.2
Income taxes:				
Current	2,281	0.1	15,945	0.8
Deferred	29,995	1.3	8,053	0.4
Other current liabilities	296	0.0	31,994	1.5
Total current liabilities	918,063	38.5	969,890	46.6
Noncurrent liabilities				
Long-term debt, excluding current maturities	210,927	8.8	268,334	12.9
Deferred income taxes	40,544	1.7	6,128	0.3
Other	80,699	3.4	28,075	1.3
Total liabilities	1,250,234	52.4	1,272,428	61.1
Minority interests	0	0.0	0	0.0
Net worth:				
Common stock, par value	30,782	1.3	18,349	0.9
Preferred stock, stated value	0	0.0	0	0.0
Additional paid-in capital	54,399	2.3	41,815	2.0
Retained earnings	1,064,495	44.6	811,432	39.0
Treasury stock	(6,975)	(0.3)	(38,881)	(1.9)
Excess value of marketable securities	(4,855)	(0.2)	(3,112)	(0.1)
Other equity	(1,794)	(0.1)	(19,912)	(1.0)
Total net worth	1,136,051	47.6	809,690	38.9
Total liabilities and net worth	$ 2,386,285	100.0 %	$ 2,082,118	100.0 %

Statement of Earnings

	2004 Participants Amount	2004 Participants Percent	2003 Participants Amount	2003 Participants Percent
Contract revenue	$ 5,643,285	99.5 %	$ 5,006,416	93.5 %
Other revenue	27,927	0.5	346,743	6.5
Total Revenue	5,671,212	100.0	5,353,159	100.0
Contract cost	(4,563,192)	(80.5)	(3,995,258)	(74.6)
Other cost	(23,581)	(0.4)	(225,935)	(4.2)
Total cost	(4,586,773)	(80.9)	(4,221,193)	(78.9)
Gross Profit	1,084,439	19.1	1,131,965	21.1
Selling, general & administrative expenses:				
Payroll	(624,621)	(11.0)	(476,920)	(8.9)
Professional fees	(32,457)	(0.6)	(27,018)	(0.5)
Sales & marketing costs	(22,321)	(0.4)	(35,830)	(0.7)
Technology costs	(14,287)	(0.3)	(5,520)	(0.1)
Administrative bonuses	(37,640)	(0.7)	(80,597)	(1.5)
Other	(336,082)	(5.9)	(322,099)	(6.0)
Total SG&A expenses	(1,067,409)	(18.8)	(947,983)	(17.7)
Income from operations	17,030	0.3	183,982	3.4
Interest income	3,886	0.1	6,986	0.1
Interest expense	(21,206)	(0.4)	(22,083)	(0.4)
Other income / (expense), net	11,333	0.2	(1,637)	(0.0)
Net earnings / (loss) before income taxes	11,043	0.2	167,248	3.1
Income tax (expense) / benefit	24,020	0.4	(7,922)	(0.1)
Net earnings	$ 35,064	0.6 %	$ 159,326	3.0 %

Number of Participants

	Number
2004	25
2003	34

Financial Ratios

	2004 Participants Average	2004 Participants Median	2003 Participants Average	2003 Participants Median
Liquidity Ratios				
Current Ratio	1.9	1.8	1.7	1.7
Quick Ratio	1.7	1.4	1.4	1.4
Days of Cash	20.9	17.1	15.0	6.4
Working Capital Turnover	6.5	9.5	8.0	8.1
Profitability Ratios				
Return on Assets	1.5 %	2.5 %	7.7 %	6.9 %
Return on Equity	3.1 %	5.5 %	19.7 %	23.2 %
Times Interest Earned	1.5	5.9	8.6	4.2
Leverage Ratios				
Debt to Equity	1.1	1.5	1.6	1.2
Revenue to Equity	5.0	7.3	6.6	5.7
Asset Turnover	2.4	2.7	2.6	2.7
Fixed Asset Ratio	47.8 %	42.6 %	45.4 %	44.6 %
Equity to SG&A Expense	1.1	0.8	0.9	0.6
Underbillings to Equity	10.1 %	12.3 %	19.3 %	11.1 %
Backlog to Equity	3.9	2.3	1.5	1.1
Efficiency Ratios				
Backlog to Working Capital	3.5	2.4	1.8	1.4
Months in Backlog	3.9	3.5	2.7	2.4
Days in Accounts Receivable	62.0	56.6	62.5	61.9
Days in Inventory	5.4	4.2	3.8	2.3
Days in Accounts Payable	27.6	24.2	28.6	25.9
Operating Cycle	60.8	60.5	52.7	47.3

Figure 4.10 Specialty trade contractors with less than $10 million revenue.

Specialty Trade Contractors

$10 - $25 Million Revenue

Balance Sheet

	2004 Participants Amount	2004 Participants Percent	2003 Participants Amount	2003 Participants Percent
Current assets:				
Cash and cash equivalents	$ 569,840	9.8 %	$ 478,313	7.3 %
Marketable securities & short-term investments	70,023	1.2	40,971	0.6
Receivables:				
Contract receivables currently due	2,943,558	50.6	3,437,356	52.7
Retainages on contracts	527,602	9.1	614,922	9.4
Unbilled work	61,758	1.1	25,501	0.4
Other receivables	158,891	2.7	91,918	1.4
Less allowance for doubtful accounts	(18,707)	(0.3)	(20,147)	(0.3)
Total receivables, net:	3,673,101	63.1	4,149,549	63.7
Inventories	144,234	2.5	227,711	3.5
Costs and recognized earnings in excess of billings on uncompleted contracts	248,736	4.3	363,607	5.6
Investments in and advances to construction joint ventures	5,287	0.1	4,929	0.1
Income taxes:				
Current/refundable	14,008	0.2	17,828	0.3
Deferred	12,792	0.2	1,967	0.0
Other current assets	171,723	3.0	221,621	3.4
Total current assets	4,909,744	84.4	5,506,497	84.5
Property, plant and equipment	1,940,162	33.4	2,382,681	36.6
Less accumulated depreciation	(1,265,778)	(21.8)	(1,559,866)	(23.9)
Property, plant and equipment, net	674,384	11.6	822,815	12.6
Noncurrent assets:				
Long-term investments	50,050	0.9	28,177	0.4
Deferred income taxes	5,467	0.1	31,526	0.5
Other assets	177,869	3.1	127,495	2.0
Total noncurrent assets	233,387	4.0	187,197	2.9
Total assets	$ 5,817,515	100.0 %	$ 6,516,510	100.0 %

	2004 Participants Amount	2004 Participants Percent	2003 Participants Amount	2003 Participants Percent
Current liabilities:				
Current maturity on long-term debt	$ 111,737	1.9 %	$ 151,628	2.3 %
Notes payable and lines of credit	359,653	6.2	563,377	8.6
Accounts payable:				
Trade, including currently due to subcontractors	1,053,994	18.1	1,018,788	15.6
Subcontracts retainages	95,894	1.6	21,954	0.3
Other	135,540	2.3	59,625	0.9
Total accounts payable	1,285,428	22.1	1,100,367	16.9
Accrued expenses	407,501	7.0	635,194	9.7
Billings in excess of costs and recognized earnings on uncompleted contracts	689,711	11.9	735,501	11.3
Income taxes:				
Current	8,907	0.2	16,029	0.2
Deferred	17,045	0.3	1,304	0.0
Other current liabilities	54,628	0.9	95,004	1.5
Total current liabilities	2,934,611	50.4	3,298,403	50.6
Noncurrent liabilities				
Long-term debt, excluding current maturities	401,797	6.9	551,047	8.5
Deferred income taxes	8,019	0.1	9,645	0.1
Other	58,257	1.0	6,088	0.1
Total liabilities	3,402,685	58.5	3,865,182	59.3
Minority interests	0	0.0	15,229	0.2
Net worth:				
Common stock, par value	104,753	1.8	98,175	1.5
Preferred stock, stated value	9,643	0.2	6,283	0.1
Additional paid-in capital	330,252	5.7	138,482	2.1
Retained earnings	2,036,379	35.0	2,534,427	38.9
Treasury stock	(153,137)	(2.6)	(238,165)	(3.7)
Excess value of marketable securities	4,227	0.1	(544)	(0.0)
Other equity	82,714	1.4	97,442	1.5
Total net worth	2,414,831	41.5	2,636,099	40.5
Total liabilities and net worth	$ 5,817,515	100.0 %	$ 6,516,510	100.0 %

Statement of Earnings

	2004 Participants Amount	2004 Participants Percent	2003 Participants Amount	2003 Participants Percent
Contract revenue	$ 15,877,980	96.6 %	$ 16,431,964	99.3 %
Other revenue	553,307	3.4	114,634	0.7
Total Revenue	16,431,287	100.0	16,546,598	100.0
Contract cost	(13,539,222)	(82.4)	(13,515,601)	(81.7)
Other cost	(375,464)	(2.3)	(88,914)	(0.5)
Total cost	(13,914,687)	(84.7)	(13,604,515)	(82.2)
Gross Profit	2,516,600	15.3	2,942,083	17.8
Selling, general & administrative expenses:				
Payroll	(1,102,686)	(6.7)	(1,250,566)	(7.6)
Professional fees	(62,859)	(0.4)	(70,100)	(0.4)
Sales & marketing costs	(45,408)	(0.3)	(60,899)	(0.4)
Technology costs	(31,962)	(0.2)	(37,382)	(0.2)
Administrative bonuses	(116,567)	(0.7)	(123,284)	(0.7)
Other	(830,417)	(5.1)	(831,845)	(5.0)
Total SG&A expenses	(2,189,900)	(13.3)	(2,374,075)	(14.3)
Income from operations	326,700	2.0	568,008	3.4
Interest income	14,636	0.1	11,722	0.1
Interest expense	(50,573)	(0.3)	(40,436)	(0.2)
Other income / (expense), net	(7,413)	(0.0)	(2,342)	(0.0)
Net earnings / (loss) before income taxes	283,349	1.7	536,953	3.2
Income tax (expense) / benefit	(28,323)	(0.2)	(26,062)	(0.2)
Net earnings	$ 255,026	1.6 %	$ 510,891	3.1 %

Number of Participants

	Number
2004	56
2003	46

Financial Ratios

	2004 Participants Average	2004 Participants Median	2003 Participants Average	2003 Participants Median
Liquidity Ratios				
Current Ratio	1.7	1.6	1.7	1.6
Quick Ratio	1.5	1.4	1.4	1.4
Days of Cash	12.5	4.8	10.4	4.5
Working Capital Turnover	8.3	9.8	7.5	10.3
Profitability Ratios				
Return on Assets	4.4 %	5.2 %	7.8 %	5.8 %
Return on Equity	10.6 %	11.6 %	19.4 %	14.1 %
Times Interest Earned	6.6	6.3	14.3	4.0
Leverage Ratios				
Debt to Equity	1.4	1.8	1.5	1.6
Revenue to Equity	6.8	8.0	6.3	7.8
Asset Turnover	2.8	3.0	2.5	3.1
Fixed Asset Ratio	27.9 %	25.4 %	31.2 %	31.5 %
Equity to SG&A Expense	1.1	1.0	1.1	0.9
Underbillings to Equity	12.9 %	12.5 %	14.8 %	14.2 %
Backlog to Equity	5.6	4.0	2.6	2.2
Efficiency Ratios				
Backlog to Working Capital	7.7	4.4	3.1	3.1
Months in Backlog	6.1	5.6	5.0	4.3
Days in Accounts Receivable	67.6	63.5	76.3	62.3
Days in Inventory	3.7	2.9	6.0	2.4
Days in Accounts Payable	30.8	26.8	28.5	21.6
Operating Cycle	53.0	52.8	64.2	46.4

Figure 4.11 Specialty trade contractors with $10–$25 million revenue.

Selected financial data by revenue best-in-class contractors

	All industrial & nonresidential	Best in class	Annual revenue: All $0–$50M	Best in class $0–$50M	All >$50M	Best in class >$50M
Number of companies	250	82	147	56	103	26
Assets ($)	31,644	28,929	6,493	6,055	67,539	78,195
Liabilities ($)	24,169	21,127	4,474	4,000	52,278	58,016
Net worth ($)	7,447	7,773	2,018	2,051	15,195	20,098
Net worth to assets	23.5%	26.9%	31.1%	33.9%	22.5%	25.7%
Revenues ($)	105,456	104,398	22,388	24,578	224,009	276,320
Gross profit ($)	6,339	6,469	1,958	2,237	12,592	15,585
Gross profit margin	6.0%	6.2%	8.7%	9.1%	5.6%	5.6%
SG&A expense ($)	5,113	4,106	1,673	1,693	10,022	9,305
SG&A expense margin	4.8%	3.9%	7.5%	6.9%	4.5%	3.4%
Net income ($)	1,259	2,433	317	572	2,603	6,441
Net income margin	1.2%	2.3%	1.4%	2.3%	1.2%	2.3%
Current ratio	1.3	1.3	1.4	1.4	1.2	1.3
Return on assets	4.0%	8.4%	4.9%	9.4%	3.9%	8.2%
Return on equity	16.9 %	31.3 %	15.7 %	27.9 %	17.1 %	32.0 %

Note: All $ amounts are in thousands.

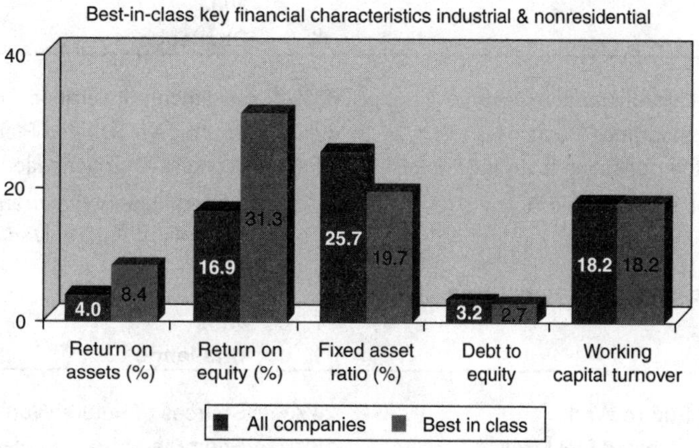

Figure 4.12 Selected financial data by revenue "best-in-class" contractors.

Selected Financial Data by Region

The composite regional financial statements are presented on the following pages. Selected financial data from these regions and the overall composite financial data are presented in the following chart.

	All Companies	REGION NE	SE	MW	SW	W	FW
Number of Companies	537	89	94	154	68	45	87
Assets ($)	28,165	26,248	31,544	22,857	41,532	34,536	22,128
Liabilities ($)	19,264	16,786	22,237	14,975	27,903	25,184	16,360
Net Worth ($)	8,876	9,460	9,193	7,878	13,621	9,347	5,750
Net Worth to Assets	31.5 %	36.0 %	29.1 %	34.5 %	32.8 %	27.1 %	26.0 %
Revenues ($)	82,160	67,584	94,057	66,423	114,282	119,314	67,751
Gross Profit ($)	6,689	6,415	7,855	5,771	8,091	7,783	5,676
Gross Profit Margin	8.1 %	9.5 %	8.4 %	8.7 %	7.1 %	6.5 %	8.4 %
SG&A Expense ($)	5,286	5,242	6,526	4,472	6,318	5,587	4,470
SG&A Expense Margin	6.4 %	7.8 %	6.9 %	6.7 %	5.5 %	4.7 %	6.6 %
Net Income ($)	1,323	1,078	1,248	1,243	2,124	1,480	1,091
Net Income Margin	1.6 %	1.6 %	1.3 %	1.9 %	1.9 %	1.2 %	1.6 %
Current Ratio	1.4	1.4	1.4	1.4	1.4	1.3	1.3
Return on Assets	4.7 %	4.1 %	4.0 %	5.4 %	5.1 %	4.3 %	4.9 %
Return on Equity	14.9 %	11.4 %	13.6 %	15.8 %	15.6 %	15.8 %	19.0 %

Note: All $ amounts are in thousands.

Figure 4.13 Selected financial data by region.

TOP FIVE CHALLENGES IN THE NEXT FIVE YEARS

Percent Selected Within Top Five

2003 Challenge	%	2004 Challenge	%
General Liability Insurance Costs	63%	Healthcare Insurance Costs	65%
Healthcare Insurance Costs	62%	Shortage of Trained Field Help	61%
Workers' Compensation Insurance Costs	61%	Workers' Compensation Insurance Costs	54%
Shortage of Trained Field Help	57%	General Liability Insurance Costs	54%
Sources of Future Work	54%	Sources of Future Work	50%

Percent Selected as Number One Challenge

2003 Challenge	%	2004 Challenge	%
Sources of Future Work	24%	Sources of Future Work	22%
Shortage of Trained Field Help	18%	Shortage of Trained Field Help	20%
Healthcare Insurance Costs	12%	Healthcare Insurance Costs	11%
General Liability Insurance Costs	12%	General Liability Insurance Costs	10%
Workers' Compensation Insurance Costs	9%	Workers' Compensation Insurance Costs	9%

Figure 4.14 Top five challenges in the next five years.

that the 2004 PSMJ survey of the A/E industry can be viewed. The survey reveals that

- Operating profits reversed their downward trend and increased to 9.67%.
- Gross Revenues increased 5%.
- Backlogs increased at a higher rate (5%) as opposed to (3%) in 2003.
- Overhead rates increased to a new 20-year high, primarily due to the increase in insurance costs.
- Direct labor increased by 6% to $25.71.

As with the construction industry, strategic planning was deemed a crucial activity for future financial health. The Executive Summary of the 2004 PSMJ report is included in App. 4.1.

ZweigWhite

ZweigWhite is a leader in management consulting, information and education for the construction and design industries, providing a wide range of services to these industries, from strategic business planning to finance and administration, marketing, project management, and delivery methods. Their yearly survey is looked upon as a bellwether of market conditions for contractors and Architect/Engineers.

Respondents to their 2004 survey were asked to rank 25 markets in terms of expected strength in 2005 and they named health care, K-12 schools, and higher education as having the strongest outlook—markets that are expected to spill over into 2006.

Health Care. The U.S. Department of Commerce statistics confirmed the opinion of the Zweig respondents that health care related construction activities will grow. They estimate that building in this field will have expanded by 7.9% in 2005. Not only are the needs there, but the money also appears to be there. Double digit increases in health insurance costs provide the capital, and an aging population provides the market.

K-12. Demographics combined with approved bond issues and court mandated programs such as "No Child Left Behind" provide the impetus for school construction growth in 2005 and beyond. School systems in the south and west will provide most of the opportunities as rising populations in those areas will rapidly fill existing facilities.

Higher Education. As Baby Boom Echos expand the colleges and universities, new and upgraded facilities will be required to make way for this surge in higher education population.

The ZweigWhite survey reports that, in the engineering design sector, air pollution remediation costs will show no growth unless current administration funding policies change. If any changes to the Clean Air Act are authorized, and if they favor industry, power plant construction will face a downturn.

These ZweigWhite annual surveys pertaining to the design and construction industries along with their design-build survey should be of interest to all firms currently engaged in that process or companies contemplating design-build as part of a company or inclusion in a strategic plan. The 2005 survey that included design-build focused on two major issues—profitability and growth—as discussed in Chap. 3 and revealed that respondents overwhelmingly found design-build more profitable and expected more growth in that area.

Chapter

5

Developing a Design-Build Program

Chapter 3 discussed the different ways in which a design-build team can be created: contractor-led, architect-led, prime contractor, and subcontractor, and the joint venture business structure.

Under the design-bid-build system, the architect's goal was to search out clients desirous of their services and produce a set of construction documents. They marketed their services accordingly. The contractor's goal was to seek contract awards for construction by either hard bid or negotiated work; both expected to make a profit from their efforts. The new design-build team must now focus on somewhat different approaches to their sales development and marketing efforts.

Back in the days when design-build was a relatively new concept, the marketing of a design-build firm would be heavily focused on explaining that approach, advising the client on the potential advantages and disadvantages of switching from a design-bid-build project to this new delivery system. Now that design-build has received its fair share of notoriety and has several published studies substantiating its efficacy, emphasis has now shifted from the system to the experience of the practitioners of the system. A string of successful design-build projects will be instrumental in obtaining new work.

The key to developing a design-build capability depends as much on the creation of the proper business entity as it does on the business development end of the venture. In effect, it is not only a good idea to have the right tool for the job but also to know how to use that tool effectively, otherwise success will be elusive.

Developing a design-build program will surely be different from the way a company that was engaged in only construction or design would have evolved, but embarking on a new business model is not that much different from the previous one.

Building on What You Have

The new team partners already have a base from which they can continue to build. If the teaming partners, architects, engineers, and contractors have been successful in their prior endeavors of either design or construction, they obviously will have created a client list that can be tapped for this new work. Clients like to deal with firms that have treated them fairly in the past and where individuals in both companies felt comfortable working together.

First, one has to decide: What market do I want to reach? Although an architect or a contractor may have engaged in a general practice or general contracting, it is most likely that they have developed some expertise in one or more segments of their respective markets. If a design firm has developed expertise in commercial structures and the contractor in midrise residential buildings, the synergistic effect of both experiences would make them a natural on a mixed-use project.

So, the first order of business might be to discuss how the experiences of each team member, and possibly some in-depth experience of a key member of either firm, could point the way to defining a targeted market segment.

The niche market approach

Design-build, like any other business, has many different ways to view potential markets or market niches, ways that are probably not too dissimilar to those already successfully practiced by the builder or design consultant. Specialization in the capital facilities field is little different from specialization in many other types of businesses or professions. Ones that come readily to mind are the legal profession—where firms specialize in construction litigation, corporate law, accident, medical malpractice, and so on, and the medical field which is rife with specialists.

Niche marketing is not unique to either contractors or architects; design-build can be a niche within a niche.

In Chap. 3 on the design-build team, the various ways a contractor and a design consultant can work together in a design-build venture have been investigated. In Chap. 6, working in the public sector and the expanding market for design-build in both horizontal and vertical construction are discussed. Chapter 8 deals with green buildings and sustainable construction—another rapidly growing field for design-builders.

But why stop there? The basic concepts that warrant design-build are being employed in a wide variety of project types and for a firm that wishes to develop a design-build capability, they need to look at what they have been doing and how it might fit into this new arena.

Let's take a look at some old and new niches.

Sports and Recreational Facilities

Back in the 1960s, when tennis left the country club to become a suburban necessity, operators could hire a preengineered building contractor to erect the structure and either install the court with all its amenities or leave it to the owner to do so. After doing one or two of these projects, the preengineered building contractor was suddenly viewed as a design-build indoor tennis court contractor.

Since the most basic sports facility requires specific heating, cooling, and ventilating performance systems, and the furnishing and installation of specialty equipment for the activity relating to the sport, design-build appears to fit perfectly into several distinct niche markets.

There are single-use and multiuse facilities; a single source may well be the tennis court, however, even those facilities often have a health and fitness club offered to other than tennis buffs. An ice skating rink may be a good example of a single source sports facility while a tennis/basketball/fitness club/spa may qualify as a multiuse facility. Many of these types of structures are single-story affairs with modest budgets but they also require a high degree of expertise to make them function efficiently. As colleges and universities expand their sports programs, there should be a steady demand for this type of product, and experienced design-builders may find this field lucrative.

Large municipal stadiums for professional sports like baseball and football begin to age, owners begin to upgrade not only to avoid that seedy look, but also to remodel areas to become profit centers, adding more suites or upgrading existing ones, installing retail clothing and memorabilia stores, or inviting an upscale restaurant to open in an underused part of the ball park. All of these situations are ripe for design-builders with sports facilities experience.

A good example is the transformation of the college level Gator Bowl stadium in Jacksonville into the Alltel Stadium back in 1995. Seven years later, with the added impetus of being awarded Super Bowl XXXIX, HOK and The Haskell Company formed a design-build partnership project to proceed with a $200 million renovation and expansion program. Not all football and baseball design and construction work is in the multimillion dollar range. Many of the older stadiums throughout the country embark on lesser value projects every year as part of an overall master plan. Some projects may be as little as $500,000, and many range from $2 million to less than $10 million.

Commercial office space

In a hot commercial real estate market, getting a product to market rapidly is what developers seek, and what better way to do this than through the proven track record of quick delivery of design-build projects.

Many commercial structures are divided into two basic components, core and shell (base building), and tenant improvements. Familiarity with the scope of work in each component is necessary to "talk the language" of the developer. Let us take a look at the scope of a generic commercial building divided into these two base components, setting aside the building's structural system.

Core and shell design

Core design. The building's core is symmetrical around one elevator shaft, providing equal access to the core element for building tenants in the north and south of each floor plate. Doors into the core elements (bathrooms, telecommunications closets, electrical closets, janitor's closets, and the like) are located in a midcore cross corridor to allow maximum flexibility for tenants to place offices tight against the core and push the access corridor into their space.

Two male and two female toilet rooms are provided on each floor to minimize the travel distance from a tenant's desk to the toilet room, which is key to a building 200-feet long. Three lavatories are to be provided in each toilet room (12 per floor) to provide equal access within each of the toilet rooms. Service sinks are provided in the janitor's closet just outside the toilet rooms.

Mechanical. Each of the typical office floors is served by two air conditioning units located in the mechanical rooms at opposite ends of the corridor. The units supply conditioned air through ductwork to the space under the raised access floor. Air must be ducted to within 60 feet of the building perimeter to effectively pressurize the plenum space. The mechanical rooms are located within the core so that minimum ductwork is required for the systems to function properly. Fresh air intake shafts and piping are provided in each mechanical room. Under-floor air distribution removes all ductwork from the ceiling spaces, affording greater ceiling height and minimal overhead systems within the tenant space.

This narrative is followed by one for each of the other building components:

Moisture protection
Doors, windows, and glass
Finishes
Specialties
Conveying systems
Mechanical
Electrical
Telecommunications
Sprinklers and fire alarm

Tenant improvements. The landlord leasing space will provide potential tenants with a "work letter," in effect, advising them about the improvements in their space included in the lease rate. The landlord will give the tenant the following improvements:

Partitions. Corridor partitions, partitions between suites full height (slab to slab) with sound insulation. Exterior wall and (unfinished) corridors are not included when measuring partitioning, but 50% of partitions between suites are included. Maximum partition allowance is one linear foot for every 15 square feet of rentable area.

Floor finish. Twenty-eight ounce direct glue-down carpet, one color from the landlord's standard colors. At the tenant's option, vinyl composition tiles or vinyl sheet goods selected from the landlord's standard colors may be substituted in selected areas.

Doors. Suite entrance doors—3′0″ × 7′0″ solid core wood, 1¾ inch thick, stained or painted from the landlord's standard colors.

Wall finishes, casework, shelving, heating and cooling terminal devices, plumbing, sprinklers, fire alarms, and electrical items provided to the tenant are spelled out in much the same detail.

Once this type of successful project has been added to the design-builder's portfolio, with all the unique qualities and demands they create, the real estate market may become a lucrative one.

Security and Design-Build

Since September 11, 2001, security considerations have been at the forefront for many building owners, spawning a whole new area of design and construction. Leading the way naturally is the federal government. The U.S. Department of State's new high-security worldwide construction and upgrade program for U.S. embassies valued at approximately $19 billion includes design-build for 25 out of its 28 projects.

Much has been said about the security of some American industrial sites, the chemical industry being more than a point of conversation for the Department of Homeland Security. The chemical industry has stated their goal to voluntarily provide more security at their high-risk sites, but if they don't show significant progress, the federal government will step in with a whole list of "to do" items, and the value of this kind of work must surely be in the hundreds of billions of dollars.

On a lesser scale, some high-security industries are installing blast resistant windows or renovating an area within an existing building to make it more secure.

Design-build in this type of work is a little more trying than usual. Because of the security associated with the work, access to knowledge of the owner's design requirements will be limited, but by its very nature, this should provide opportunities.

An owner desirous of designing and building a secure structure or enhancing an existing one wishes to limit their program to the least number of people. With a conventional design-bid-build program, even if the project is negotiated, the owner must spread the information base over a wider field. With a sole source commitment, an owner needs fewer meetings with less people under a more controlled atmosphere, thereby maintaining tighter security. Sensitive information can be placed in the hands of one or two individuals who are then charged with the responsibility of protecting and preserving that information. The lesser need for reams of electronic and paper-based communications, and the inherent advantage in the design-build process lends itself perfectly to security. The fewer the documents produced, the lesser the requirement for security.

Lessons learned as the design-build team progresses from client to client make them invaluable as a one-stop shop for security.

Green Buildings and Sustainable Construction

This topic is discussed in detail in Chap. 8 and the reader will acquire information on the green building movement and the opportunities sustainable construction can provide to those design-builds firms willing to learn about this method of design and construction.

Design-Build in the Process and Biotech Industries

The process, petrochemical, power generation, pharmaceutical, and other types of manufacturing industries have frequently employed design-build, often in turnkey projects, where not only were the building and building systems provided, but process machinery was also purchased, installed, and commissioned.

The writer was involved in one such project as the project executive. His firm was responsible for design and construction of the base building that contained general office space, research laboratories, and a chemical processing plant that would produce a monomer to be used in the production of dissolvable sutures in a contiguous part of the building. The product and building had to meet the Food and Drug Administration standards, and a part of the job was to construct a building that would conform to those standards, which included quality of inside air and rigid restrictions on air temperature and specifically on humidity. The project was not considered complete until the FDA's approval was obtained. The lessons learned would prove invaluable should other such design-build projects come the firm's way.

According to an article written by Mr. Mark P. Shambaugh, P.E., Shambaugh & Son, L.P., an MEP specialty contractor, in an issue of *Design-Build Dateline* magazine, selecting a design-builder for a process project is most frequently achieved by one of these three methods:

1. For small projects with a fixed scope, proposals from design-build firms will be solicited with little upfront engineering.
2. When a quick response to market issues is required, a qualification-based selection (QBS) method is employed. Owners will develop their own criteria for evaluation and arrange personal interviews to examine several key team capability issues such as process experience, proposed organizational team, financial strength, innovation, safety records, and of course, fee structure. An in-depth review of similar past projects is also a key factor in the selection process.
3. A third method involves the issuance of an RFP for a Part 1 submission, which would include a 20% engineering study and a GMP proposal from respondents which will evolve into an 8- to 14-week evaluation process before a Part II construction contract can be awarded.

Add in college dormitory design-build work, a promising field reported by one major New England design-build firm, and mixed-use projects consisting of high-rise residential condominium buildings surrounded by low-rise commercial and retail space, addressed by a major residential builder, and the market for this form of project delivery system appears to know no bounds.

Analyzing the Market

Knowledge of a market can be gained from many sources—newspapers, magazines, and the Internet, and sometimes research of markets can point the way to new opportunities for the design-build firm.

Government projects are advertised publicly and are also picked up by subscribing to the services of various project reporting companies. Announcement of bidding dates and program requirements are readily available.

Business Development and Design-Build

The ability to develop a client base, bring in work, and keep those clients happy is the goal of any firm, be it design-build or others, and strong business development is the route to take to achieve that goal.

Business development consultants are often called upon to look at a company's present method of attracting clients and can provide many detailed plans and advice based on their experience in a particular field. However certain basic elements seem to jump out that certainly apply to design-build.

Building client relations

Depending on the niche that has been selected for business development, past relationships with key personnel of those types of firms is a good way to start, often with a call to renew acquaintances and advise them of your company's new business plan. Clients tend to prefer working with people they know, so it stands to reason that developing new contacts in the selected field of endeavor is a foregone conclusion. Taking part in activities of professional and trade organizations, where owners, facility managers, lenders, design consultants, and contractors can meet informally, is a way to begin to establish relationships. Isn't this what networking is all about? Other than the hard bid process, look at the past projects, either from a design or construction standpoint, and list those contracts where an introduction to the client or someone with a relationship with the client produced that job.

Attending trade or professional conferences and seminars is another significant networking opportunity and although it may be difficult at times to leave a busy office, this is as much an educational experience as it is a chance to meet someone who may introduce you to someone who is considering a capital building program.

Developing a Follow-Up Plan

A common phrase uttered around the office is, "We're all salesmen (or salespersons?)." From the friendly voice of the receptionist to the outstretched hand of the CEO, everyone in the organization should be made aware of the "dos" and "don'ts" when prospective clients visit the office, call, or email individuals within the office. Don't be rude or curt; be pleasant and helpful. If phone calls or emails require a response, do so promptly, and efficiently. If the client has a question, find the answer. We all have our lists of "dos" and "don'ts" and all employees should be made aware of them.

At a CMAA 2004 National Conference in Texas, Cinda Bond, vice president, Carter & Burgess, Inc., a business development company, outlined a blueprint for a successful pursuit. Pursuit in this case, simply means a method to pursue prospective work.

- *Pursuit sponsor.* The individual responsible for the overall pursuit
- *Pursuit information.* The client, project, and services targeted
- *Background.* Project size, scope, and history; names of consultants involved in the previous phases; schedule; key concerns; and "hot buttons"
- *Schedule.* Key solicitation dates and milestones
- *Client contacts.* Names and contact information of key decision makers for the targeted project (include the name of the in-house primary contact for each person) may also include date of last contact and date for next contact
- *Teaming partners.* Names and strengths of teaming partners and subconsultants along with the contact information for each firm and the in-house primary contact for each
- *Competitors.* Names of known or likely competitors for the project, including a summary of their strengths and weaknesses
- *Proposed team.* Names of the individuals who will or may be proposed for the assignment, includes in-house and teaming partners
- *Strengths and weaknesses.* An honest assessment of the strengths and weaknesses of your team and the strategies to minimize the weaknesses
- *Action items.* A rolling list of actions to be taken to move the prospect into a "win" including client meetings, strategy sessions, meetings with teaming members, collateral material development, and other key issues; identify an individual for each item along with an anticipated completion date

Producing Effective Presentations

We have all anguished, to some degree or other, prior to and during an oral presentation to a prospective client or a panel of their managers. Did that raised eyebrow mean surprise or skepticism or approval? Most presentations are a combination of written and oral, and a well-planned-out approach to both is necessary to produce a winning one.

The written proposal:

- Don't expect your first draft to be the final version. Expect to edit and rewrite a couple of times, which means allotting plenty of time to produce the finished product. Checking for grammar, punctuation, and spelling goes without saying.
- With word processing, corrections come easy. Look not only for misspellings, but also for margin spacing and other aesthetics. Does it look professional? Is the font size and type appropriate?
- Reread the RFP to ensure that all bases are covered and all points requested by the client are appropriately, precisely, and clearly covered. Then give it to someone else to read and verify.
- Brevity, without being abrupt is key. Various aids, such as graphs, charts, and photos lead your reader to the salient points of the proposal.

The Oral Presentation

- Look at the oral portion of the presentation as a way to illustrate your team's ability to communicate effectively with the client. A presenter devoid of personality will cause a client's eyes to glaze over.
- If graphics are used, keep them simple. Most people tend to focus on one or two items during a presentation; too many items may tend to get lost in the process.
- Continue the themes set forth in the written presentation.
- If other team members are going to participate in the oral presentation, have an in-house run-through (or two) to make sure everyone is confident in their portion of the presentation and is familiar with their place in the presentation (whom do they follow).
- Anticipate the questions that your written and oral presentation will generate and prepare answers, not only for your portion, but also so that the other team members are prepared if they are called upon to respond to questions. Having several members of your team participate in the oral presentation gives the client an impression of depth of staff.
- Whether won or lost, follow up with a written thank-you note. You may have lost by a very small margin, but a well-thought-out thank-you letter may be the key to winning the next one.

The day of the master builder has been resurrected and reappears under the new name of design-build. But a master builder was much more; someone who put the interests of the client front and center, someone who would not compromise on quality, and someone whose words really meant something. Maybe that's what ought to go into that sales brochure.

Chapter

6

Design-Build in the Public Sector

All across this nation the federal government along with the local and state governments are being challenged to perform more efficiently. They have all begun to realize that changes need to be made to existing laws to permit more flexibility in a variety of fields. As they relate to the design and construction industries, ways to create public-private partnerships would be a major step in that direction.

Public capital project development has often been a long and drawn out process governed by budget limitations and multiple review and evaluation procedures, all in the name of maintaining transparency and accountability. The private sector, by its very nature, is willing and able to take risks that are not an option in public work, so the blending of the two environments into a public-private partnership was seen by many as a further step in operating more cost-effectively.

Legislators at all levels of government put their heads together and arrived at many innovative ways to involve the private sector in their quest for efficiency, and design-build was at the forefront of their capital buildings agenda.

Various ways were pursued to create this pubic-private partnership while preserving the public's interests and still allowing for the innovation that the private sector would bring to the process.

Direct selection. A competitive process where a design-builder is selected based on definable, objective criteria, prior experience, complete scope of work, terms, and price.

Best value. Award based on the combination of price and qualitative evaluations.

Equivalent design/low bid. A best value selection where technical submissions are followed by a critique of the proposal and respondents are afforded an opportunity to change their design and adjust their bid accordingly.

Fixed-price design. The owner's Request For Proposal (RFP) contains the maximum cost of the project, with the award based on the best qualitative design proposal.

Adjusted low bid. On selection of the qualified low bidder, the price may be adjusted by further negotiations.

Although every public agency may have their own particular method of initiating a design-build project, they more or less follow the same procedures as those in the private sector:

- Program definition by the owner
- Request For Qualifications (RFQ)
- RFP
- Preproposal Q&A conference followed by the issuance of Addendas, if necessary
- Proposal submission and evaluation
- Postproposal interview
- Contract award
- Start of design and construction postaward process

The Federal Government and Design-Build

The federal government via passage of the Federal Acquisition Regulations (FAR) puts federal agencies in design-build work.

The Army Corps of Engineers doubled their design-build work from $1.1 billion to $2 billion from 1999 to 2002.

Richard C. Viohl, Jr., spokesman for the Naval Facilities Engineering Command, in 2003 said, "Now design-build is the dominant method for procuring Navy construction."

In May 2005, the federal government announced that 25 of their 28 worldwide U.S. Embassy new construction and upgrade programs would use design-build instead of design-bid-build.

The States Use Design-Build

Design-build for state transportation projects where federal funds were involved was not allowed prior to 2003. The Federal Highway Administration (FHWA) regulations prior to that time did not permit design-build for highway construction, except for the Special Experimental Project (SEP) No. 14—"Innovative Contracting." SEP-14 allowed 25 participants to conduct a limited number of design-build projects to evaluate that process.

No longer is design-build road building an "experiment," and effective January 2003 the SEP-14 provision, minus the experimental qualifier, removed the requirements that design-build road projects be defensible as cost-effective. Now *value engineering* principles could also be used to evaluate contract bids. Today, over one-half of all state transportation agencies employ design-build in some form or other.

Subsequently, an American Association of State Highway and Transportation Officials (ASHTO) report verified the federal government's decision to allow design-build where federal funds were included in state projects. But even with some previous reports of good results with the design-build delivery system, ASHTO's report was an eye-opener:

- The original 11 projects in the SEP-14 experimental program, worth about $30.5 million, showed a 36% decrease in design and construction time, albeit at a 5% increase in cost.
- These projects had a higher quality level.
- Local governments experienced less than half the legal claims and litigation associated with design-bid-build projects.
- Design-build projects experienced no contract growth.

The Wisconsin Department of Transportation (WisDOT), after reviewing the ASHTO report about design-build transportation projects in Florida, Utah, Arizona, North Carolina, Minnesota, Colorado, and Washington, found several important options to consider when they planned to embark on a design-build program.

Incentives and disincentives. Include a contingency in the program and offer a certain portion of it to the contractor as payment at the end of the project, depending on the amount of funds used during design and construction. Arizona linked incentive programs to provisions keyed to allowing more traffic sooner on the new road. Other agencies have awarded bonuses for early completion.

Low value or best value as the determinant. Low bid often diminishes innovation and quality levels. Best value seems to be the more advantageous approach.

Miscellaneous contract provisions. Dealing with bonds—100% payment and performance bonds on smaller projects may be justified, but relaxing the 100% standard on larger projects may open up the list of bidders.

Performance versus prescriptive specifications. Some states use prescriptive specifications that permit fewer options to firms submitting bids. Performance specifications, on the other hand, allow a smart firm more freedom in specifying materials, equipment, and even design components that benefit both owner and design-builder in cost, speed of work, and technology transfer. On the other hand, prescriptive specifications may afford an owner more control, particularly when it comes to quality control and quality assurance.

Planning and design input. The more the design input provided by the agency, the less will be the innovation permitted by the design-build firm. Thirty percent design presented by an agency will have the effect of fast tracking a project, but some states, particularly Utah, on its I-15 Project, found that too fast a start did not leave enough time for complete review. Indiana has found a way around that problem by awarding design-build projects in late fall allowing for adequate design progress and review before construction can start in the spring.

Pre-let permitting. Environmental permits and right-of-way agreements are generally the responsibility of the agency, but not in all cases. Timing of securing these permits can significantly alter the progress of a project.

What Other States Are Doing

In July 2004, the Commonwealth of Massachusetts became the 46th state to adopt the design-build project delivery system. Massachusetts spends about $3 billion per year on public-related construction and has accepted design-build for road projects, but adopts a "wait-and-see" attitude before allowing vertical construction in that same mode. The Commonwealth of Virginia, with the enactment of the Public-Private Educational Facilities and Infrastructure Act of 2002, now allows design-build on just about any public building project.

Just looking at the various state legislative action in 2004 provides a window on the acceptance of design-build in public works projects:

California. Legislation permitting transit operators to use design-build.

Florida. Previous approval of the state's design-build high-speed rail project was deemed tax exempt. New laws permit toll road construction via design-build.

Georgia. Design-build use on buildings, bridges, and "other projects" not over $10 million allowed.

Louisiana. New Mississippi River bridge okayed for design-build, plus one other project whose value cannot exceed $45 million.

Maryland. Permits design-build on county public school projects.

Mississippi. Design-build allowed on a pilot project not over $10 million and three projects per year not over $50 million.

Minnesota. Design-build approved for highway projects.

New Hampshire. Addition to Department of Motor Vehicles (DMV) building via design-build, $3.9 million approved.

New Mexico. Design-build approved for public school projects.

Ohio. Design-build pilot project for lounge and convention center at Geneva State Park.

Utah. Adds a very interesting amendment to the existing design-build law stating that an agency can offer an award to *a responsible* bidder that offers design-build services rather than *the lowest* responsible bidder.

As the benefits of design-build projects continue to be recognized, its use by government will obviously grow.

The State of Arizona, a leader in design-build road projects, recently reported on some of their experiences. In a report to the Acting Deputy State Engineer in January 2004, Mr. Julio Alvarado, Assistant State Engineer, said that the widening of a portion of U.S. 60, using design-build, was completed 450 days ahead of schedule, resulting in a savings of lost time to motorists of $22.5 million. Although the final cost of the project was 8.5% over the bid amount, this was significantly lower than the conventional design-bid-build project, which historically ran over budget by 10.5%.

A Phoenix project, adding HOV lanes to State Route 51, solicited bids from a short list of three bidders with Ames Construction/Edward Kraemer, JV the successful bidder at $75,685.000, approximately $6.8 million under budget. In March 2004, when the lanes opened, construction was 5 months ahead of schedule.

Public highway officials in Arizona site four benefits of design-build work:

- Speed of construction
- Savings in total construction costs
- Savings in contract administration costs
- Tremendous reductions in motorist delays

The list of state governments enacting legislation to permit design-build work has been growing rapidly. Between 2000 and 2003, 429 design-build-related bills were introduced in various states and 152 of these bills were passed.

Quicker Delivery of Design-Build Projects

One universal benefit from design-build, as reported by the states in a survey conducted by ASHTO in 2002, is the shorter delivery time for a design-build project.

In Florida, a review of 11 completed design-build projects revealed a 36% decrease in design and construction time. Projects have been completed, on an average, 33% faster than conventional projects.

Utah's $1.56 billion replacement of Interstate 15 with an eight lane highway, HOV lanes, and reconstruction of 142 bridges in preparation for the 2002 Winter Olympics was extremely successful. This project that was estimated to take 7 years if constructed in a conventional manner, was completed in $4^1/_2$ years.

In North Carolina, NCDOT reported that the speed and innovations provided by design-build can shorten the entire course of some projects by 3 years.

A Look at Cost Savings

Cost savings, as reported by some agencies referred to time savings by commuters when compared to lengthy traffic delays occurring when conventional design-bid-build projects are employed. Cost savings in total project costs were minimal in most cases, or in some others nonexistent. In Florida, early in the pilot process, costs were higher than conventional projects, approximately 10% over traditional project delivery systems.

On the other hand Utah claimed that they saved $30 million on the I-15 project.

Licensing Laws Affecting Design-Build

Licensing laws had to be revised to allow architects to practice construction, and laws were required to define what design-build is and how the states can solicit proposals for this type of work.

The State of Pennsylvania's recent passage of an amendment to the Architect's Licensure Law is typical of the approach many states are taking to allow design-build projects to be built. This licensure law had prohibited any party from providing architectural services, unless they were authorized under existing Pennsylvania law to practice architecture. This law essentially prohibited contractor-led design-build in that state, and because architects were reluctant to form architect-led design-build entities due to bonding, insurance, and capital concerns, the design-build delivery system was stymied.

The amendments to the Architect's Licensure Law passed by the state legislature defines design-build as a project delivery system whereby a single contract is issued to provide a combination of architectural and construction services to a client.

The law also authorizes firms that practice architecture to provide design-build services, and lastly, this new amendment eliminates the requirement that only a firm practicing architecture can offer architectural services.

Some states are addressing the practice known as "bridging" as a modification to the design-build process. A $96 million federal courthouse in Las Vegas, Nevada, was recently completed by Chanen Construction, headquartered in Phoenix, utilizing bridging. The owner, in this case the General Service Agency, hired an architectural firm to produce schematic drawings and outline specifications that were used to solicit bids from design-build contractors. The architectural firms hired during the bridging phase were precluded from entering into the subsequent design-build competition since the first phase is acknowledgment of a design (they created) that must meet the agencies program.

Bidders point out some shortfalls in the process

Some design-build firms point out the cost to prepare and submit an RFP responding to a design-build project with costs ranging from low five figures to high six figures. These costs can deter a number of firms from entering into, what is still, in effect, a hard bid in its initial phase. In some locales design requirements are very high; one firm said they are required to submit a full schematic that really represented a 30% design document, and they were also required to include a rendering and some engineering drawings. Another design-build firm said that over a period of 4 years they have been awarded nearly $200 million, but this required responding to about 60 RFPs at a cost of 5% to 10% of their proposed fee. This same firm said that typically the military only prequalifies three to five firms, which means that a lot of design-build firms are contributing a lot of design time and money hoping to snag a winner. Some government agencies offer a stipend to proposers that don't make the short list and even though this helps to soften the blow, it is not sufficient compensation to cover a significant portion of the cost of the proposal.

A stipulated fee (stipend). State governments have realized for some time now that offering to make a payment or stipend, albeit a small one, to short-listed firms makes their interest in submitting design-build proposals a little more palatable.

On the federal level, several agencies are responding to concerns voiced by many design-build companies and are reviewing the amount of the stipend, minimizing some submissions, eliminating nonrelevant specification details, and standardizing RFPs in an attempt to reduce the cost of the RFP.

The Experience Factor

Another limiting factor in public sector RFPs is the inclusion of experience requirements. Points are awarded for experience in projects of similar scope, experience of the design-build team working together on a previous project, and on-budget-on-time completion of other design-build projects.

This is as it should be, with officials in the pubic sector seeking out only those who have a proven track record in design-build. However, if other things are equal in the response to an RFP, a firm, possibly with a superior understanding of the project or a much better approach to the owner's program, may be disqualified from the short list based solely on their lack of experience. Perhaps government agencies awarding these types of projects may find a way to equate experience required with project size and complexity, thereby allowing design-build firms with little or no experience, but highly qualified otherwise, to participate on smaller, less complex projects giving them an opportunity to work their way up the experience ladder.

Private Sector Teams Learning from Public Sector Procedures

Design-builders interested only in the private sector can learn a lot from the proposal and evaluation process that has evolved through several iterations of public policies as reflected in their RFPs. By reviewing and evaluating what states deem most important in design-build proposals, firms operating in the private sector may gain more knowledge of what owners are looking for and be able to stress these points in their written and oral presentation.

The two-part RFP on the federal and state levels

Kansas City, Missouri recently adopted design-build for eight projects: four fire houses, a police station, two bridges, and a street improvement project, four of which were estimated to cost $7 million and four to exceed $7 million. Their approach to this process, similar to that of many other public agencies, requires eight steps:

1. The city will issue an RFQ.
2. Interested firms will submit their statement of qualification (SOQ).

3. The city selection committee will evaluate and score each SOQ.
4. The city will invite the highest scoring firms (short list) to submit bids.
5. The firms will provide a two-package proposal—technical submission and a cost submission.
6. The city will score the technical submission based on selected criteria.
7. The city will score the cost submission also based on selected criteria.
8. The best combined scorer will be recommended for the project.

The federal two-part RFP. At the federal government level, FAR 48FAR, Chapter 1, Part 15 is representative of the way in which a two-part design-build proposal is offered. The contracting officer for the agency is charged with certain responsibilities that require them to look inside their department to ensure that they have a need to initiate a design-build proposal.

They must consider:

- The extent to which the project definition and requirements can be established and incorporated into an RFP.
- Whether there is sufficient time allowed for a two-part proposal in which the bidder's qualifications will be evaluated and result in a preselection process before they will be requested to submit technical and cost information.
- Whether the agency has considered the capability and performance of the proposed bidders.
- Whether the project is actually compatible with the planned two-stage proposal process.
- Whether the agency on receipt of both parts of the proposed proposal can properly evaluate them with the personnel currently on staff, or if additional staff is required.

The two-part RFP divided into a Part I or Part A phase and a Part II or Part B phase are generically similar throughout the public sector, some with more details than others.

Part I or Part A. This portion of the RFP is devoted to establishing the bidder's qualifications, which will be evaluated before short listing and proceeding on to the next phase. This questionnaire will invite responses to

1. Verify the bidder's technical competence and experience in the type of project being considered.
2. Document past performance of the proposed design-build team—the contractor and the design consultants.
3. Detail the capacity of the team to meet the criteria included in the RFP.
4. Answer to other factors that may be appropriate to the specific situation at hand.

Part II or Part B. This phase of the RFP will require bidders to

1. Provide a technical proposal to meet the goals established by the agency.
2. Provide cost and pricing information commensurate with the technical data they submit.

Requirements for Complying with the Prequalification Phase of a Design-Build RFP

There are several factors that a public agency, at minimum, will consider when determining whether a responder to an RFP will be qualified to fulfill the objectives of the project. These are

- Design and construction experience in the facility type under consideration by the agency
- Experience in design-build method of project delivery and the experience of the team being proposed
- Limits of bonding capacity and proof in the form of a letter from the bonding company (not the agent) attesting to available limits
- The ability to provide insurance in the type and amounts required as documented by a letter from the insurance provider
- Operative geographic area of the design-build team—contractor and design consultants
- Proposed composition of design-build team including all subconsultants and specialty consultants required for the project
- Proposed participation by any disadvantaged business enterprises—minority or women-based enterprises
- Status of all professional and government licenses/registrations, as applicable, for each member of the design-build team

The project owner may also include a more detailed list requesting information that would aid their selection committee in evaluating the qualifications of the bidders:

- Owner and industry references attesting to the bidder's high standards of both design and construction
- Specific experience of the key members of the design and construction team
- Specific design-build experience in projects similar to the one in the RFP
- Experience of the design-build team that had successfully worked together on a design-build project
- Financial strength of both the design and construction team members to include any lines of credit available

- Performance records of the design-build team members including fee structure, validity of cost databases, and valued engineering experience
- Quality of the proposed technical and managerial staff
- Quality of key individuals and their proposed positions in the design-build team
- Design approach, philosophy, and preliminary design concept
- Construction project management plan to include a summary of schedule, cost, and quality control plans
- Participation of local labor, business firms, and DBE, MBE, and WBE involvement

The Evaluation Process

The Minnesota Department of Transportation issued a document outlining the methodology and criteria for evaluation in 2001 (App. 6.1). This well-thought-out paper sets out the procedures to ensure consistency and fairness in the evaluation of design-build respondents to their RFPs. In the Scoring Allocations section (Table 1), note that experience and capabilities of the team are worth 35 out of a possible 100 points.

As part of the selection process, many government agencies schedule face-to-face interviews with each respondent. The format will generally consist of a Q&A session followed by an oral presentation by the proposer. Each member of the interviewing team will grade the responses to each question individually.

State of Maryland weighted evaluation approach

Often, each part of the RFP evaluation is weighted, similar to a proposal for an elementary school put out by the Prince George's County Public School system in Maryland. Their breakdown of points for each category was:

Appropriate project experience. Total 35 points (21 points required as minimum). Qualifications and experience of the design-build firm, including

1. Experience with similar design-build projects
2. Experience with other types of design-build projects
3. Management approach for the design-build delivery method
4. Experience with public school design and construction issues in Maryland
5. Design and construction quality as evidenced by industry awards and recommendations
6. Related project experience
7. Specifically, State of Maryland and Prince George's County major project experience

Team resources and capacity. 30 points (18 points required as minimum)

1. Team resources, ability, and capacity to meet this project's design and construction requirements and to complete the project within the schedule and budget
2. Key personnel's experience with similar design-build projects and with the design-build delivery system
3. Design-build team history of working together

MBE compliance. 30 points (18 points required as minimum)

1. Past record of MBE participation
2. Proposed plan to achieve MBE participation in design and construction

Educational support. 5 points (3 points required as minimum). Experience with and willingness to develop and participate in graduate school internship, mentoring, and apprentice programs with local architectural, engineering, and business science student residents. (This is rather an unusual requirement that appears to represent an attempt by the local school board to have the DB team transfer some skills and technology to local students.)

Maryland adds flexibility to the evaluation process

A recent bill passed by the Maryland legislature in 2004, Senate Bill 787, refers to the financing of public schools in the state. Recognizing that accepting the low bid is not always in the public interest, SR 787 allows prospective bidders to use their knowledge and experience to include other components in their design-build proposal package. Those sections of this bill that have applicability to the design-build process are as follows:

> *Section A-(5).* Design-build arrangements that permit a county board to contract with a design-build business entity for the combined design and construction of qualified facilities, including *financing mechanisms where the business entity assists the local governing body in obtaining project financing.*
>
> *Section B-(2).* Engage in competitive negotiation rather than competitive bidding, in limited circumstances including construction management at-risk arrangements, and other alternative project delivery arrangements, as provided in regulations adopted by the Board of Public Works.
>
> *Part (4).* Use quality-based selection, in which selection is based on a combination of qualifications and cost factors, to select developers and builders, as provided in regulations adopted by the Board of Public Works.

An official in the Public School Construction Program elaborated on this competitive negotiation process by referring to one school project that was awarded based on a unique design feature that elevated one design-build proposal above another. One team included in their design an exposed solar panel installation behind a glass enclosure complete with plumbing and valves to show the elementary level student how solar power works—it won them the competition.

ADOT's Short Bidder's Compensation Provision

The Arizona Department of Transportation (ADOT) in their *Design-Build Procurement and Administration Guide* includes a provision requiring the agency to pay a stipend to short-listed bidders that were not selected as the successful proposers. This provision states:

> Stipulated Fee (Stipend)
>
> The Department is required to pay all short-listed firms a stipulated fee (or stipend) equal to 0.2% of the engineer's estimate for the project. The selected Design-Builder does not receive the stipend. Only short-listed design-build firms that are not selected but submitted responsive proposals are allowed to receive the stipend though upon request, a firm may elect not to receive the stipend. This election prevents the Department from using any of the ideas and information contained in the firm's technical proposal. If the Department cancels the contract, all short-listed firms including the selected Design-Builder will receive the stipend. The stipend must be paid within 90 days from the award of the contract or from the day the decision is made not to award.

This is a smart move by the agency. It encourages responsible bidders to submit proposals, partially rewarding them for their efforts if they are not successful, and also allowing the agency to evaluate any of the ideas included in those proposals for future work. The stipend of 0.2% of a $1.5 million project would be $30,000, offsetting some of the contractor's expenses and providing the agency with some reasonably inexpensive design ideas.

State of California as Innovator

California has been the birthplace of new movements from the Hip Generation of Haight-Asbury to requiring automakers to manufacture automobiles with more stringent pollution controls for sale in their state. The movement to embrace design-build is no exception. Seventeen statutes (Fig. 6.1) have been enacted since 1993 to permit design-build by local and state authorities. Since 1995 several design-build projects have been completed (Fig. 6.2) and the counties and cities where these projects were built have generally expressed favorable opinions of the process.

The Legislative Analyst's Office, a California nonpartisan fiscal and advisory agency, took a look at two prevalent means of construction project delivery systems, design-bid-build and design-build, and compared the advantages and disadvantages of each system (Fig. 6.3).

They also compared the advantages and disadvantages of design-build, where a stipulated sum contract would be awarded and also design-build via the CM approach. Their findings are discussed in the following sections.

Recent State Laws Authorizing Design-Build

Authorization	Facilities	Comments
State		
Ch 429/93 (AB 896 Brown)	Junipero Serra (Los Angeles) and Civic Center (San Francisco) buildings	
Ch 430/93 (SB 772, Petris)	Elihu Harris (Oakland) building	
Ch 761/97 (SB 1270, Johnston)	East End Project (Sacramento)	
Ch 252/98 (SB 776, Johannessen)	Permits Department of General Services to use design-build on at least five projects authorized by Legislature	• Used for CalTrans District 7 building (Los Angeles) • Expires 1/1/06
Ch 782/98 (SB 1934, Johnston)	Department of Corrections headquarters (Sacramento)	• Not used
Ch 733/99 (AB 290, Steinberg)[a]	Department of Parks and Recreation, Stanford Mansion restoration (Sacramento)	
Ch 672/01 (SB 809, Ortiz)	West End Project (Sacramento)	• In planning stages
Local		
Ch 663/95 (AB 1717, Cortese)	Four specified counties	• Projects not exceeding $50 million • Expired 1/1/01
Ch 1040/96 (AB 2660, Aguiar)	Authorized local agencies to enter into agreements for private funding and development of revenue producing facilities	
Ch 258/99 (AB 755, Corbett)	Alameda County, juvenile justice facility	
Ch 541/00 (AB 958, Scott)[a]	Transit operators	• Projects exceeding $10 million • Expired 1/1/05
Ch 594/00 (AB 2296, Dutra)[ab]	Seven specified counties	• Projects exceeding $10 million • Expires 1/1/06
Ch 767/00 (SB 1144, Johannessen)[a]	Two specified cities	• Projects not exceeding $50 million
Ch 421/01 (AB 1402, Simitian)[a]	School districts	• Projects exceeding $10 million • Expires 1/1/07
Ch 637/02 (AB 1000, Simitian)[ab]	Three specified community college districts, and five additional as selected by the community colleges chancellor	• Expires1/1/08
Ch 976/02 (SB 1759, Johannessen)[ab]	Four specified cities	• Projects exceeding $5 million • Expires 1/1/06
Ch 196/04 (SB 1130, Scott)	Transit districts	• Revised Ch. 541/00 • Expires 1/1/07

a Required to report information to Legislature.

b The LAO is required to report on local implementation.

Figure 6.1 Recent California Laws authorizing design-build. (*Source: Legislative Analysts Office, State of California.*)

Summary of Design-Build Activities by Authorized Cities and Counties

Agency	Used Design-Build	Did Not Use Design-Build	Types of Projects
Counties			
Chapter 663, Statutes of 1995			
Solano	X		• $2.3 million juvenile hall expansion • $0.4 million county recorder's office renovation
Chapter 594, Statutes of 2000			
Alameda	X		• $15 million county recorder's office building • $135 million juvenile justice center (under construction)
Contra Costa		X	
Sacramento	X		• $2.5 million branch library
Santa Clara		X	
Solano	X		• $18.4 million health and social services building (under construction) • $80 million county administration center (under construction)
Sonoma		X	
Tulare		X	
Cities			
Chapter 1040, Statutes of 1996			
Woodland	X		• $14.4 million police station
Chapter 767, Statutes of 2000			
Davis	X		• $7.3 million police station
West Sacramento	X		• $2.6 million pump station
Chapter 976, Statutes of 2002			
Brentwood		X	
Hesperia		X	
Vacaville		X	
Woodland		X	

Figure 6.2 Summary of design-build projects in California. (*Source: Legislative Analysts Office, State of California.*)

Design-build—stipulated sum contract award

Advantages

1. Affords the agency with best certainty of cost of the project at its outset. The risk is similar to all lump sum low bidder awards–quality may suffer.
2. The agency may avoid conflicts because the designer and builder are part of the same entity and the public agency is not the guarantor of the completeness and accuracy of the design—conflicts that often arise in a conventional design-bid-build project.

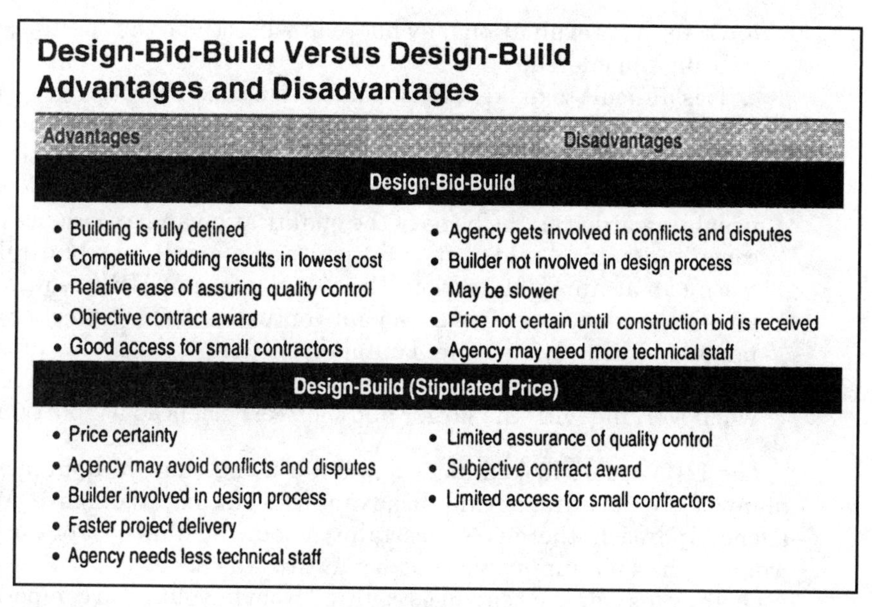

Design-Bid-Build Versus Design-Build Advantages and Disadvantages

Advantages	Disadvantages
Design-Bid-Build	
• Building is fully defined	• Agency gets involved in conflicts and disputes
• Competitive bidding results in lowest cost	• Builder not involved in design process
• Relative ease of assuring quality control	• May be slower
• Objective contract award	• Price not certain until construction bid is received
• Good access for small contractors	• Agency may need more technical staff
Design-Build (Stipulated Price)	
• Price certainty	• Limited assurance of quality control
• Agency may avoid conflicts and disputes	• Subjective contract award
• Builder involved in design process	• Limited access for small contractors
• Faster project delivery	
• Agency needs less technical staff	

Figure 6.3 Advantages-disadvantages of design-build as reported by agencies. (*Source: Legislative Analysts Office, State of California.*)

3. Because the builder is involved in the design process from the beginning, they can provide useful costing information and availability of materials and equipment to make the design more efficient and less costly.
4. By overlapping design and construction to some degree, and potentially reducing conflicts between design consultants and builder, design-build can deliver a project faster.
5. With a design-build project, the agency does not have to review the accuracy and completeness of the design consultant's work and therefore may not need more in-house technical staff assigned to the project.

Disadvantages

1. *Limited assurance of quality control.* This is because the agency's wants may not be sufficiently defined in detail when it enters into a contract. The agency may have little control over the quality of the construction work.
2. *Design-build projects.* These are generally awarded on the basis of subjective reviews such as experience, qualifications, and best value. Even though agencies develop various point systems and other processes for evaluation, drawbacks can still occur.
   a. Public managers have discretion in awarding points, and there is no objective way to determine the exact number of points that a bidder should receive for a specific activity, i.e., does one bidder's past experience in similar projects rate a 43 out of 50, while another bidder is awarded a 44?

b. Criteria for evaluation may not relate directly to the specific building type being considered.

c. It is difficult to make a reasonable comparison of alternative added value proposals. It will be difficult to compare one bidder's enhanced plumbing system with another bidder's upgraded electrical distribution system.

3. *Limited access for small contractors.* Because design-build contracts are usually awarded on the basis of the qualification and experience of the design consultants and builder, it is difficult for a small, newly established contractor to attain qualification. Even in the face of MBE requirements with points awarded, large firms via joint ventures (JVs) can meet or exceed some minority hiring goals. The JV approach does have its benefits because it allows an MBE or DBE firm to acquire more knowledge and sophistication when working with the larger and more experienced JV partner.

The LAO report looked at design-build using the CM approach and found many of the advantages and disadvantages enumerated above. With the CM agency approach, there is less certainty about the final cost of the project than would be had if a lump sum contract award was used.

On the plus side, any savings resulting from buyouts and competitive bidding will accrue to the agency rather than to the design-build contractor in a lump sum arrangement or even to the owner and contractor in a GMP contract, where any savings are shared.

Lessons Learned

The local agencies where design-build projects were built made various observations about the efficacy of a process:

- Statutory requirements regarding specified maximum or minimum project costs prevented the agencies from using design-build on certain projects. The local agencies saw no compelling reasons for imposing cost thresholds as a criterion of whether or not to employ design-build.
- Adding objectivity in the procurement process would be a plus. Using a two-step process to select a design-builder seemed to be a way to achieve more objectivity. Three counties used subjective criteria such as experience and qualifications to create a short list. These finalists then submitted design and cost proposals based on county criteria and a contract was awarded on the objective criteria of lowest cost. So a mixture of subjective criteria—experience and qualifications and proposals of best value, combined with a second objective review of cost appeared to satisfy this need for objectivity.

Good project definition is needed before awarding a design-build contract. Agencies need to use conceptual drawings, program statements, and other documentation to (1) provide bidders with a complete understanding of what is required and (2) form the basis for a contract between the agency and the design-builder.

- Design-build is best suited for straightforward projects. Where projects are less defined, such as in jail and hospital construction, there was less certainty that design-build was the best delivery system.

The Arizona Department of Transportation published a guide to design-build in December 2001 in which they succinctly described the virtues of employing design-build in the public sector and those observations hold true for private sector work as well.

> A certain amount of enlightenment and synergy occurs between the designer and constructors when they work closely side by side to solve problems. In the traditional role, where each group tends to work in isolation, many beliefs and practices are taken for granted and are rarely challenged. The enlightenment and synergy required on design-build projects cause team members to both question standards and look for cost-effective, innovative alternatives that meet the construction needs of the project. Technical leaders are often questioned about ADOT design policies and standards. Rather than rigidly applying the standards; technical leaders are encouraged to look beyond the standards and policies to identify the underlying issues the standard or policy attempts to resolve. Once these issues have been identified, the technical leaders are in a better position to decide the merits of the design-build team's innovation and help develop solutions that meet everyone's interests.

Appendix 6.1: Minnesota Department of Transportation Evaluation Procedure

T.H. 52 RFQ

SOQ Evaluation Procedure

1.0 Introduction and Purpose of the Procedure

This document provides the methodology and criteria for evaluation of the Statement of Qualifications (SOQs) received in response to the Request for Qualifications (RFQ) for the T.H. 52 Design-Build Project (Project) issued on December 17, 2001.

The purpose and goal of these procedures is to ensure consistency and fairness in the approach to determine the most qualified respondents to the RFQ for the purposes of shortlisting the most responsive design-build Proposer that will receive a Request for Proposal (RFP) for the Project. The intent is to protect the interests of the Minnesota Department of Transportation (Mn/DOT), as well as those of the respondents to the RFQ.

2.0 Evaluation Procedure

The scores will be developed using the procedure summarized in this document and detailed below:

- Mn/DOT will select technical advisors.
- HDR Engineering will evaluate each SOQ for compliance with minimum qualification criteria.
- Each Evaluator will assess and score individual SOQs passing the minimum qualifications using the overall criteria described in this document and with the assistance of technical advisors. The technical advisors will review the submitted SOQs and assess the SOQ for the Proposer's level of competence in responding to the RFQ and Project requirements. The technical advisors will support and assist the Evaluators on the Evaluation Committee in connection with their review and scoring of the SOQs but will not individually or independently score any SOQ.
- The Evaluation Committee shall meet and discuss the submitted SOQs according to the methodology outlined in this manual and feedback from the technical advisors.
- The technical advisors and Evaluation Committee may prepare written or oral questions to ask some or all of the Proposer teams before or at the oral presentations, at the option of the Evaluation Committee.
- The Evaluation Committee may request Proposer team oral presentations for the Evaluation Committee and technical advisors.
- After completion of the oral presentations, if held, the Evaluation Committee, with the assistance of the technical advisors, will have the opportunity to adjust their scores and enter them using the appropriate column on the Qualitative Matrix, along with comments discussing the basis of adjustment.

- The Evaluation Committee will examine the total adjusted scores for each SOQ and determine a logical breaking point for the shortlisting of responsive Proposers.
- The Evaluation Committee will submit its completed matrices to the Chairperson and will prepare a report documenting which summarizes the results of their evaluation. The report will be forwarded to the Mn/DOT Commissioner of Transportation (Commissioner) for approval and finalization of the shortlist.
- The Commissioner will receive and review the summarized evaluations of the RFQ Evaluation Committee. By approval of the Commissioner, a shortlist of respondents will be established. The shortlisted respondents will then be invited to respond to the RFP for the Project.
- All Proposers submitting SOQs will be notified in writing of the results of the evaluation process.

Mn/DOT may, but is not required to, debrief those Proposers that are not shortlisted. Such debriefings are at Mn/DOT's sole discretion.

3.0 Chairperson and Evaluator Responsibilities

The Chairperson shall serve as a point of contact if Evaluator or Evaluators have questions or encounter problems relative to the evaluations. The Chairperson shall coordinate and facilitate the participation of technical or other advisories as may be necessary during the course of the evaluation and selection process.

The Chairperson is responsible for ensuring the timely progress of the evaluation, coordinating any consensus meeting(s) or reevaluations, and ensuring that appropriate records of the evaluation are maintained.

To the extent the Chairperson determines it appropriate, the Chairperson may deviate from any procedure as prescribed herein as long as said deviations do not otherwise constitute violation of applicable law. The change or modification shall be documented in the RFQ Evaluation Committee's report to the Commissioner.

Each Evaluator will individually review and assess individual SOQs using the overall criteria set forth in the attached evaluation matrices. Each Evaluator shall record his/her impressions and judgments via the attached Evaluation Forms. These forms are intended to provide a record of the evaluation and will be utilized as a beginning point for further discussions and evaluations. The Evaluation Forms should be completed in a manner that adequately indicates the basis of the Evaluator's assessment, including the significant advantages, disadvantages, and risks supporting the assigned ratings. Reasoning for assigned scores or comments shall be thoroughly

documented. It is critical that the Evaluator's evaluation comment and score justification statement for each SOQ be specific and not a generalization.

Each Evaluator will review the criteria prior to assessing the submitted SOQs. If an Evaluator has any questions regarding the evaluation criteria, a clarification shall be requested from the RFQ Evaluation Committee Chairperson.

Evaluators shall comply with all applicable law, including any relating to nondisclosure of proprietary or confidential information and other source selection information.

Upon receipt of the SOQs, the technical advisors and Evaluation Committee members will deliver a written disclosure to the Commissioner identifying any conflicts of interest or relationships with individuals or entities on any Proposer's team or with any Proposer's team member.

If an Evaluator is unable to complete his/her evaluation responsibilities to the extent the Chairperson determines necessary or if additional Evaluators are necessary to evaluate the SOQs more completely, the Chairperson shall take whatever steps he/she determines appropriate to arrange for substitution and or/supplementation of evaluation personnel.

4.0 Technical Advisors

Technical advisors will submit an original copy of their assessments to the RFQ Evaluation Committee Chairperson for distribution to the Evaluators for consideration in completing the scoring matrices. The technical advisors will be available to the Evaluation Committee during the evaluation process and will participate in the oral presentations if held.

5.0 Detailed Evaluation Criteria

The RFQ specifies that each Design-Build firm is to include in its response detailed information that demonstrates the Developer's experience and qualifications in projects of a size and complexity similar to or greater than the Project. The SOQs are required to contain specific information and to elaborate on the Proposer team's specific qualifications and experience.

5.1 Pass/Fail Evaluation Portion

The pass/fail section of the evaluation requires that each Evaluator assess the SOQ for meeting the general submittal requirements of the RFQ as well as legal and financial issues and assign a pass/fail score. The pass/fail ratings are based on the following general RFQ evaluation criteria as it relates to the Design-Build firms proposal:

- General submittal requirements
- Financial stability and capability
- Legal implications of Proposer structure
- Ability to obtain a performance bond

The T.H.52 RFQ Pass/Fail Criteria Worksheet is a listing of required information and can be found in Appendix A. The Design-Build firms who substantially comply with the requirements of the RFQ will be given a passing score in this portion of the evaluation. Failure to address a particular requirement or failure to include or deliver an important item of information that is required by the RFQ may be grounds for failing the Proposer on that item.

A failing score in one or more of the items listed in the pass/fail portion of the evaluation process may be grounds for a determination that a particular Proposer is noncompliant and may not be shortlisted for the Project. In addition, proposals must substantially meet the pass/fail criteria to be advanced to the qualitative evaluation process. The RFQ Evaluation Committee Chairperson may correspond with a Proposer to request information to correct a failing category.

5.2 Qualitative Evaluation Portion

The qualitative section of the evaluation requires that each Evaluator assess the SOQ in the categories listed below and assign a qualitative score from Excellent to Poor:

a) Organization Issues.
b) Project Team Experience and Capabilities.
c) Project Understanding.
d) Design-Build Project Approach.

The Excellent to Poor ratings are based on the following four general RFQ evaluation criteria:

a) Organization Issues
  - Effective project management authority and structure
  - Realistic and efficient design and construction management structure
  - Effective utilization of personnel and equipment

H

- Key management/staff experience, capabilities and functions on similar projects.
- Owner/client references

b) Project Team Experience and Capabilities
   - Experience on projects of similar scope and complexity
   - Experience with timely completion of comparable projects
   - Experience with on-budget completion of comparable projects
   - Experience of Design-Build team members working together
   - Team members with experience and qualifications that cover Project scope

c) Project Understanding
   - Understanding of Project scope
   - Understanding of Mn/DOT's goals for the Project
   - Understanding and inclusion of expertise necessary to develop Project
   - Understanding of impacts on the community
   - Understanding of required interaction with local governments, municipalities, property owners and utility entities/companies
   - Understanding of permitting needs and strategy

d) Design-Build Project Approach
   - Completing Project on time and within budget
   - Delivery of high-quality, safe, durable Project
   - The significance of creating/maintaining positive public image and effective response plan
   - The importance of effectively managing community interests, local/state government concerns and political focus on the Project
   - Effective management plan
   - Flexibility and ability to handle conflicts/issues
   - Ability to meet DBE project goals

The five assessment levels of general competency of the Proposer qualifications as related to the stated evaluation criteria are:

Excellent (E): The Proposer demonstrates an approach that is considered to significantly exceed stated requirements/objectives in a beneficial way and provides a consistently outstanding level of quality. There is very little or no risk that the Proposer would fail to satisfy the requirements of the Design-Build contract.

Very Good (VG): The Proposer demonstrates an approach that is considered to exceed the stated requirements/objectives and offers generally better-than-acceptable quality. There is a very small risk that the Proposer would fail to satisfy the requirements of the Design-Build contract. Weaknesses, if any, are very minor.

Good (G): The Proposer demonstrates an approach that meets the stated requirements/objectives and offers acceptable quality. There is a very small risk that the Proposer would fail to satisfy the requirements of the Design-Build contract. Weaknesses are minor and can be readily corrected.

Fair (F): The Proposer demonstrates an approach that is considered to marginally meet the stated requirements/objectives and has a marginal level of quality. There are questions about the likelihood of success and tangible risk that the Proposer would fail to satisfy the requirements of the Design-Build contract. Weaknesses are prevalent, and may or may not be readily correctable or acceptable in accordance with standards.

Poor (P): The Proposer demonstrates an approach that contains significant weaknesses, deficiencies and/or unacceptable quality. The SOQ proposal failed to meet the stated requirements/objectives and/or lacked essential information and is conflicting and/or ineffective. There is not a reasonable likelihood of success, and there is a high risk that the Proposer would fail to satisfy the requirements of the Design-Build contract.

The T.H. 52 RFQ Qualitative Criteria Worksheet for each individual evaluation criteria listed above can be found in Appendix B.

For evaluation of the SOQs, comparable projects are defined as projects that include one or more of the following components:

H

- Design, permitting, and construction of large highway projects.
- Design-Build Agreements, innovative contracting procurements, and public/private partnerships.
- Federally funded highway projects.
- Projects with construction warranties by the design-build firms.
- Projects with guaranteed maximum/fixed prices.
- Projects with guaranteed completion dates.
- Extensive community outreach and relations program.
- Construction value of at least $50 million.
- Environmentally sensitive activities.

The RFQ Evaluation Criteria listed above will be used as a guide once the Evaluators begin evaluating the submittals. The term "team members" is used to refer to companies that are identified in the SOQ as having an equity position in the Project and/or are given a substantial role in the performance of the terms of the Design-Build Agreement. Proposers that present substantial relevant experience and positive references should score higher than those with less relevant experience or weaker references.

Following the qualitative evaluation, the Evaluator will determine a numerical score for each major selection category based upon the overall category adjectival rating. Numeric scores will be assigned to the four major evaluation categories. The numerical range associated with each qualitative response is listed on the SOQ Team Summary Sheet found in Appendix B. A maximum of 100 points will be assigned based on the allocations shown in the Table 1 below.

Table 1
SCORING ALLOCATIONS

Evaluation Criteria	Maximum Score
Legal and Financial Issues	Pass/Fail
Organization Issues	15
Project Team Experience & Capabilities	35
Project Understanding	20
Design-Build Project Approach	30
TOTAL	**100**

6.0 Scoring /EvaluationForms

There are two sets of forms to be used during the evaluation process: i) the pass/fail forms and ii) the qualitative forms. Each of the four qualitative evaluation categories listed above have a set of forms that will allow the Evaluators to rate the Proposer based upon the criteria described in the RFQ. Once the qualitative rating have been assigned for each criteria, the Evaluator will then determine a numerical scope of the Proposer for each evaluation criteria listed above in Table 1. The numerical scopes for each evaluation criteria will be shown on the SOQ Team Summary Sheet found in Appendix B. The composite total score is then at the bottom of the SOQ Team Summary Sheet by summing the individual categorical scores. After completion of all SOQ evaluations and scoring, a T.H. 52 Project Summary Sheet (found in Appendix C) will be generated from all individual Evaluator's SOQ Team Summary Sheets, allowing for the total SOQ scores for all Proposers to be compared side-by-side and a final ranking to be assessed.

7.0 Information Release

No information regarding the contents of SOQs, the deliberations by advisors or the Evaluation Committee, recommendations to the Commissioner or other information relating to the evaluation process will be released except to authorized Mn/DOT persons or will be made without the authorization of the Chairperson or his/her designated representative.

8.0 Notification and Debriefing

All Proposers submitting SOQs will be notified in writing of the results of the evaluation process within a time specified by the Chairperson.

Those prospective design-build firms that do not appear on the most highly qualified list (shortlist) will be contacted by the Chairperson or his/her designee and given the opportunity to request a debriefing which may be conducted by a designee of the Chairperson, at the discretion of Mn/DOT. The Commissioner or his designee will coordinate with the Chairperson of the Evaluation Committee to schedule the debriefings. Participants in a debriefing may include the RFQ Evaluation Committee Chairperson and any other person designated by the Chairperson. Only information pertaining to the SOQ submitted by the team attending a debriefing will be shared with that team. Discussions regarding the qualifications of other Proposer teams will not occur. No scoring information will be disclosed.

CONFIDENTIAL ***T.H. 52 RFQ***

SOQ Evaluation Procedure

Name of SOQ Evaluator: Proposer:

T.H. 52 Pass/Fail Checklist

PASS/FAIL TASK	PASS	FAIL
Provide 20 copies of SOQ in loose-leaf 3-ring binder in sealed packages		
Submittals prepared on letter-size, white paper and bound with pages sequentially numbered and not to exceed 60 pages		
Employee resumes of key individuals submitted and provided as an appendix to the proposal		
Previous client references, location and address summaries of for each member of the D-B team. Awards, licenses and certifications included in a separate volume		
Include transmittal letter, submitted on D-B team lead firm stationery		
Statement that representations made by lead firm on behalf of the signer's principal firm have been authorized by, are correct and accurately represent the role of the signer's principal firm in the D-B team		
Identify D-B and its owners or the lead or managing entity		
Identify legal nature of the entity and state of organization		
Identify name, title, address, telephone and fax numbers and e-mail address of principal contact(s)		
Explanation of legal relationship among and role played by each member entity and involvement in D-B team		
Identify projects of construction which includes: ▪ Project name and contact number ▪ Owner's name, address, contact and current phone and fax numbers ▪ Dates of work performed ▪ Project description ▪ Description of work and percentage actually performed by such entity ▪ Initial contract price ▪ Final contract price (include number and value of contract modifications and claims) ▪ Explanation regarding the causes of contract value adjustments ▪ Initial contract completion date ▪ Final completion date ▪ Number of time extensions sought, explanation regarding time extension clauses		

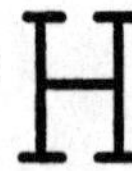

CONFIDENTIAL ***T.H. 52 RFQ***

SOQ Evaluation Procedure

Name of SOQ Evaluator: Proposer:

PASS/FAIL TASK	PASS	FAIL
Identify past joint owner/contractor D-B agreements which includes: ▪ Description of the nature, type, location ▪ Value ▪ Projected/actual completion dates ▪ Description of Proposer team participant role ▪ Provide manager name of each participating firm ▪ Owner's name, address, contact and current phone and fax numbers		
Separate resumes included for the key management staff: ▪ Design-Build Project Manager ▪ Construction Quality Control Manager ▪ Design Manager ▪ Design Quality Control Manager ▪ Geotechnical Engineer ▪ Design Project Engineer – Structures ▪ Design Project Engineer - Roadway ▪ Traffic Engineer ▪ Traffic Control Supervisor ▪ Project Utilities Coordinator ▪ Construction Project Engineer – Roadway ▪ Construction Project Engineer – Structures ▪ ITS Manager ▪ Survey Manager ▪ Landscape/Aesthetics Manager ▪ Safety Manager ▪ Public Affairs Coordinator ▪ Hydraulics Engineer ▪ Civil/Utilities Design Engineer ▪ Environmental Compliance Manager ▪ Project Superintendent ▪ Structures Superintendent ▪ Paving Superintendent ▪ Grading Superintendent		
List D-B team member firms proposed for the Project classified or specialty designer or specialty contractor		
Provide financial statements (income statement, balance sheet and cash flow statement) for any identified member of the D-B teams	Not Required by RFP	
Provide financial ratings for all rated team participants	Not Required by RFP	

CONFIDENTIAL ***T.H. 52 RFQ***

SOQ Evaluation Procedure

Name of SOQ Evaluator: Proposer:

PASS/FAIL TASK	PASS	FAIL
Submit parent or affiliate financial statements and information for D-B team member or Major Participant or statement of non-existence of such information	Not Required by RFP	
Provide information on material changes or letter from chief financial officer or treasurer certifying non-existence of such changes	Not Required by RFP	
Surety has an A.M. Best and Company rating level of A- or better and Class VIII or better		
Surety letter states that participant's backlog and work-in-progress has been evaluated in determining bonding capacity		
Surety letter states that surety has read RFQ and understands general obligations of the D-B as defined, any potential guarantees as outlined in the RFQ	Not Required by RFP	
Surety letter states recognition of joint and several liability obligations in the teaming/joint venture agreement for JVs		
Surety letter states surety's analysis of team member's financial condition for those that anticipate material change to financial condition		
Significant anticipated legal issues that must be resolved in order to carry out the Project and its obligations under a D-B contract are identified and explained		
Describe proposed insurance coverage for development of the Project	Not Required by RFP	
Describe goals/expectations of the Developer team relative to use of subcontractors, suppliers and vendors		
Provide summary of disputes or claims (including litigation, arbitration or other alternative dispute resolution procedures) to which each member of D-B team has been party to with respect to capital projects		
Describe outcome of the dispute/claim		
Provide details of any fines or enforcement penalties levied on each Developer team member		
Provide an explanation of the reasons stated by the court or administrative agency for the levying of fines or enforcement penalties		
Identify owner's representative who can verify the resolution of the dispute or claim with current phone or fax number, case or docket number (if applicable), case style and other identifying information		
Describe any project which resulted in assessment of liquidated damages during the last five (5) years for each D-B team member		
Describe causes of delays and amount assessed		
Describe any outstanding damage claims by owner, subcontractor, vendor or supplier		
Describe any amounts currently being withheld by any owner pending claim resolution including owner's representative with current phone and fax numbers		

CONFIDENTIAL ***T.H. 52 RFQ***

SOQ Evaluation Procedure

Name of SOQ Evaluator: Proposer:

PASS/FAIL TASK	PASS	FAIL
Indicate whether any Developer team member or firm which then employed any personnel proposed to be used on the Project has been debarred or similarly been denied its right to pursue business in any jurisdiction	Not Required by RFP	
Identify the nature and cause of the debarment, suspension or other action and the jurisdiction	Not Required by RFP	
Identify any contract entered into by a team member during the last five years has been terminated for cause or required completion by another party		
Describe the reasons for termination and amounts involved		
Identify any capital project exceeding $500,000 where following completion, material post completion corrective and/or repair work was required for each D-B team member	Not Required by RFP	
Indicate whether any team member has ever filed for bankruptcy or other types of receivership under similar state or Federal law	Not Required by RFP	
Identify caption, court or docket number, if applicable	Not Required by RFP	
Include original of good standing certificate for each D-B team member in the state of their organization or formation	Not Required by RFP	
Provide evidence to do business in the State of Minnesota		
Acknowledge receipt of all addenda issued to the RFQ and all responses issued to questions and requests for clarification	Not Required by RFP	

APPENDIX B

SOQ QUALITATIVE FORMS

H

Name of SOQ Evaluator: Proposer:

Evaluation Grade:

☐ **Excellent** ☐ **Very Good** ☐ **Good** ☐ **Fair** ☐ **Poor**

Organization -- Effective project management authority and structure

a) **Detailed description and establishment of the authority of the Project Manager to effectively manage project during the design and construction phases and maintain a continuous flow of development and operations.**

Excellent:

- Project manager authority and enforcement of such authority is clearly defined for the design and construction phases. The role of the project manager will most definitely contribute to the successful execution of the Project responsibilities.
- Project manager authority clearly allows for maintaining communication and fostering dispute resolution issues throughout the Project.

Very Good:

- Project manager authority and enforcement of such authority is well defined for the design and construction phases. The role of the project manager will most likely contribute to the successful execution of the Project responsibilities.
- Project manager authority generally allows for maintaining communication and fostering conflict/issue resolution throughout the Project.

Good:

- Project manager authority and enforcement of such authority is adequately defined for the design and construction phases. The role of the project manager will probably contribute to the successful execution of the Project responsibilities.
- Project manager authority adequately allows for maintaining communication and fostering conflict/issue resolution throughout the Project.

Fair:

- Project manager authority is not clearly defined for t

Chapter

7

The Construction Manager Approach to Design-Build

It would appear that a construction manager (CM) functioning in a design-build mode is merely adding another layer of professional involvement, and additional costs, to project delivery system that was designed to simplify matters. On the other hand, the CM makes their services so important to an owner during the preconstruction phase of a project through their expertise in matters of construction, their up-to-date information about local labor markets and availability of materials, and a current database of costs. The CM fulfills a critical role in the design-build process, especially when an owner lacks the professional staff to guide them through the process or when an owner elects not to increase their existing staff when they have multiple construction projects in the pipeline.

In 2005, the Legislative Analyst's Office within the State of California, in their report, Design-Build: an Alternative Construction System, had this to say about design-build and construction management:

> The advantages and disadvantages of design-build construction delivery systems using construction management methods are similar to those for design-build using a stipulated price with two main exceptions:
>
> *Price.* The public agency has far less price certainty under this method, if the stipulated price approach is used. Even so, construction management still provides more certainty than design-bid-build, where the total price is not known with reasonable certainty until design is finished and bids have been received. With construction management, a series of trade contracts is bid over time. This provides partial cost information earlier, and allows design changes to be made in subsequent trade packages to control costs and keep the project within budget.
>
> *Benefit of competitive bidding flows to agency.* With the construction management approach to design-build delivery, the savings resulting from competitive bidding for subcontracts and supplies benefits the public agency rather than the

design-build contractor. This is an important advantage that construction management has over stipulated price.

Mr. Rick Thorpe, Executive Officer, Construction Project Management, Los Angeles County Metropolitan Transportation Authority, states the case for a CM rather succinctly.

> I do contract with design-build firms. I only use CM as an oversight role. CMs that work for us are essentially an extension of our staff. Typically we staff using our own CM staff and supplement with consultants as needed. We don't have a big CM staff so almost always we wind up needing supplemental CMs from outside sources. We don't typically use CM during design except for constructability reviews which are toward the end of the design. Again, the CM works together with our staff as if they were part of the same organization. On my last job the CM consultants were given owner cards with their name and title on them.

So, one of the major roles the CM plays in design-build is acting as the owner's representative to supplement their existing staff. Owners who do not have qualified professionals on staff, will engage a CM to assist in guiding a design-build project through design and construction, acting as the owner's representative.

Both instances provide a sound rationale for employing a CM.

CM Defined

The Construction Management Association of America (CMAA) defines construction management as follows:

- A project delivery system comprising a program of management services
- Defined in scope by the specific needs of the project and the owner
- Applied to a construction project from conception to completion, in order to control time and cost, and to maintain project quality
- Performed as a professional service under contract to the owner by a CM
- Selected on the basis of the experience and qualification of the CM firm or consultant
- Compensated on the basis of a negotiated fee for the scope of services rendered

Agency versus GMP CM

A further distinction of CMs deals with the two basic contractual relationships they can have with an owner.

CM-agency. The CM will provide services to an owner during preconstruction and/or construction as an agent of the owner. In the case of a design-build, the owner will hold a construction contract with the design-builder, and all related payments will be made by the owner, with the CM's approval and recommendations.

CM-at-risk. In instances where a CM provides preconstruction and/or construction services for an owner, the CM may elect to guarantee the total cost of the project; this, in effect, puts them at risk. The at-risk approach can be accomplished by the issuance of a stipulated or lump sum type contract or via a guaranteed maximum price (GMP) contract with the owner, where some form of savings distribution is included if final costs are less than the GMP.

Some critics claim that the objectivity of the CM's decisions can be affected since they now serve two masters—the owner and themselves. When some decisions have to be made that affect costs, the CM's decision may waver in favor of protecting their guaranteed maximum cost.

A Snapshot Comparison of the Two Approaches

CM-agency	CM-at-risk
Acts solely as the owner's agent	Acts as the owner's representative
Fee is a percentage of the cost of the project	Fee is included in the contract sum
Owner deals with contractors and issues contracts to them, and pays them	CM deals directly with contractors and absorbs any cost overruns
Liability of CM is similar to "Standard of Care" provision of design professional	Liability of CM is similar to general contractor with lump sum/GMP contract

It is doubtful that any construction management firm planning to remain in business for long would split their loyalties.

The Selection of a Construction Management Firm by Prequalifying

The prequalification process of selecting a construction management firm requires an owner not only to seek out firms with experience in construction management as it is practiced in design-build, but also to engage one that gives the owner a strong sense that "I can work with this guy."

Most prequalification processes will be accomplished by the submission of a qualification questionnaire composed of four parts:

1. *General information.* Information supplied by the owner indicating a general description of the project, anticipated budget, proposed time for design and construction, and the type of project delivery system being contemplated (if known) or leaving that up to the successful CM candidate to recommend. The responding CM firms should have *X* number of years in the business and *Y* number of years in the management of design-build projects and, more specifically, experience in the particular type of project under consideration.

2. *Past performance and capabilities.* This section of the CM Request For Proposal (RFP) will require respondents to list specific previous design-build projects that they have successfully completed that are of a similar nature in both scope and cost. Financial statements, letters from bonding companies, and references from owners, contractors, and designers will be required.
3. *Project management plans.* The owner wants to know how the project management team will function and their specific duties and responsibilities, during both design and construction phases. If the CM team progresses to the oral interview stage, it will be required to have the actual management team present to be interviewed by the owner.
4. *Construction Manager's fee structure.* List of reimbursable expenses and how the cost of these reimbursables is to be established (referred to as multiples) and billed. For example, a multiple of 1.5 means that the actual cost of the expense will be multiplied $1^1/_2$ times for billing purposes, a multiple of 2 will result in the actual cost being doubled. CM fees run the gamut from a low of 3% to a high of 10 to 11%, depending on levels of staffing, expenses to be reimbursed, and those not to be reimbursed.

Evaluation procedures can vary, but they usually follow a point system when comparing scope, management plans, financial strength, and performance. The oral presentation will give the owner some assurance that they can work with this team or not.

Risk Management and the Role of the CM

One of the more important functions that a CM can serve is to make owners aware of the potential risks in a construction project. By doing this upfront, it could help the owner make important decisions about their approach to the work up ahead. The term EGAP (Everything Goes According to Plan) is rarely applied to a building program. Even with a rather extensive site exploration, the one area not examined can turn out to be the one area containing highly hazardous material, long buried and forgotten about. The best and most thoroughly designed projects often experience costs in the form of unanticipated change orders. Labor disputes, plant closings, and expected severe weather will impact a project's cost and delivery time.

The CM can perform a valuable service by alerting the owner to the potential risks in the construction process so that the necessary contingency accounts can be introduced into the program.

The need for a contingency account

There are actually two types of contingency accounts that a CM ought to recommend—one for the owner's account' and once the type of project delivery system is established, one in the construction budget to be used "at the discretion of the design-builder" or contractor, whichever the case may be. During a period of high inflation, future costs may be difficult to predict, but a separate set-aside fund just for that factor may be warranted. Contingency accounts can range from a low of 5% to a high of 15% depending on the type of project being undertaken, renovation and rehabilitation work requiring the higher amounts, and the CM can be of great assistance in establishing this contingency account even before design begins.

The risks most likely to be encountered in a construction project are:

- An unrealistic budget at the outset
- Site-related risks
- Severe, unanticipated weather patterns
- Lengthy and costly delays caused by an owner or design-builder
- Change in design and/or specification, at the owner's direction, with resultant cost and time impact
- Sudden inflationary spiral affecting those costs not under contract
- Failure to provide for a contingency, or including having a contingency account that is insufficient

The objectives of a CM-generated risk assessment plan would be to:

- Carefully scrutinize the owner's budget and the design-builder's budget as it develops to ensure that they are compatible.
- Identify risk in the design/construction schedule early on and maintain close scrutiny of schedule changes as design proceeds to construction.
- Perform due diligence in examining the site to eliminate unknown conditions as much as possible.
- Closely review plans and specifications to eliminate/reduce errors, omissions, and redundancies to ensure that the owner will not experience added costs.
- Continuously evaluate the potential for risk as the project develops.
- Track monthly events to alert the owner to risks, increased costs, and schedule problems, and intervene when necessary to mitigate those risks.

One of the basic tenets of a successful project involves some degree of risk sharing, and the CM can not only define and track potential risks but also intervene and offer professional advice when discussions about risk sharing occur.

The Role of the Construction Manager during Design

Some adherents to the CM approach say the most value derived from hiring a CM occurs during the design and initial planning stages of a project. One aspect of a CM's input during design development may be to simply translate the design and construction jargon and terminology to an unfamiliar owner. The basic services to be provided by a CM during the design phase are summarized as follows; the full text can be found in CMAA Document A-1 (2005 Edition), App. 7.1:

- Prepare a construction management plan to include the owner's schedule, budget, and general design requirements, and develop alternatives for the scheduling and management of the project.
- Assist the owner in designer selection by developing lists of potential firms, criteria for selection, preparing RFPs, interviewing, and evaluating candidates.
- Assist the owner in conducting designer orientation sessions.
- *Time management.* Develop a master schedule and, on acceptance by the owner, develop a milestone schedule for the design phase.
- *Cost management.* Survey the local market for labor, material, and equipment updated costs and availability, and prepare a project budget to include contingencies and review with the owner. Prepare a preliminary estimate and budget analysis.
- *Management information system (MIS).* Develop an MIS to establish communication with the owner and other parties of the design and construction team to include procedures for reporting, communication, and administration during the design phase.
- *Project management.* Conduct a project conference attended by the owner and designers to review the construction management plan, schedules, and project budget.
- Monitor the designer's compliance with the construction management and MIS, and coordinate the flow of information between the owner and designer. Conduct periodic meetings with the owner and design consultants to serve as a forum of exchange of information and review of design progress.
- Review the design documents with regard to constructability, scheduling, time of construction, clarity, consistency, and coordination among the various consultants.
- Expedite the owner's design review and convey comments to the design team.
- Coordinate transmittal of documents to regulatory agencies for review, and advise the owner of any potential problems.
- Assist in preparing supplemental conditions of the construction documents.
- Assist the owner in preparing documents for use in obtaining or reporting on project funding.
- Recommend revisions to the master schedule as required.

- Monitor compliance with the design phase milestone schedule, and prepare a prebid construction schedule for each part of the project.
- *Cost management.* Prepare an estimate for each submittal of design documents. If the budget figure is exceeded, suggest necessary steps to revise the project's general scope or modify the design requirements appropriately. Make recommendations to the owner concerning revisions to the project that may result in budget changes.
- *Value engineering studies.* Provide *value engineering* recommendations to the owner.
- *Management information systems (MIS).* Prepare and distribute schedule maintenance reports comparing actual progress for the design phase versus schedule progress. Prepare and distribute cost reports compared to the project budget and make recommendations for any corrective action required.
- Prepare periodic cashflow reports.
- Prepare and distribute design phase change reports that contain all owner-approved changes.

The responsibilities of the CM during the procurement phase as outlined in A-1 are as follows:

- *Prequalify bidders.* Assist the owner in developing lists of bidders by preparing and distributing questionnaires, interviewing potential bidders, analyzing completed questionnaires, and preparing recommendations to the owner.
- Assist the owner in soliciting bids by preparing and placing notices and advertisements to solicit bids.
- Expedite the delivery of bid documents to bidders.
- In conjunction with the owner and designers, conduct prebid conferences to explain and clarify project requirements.
- Develop procedures to provide answers to questions submitted by bidders. All such questions and answers should be in the form of an addenda.
- Assist the owner in the opening of bids and the evaluation of those bids. Make recommendations to the owner regarding acceptance or rejection of bids.
- Conduct a postbid conference to review contract award procedures.
- Assist the owner in the assembly, delivery, and execution of the contract documents.
- *Time management.* At prebid conference stress construction schedule responsibilities. Recommend any revisions to the master schedule.
- *Cost management.* Prepare an estimate for all addenda costs. Analyze bids including alternate bid prices and unit prices and make recommendations to the owner.

- *Management information system (MIS)*. Prepare and distribute schedule maintenance reports, compare actual bid and award dates, and summarize the progress of the project.
- Prepare and distribute project cost reports during the procurement phase, comparing actual contract amounts to the project construction budget.
- Prepare and distribute cashflow reports based on actual contract award prices.

The responsibilities of the CM during the construction phase

- Conduct a preconstruction conference to review the project reporting systems and other requirements for the work.
- Verify that the contractor has provided evidence that all permits, bonds, and insurance have been obtained.
- Provide an on-site management team to provide contract administration as an agent of the owner.
- Establish and implement procedures for reviewing and processing requests for clarification and interpretation of the contract documents, shop drawings, samples and other submittals, contract schedule adjustments, change-order proposals, requests for substitutions, payment applications, and maintenance of logs.
- Conduct periodic project site meetings including coordination meetings with the contractor and designer.
- Coordinate technical inspections and testing provided by others, receive copies of reports, and pass them on to appropriate parties.
- Authorize minor variations in the work that does not involve adjustment to the contract price and which is consistent with the overall intent of the contract documents.
- Establish and implement a change-order control system. A request from the contractor shall be accompanied by the drawings and specifications prepared by the designer and contain detailed information concerning the price and any time adjustments. The CM will review the proposal and verify that such a request has any validity.
- The CM shall provide the designer with all copies of change orders and make recommendations to the owner. At the owner's direction the CM will prepare and issue an appropriate change-order document.
- Whenever the contractor notifies the CM that a surface or subsurface condition is encountered that differs from what the contractor feels is at variance with the contract requirements, the CM will consult with the designer for review and, if necessary prepare a change order.

- *Quality review.* The CM shall establish and implement a program to monitor the quality of the work.
- The CM shall require each contractor to prepare and submit a safety plan for review and implementation.
- Disputes between contractor and owner shall be reviewed by the CM with the decision passed on to the owner.
- The CM shall receive all contractor operation and maintenance manuals, warranties, and guarantees, and send copies to the designer and the owner.
- The CM shall determine when substantial completion has been achieved and, in consultation with the designer, prepare a list of incomplete work or work that does not comply with the contract requirements.
- Final completion shall be determined by the CM after consultation with the designer.

The CM's postconstruction activities

- Coordinate and expedite information from the contractor that will allow the designer to prepare record drawings.
- Compile all O&Ms, warranties and guarantees, and certificates in a binder for submission to the owner.
- Assist the owner in getting an occupancy permit by coordinating final testing and submission of required documentation to all government agencies.
- Prepare an occupancy plan to include the schedule for location of furniture, fixtures, and equipment.
- Continue through the postconstruction period to provide services relating to change orders.

CMAA documents are, in some cases, complimentary, and are to be used in conjunction with other forms such as the Standard Form of Contract Between the Owner and Contractor (CMAA Document A-2), the General Conditions of the Construction Contract (CMAA Document A-3), and the Standard Form of Agreement Between Owner and Designer (CMAA Document A-4).

Other Construction Management Contracts

The American Institute of Architects publishes a series of CM type contract forms. These are as follows:

A101 tm CMa-1992—Standard Form of Agreement Between Owner and Contractor, Where the Basis for Payment Is a Stipulated Sum. This contract is an agreement between the owner and contractor, where the basis of payment is a fixed-price and the CM is assisting the owner in an advisory capacity rather than as the constructor in both the design and construction phase.

A201 tm CMa-1992—General Conditions of the Contract for Construction, Construction Manager Edition. To be used when the CM has been added as an advisor to the team of owner, architect, and contractor and the owner will enter into multiple contracts with prime trade contractors.

A511 tm CMa-1993—Guide for Supplementary Conditions, Construction Manager Adviser Edition. This can be used where the CM is employed in the capacity of an advisor to the owner and not where the CM is a constructor.

B141 tm CMa-1992—Standard Form of Agreement Between Owner and Architect, Construction Manager Adviser Edition. This is a contract between the owner and architect where the CM will provide construction management services under a separate contract with the owner.

B801 tm CMa-1992—Standard Form of Agreement Between Owner and Construction Manager. To be used when CM services are separate and independent of the architect and contractor and the CM will act solely as an advisor to the owner.

Other AIA CM–related contract forms are their G series.

G 701—Change Order, CM Adviser Edition

G 702—Application for Payment, CM Adviser Edition

G 704—Certificate of Substantial Completion, CM Adviser Edition

G 714—Construction Change Directive (CCD), CM Adviser Edition

G 722—Application and Project Certificate for Payment, CM Adviser Edition

AGC contracts

The Associated General Contractors of America (AGC) also publishes a series of construction manager contract forms. The AGC 400 Series includes the following standard contract forms:

AGC Document 410—Standard Form of Design-Build Agreement and General Conditions Between Owner and Design-Builder (Fig. 7.1).

AGC Document No.465—Standard Form of Agreement Between Design-Builder and Design-Build Subcontractor (Fig. 7.2).

AGC Document No.499—Standard Form of Teaming Agreement for Design-Build Project (Fig. 7.3).

THE ASSOCIATED GENERAL CONTRACTORS OF AMERICA

INSTRUCTIONS FOR COMPLETION OF
AGC DOCUMENT NO. 410
STANDARD FORM OF DESIGN-BUILD AGREEMENT AND GENERAL CONDITIONS BETWEEN OWNER AND DESIGN-BUILDER
(Where the Basis of Payment is the Cost of the Work Plus a Fee with a Guaranteed Maximum Price)

1999 EDITION

This edition of the Standard Form of Design-Build Agreement and General Conditions Between Owner and Design-Builder (Where the Basis of Payment is the Cost of the Work Plus a Fee with a Guaranteed Maximum Price), AGC Document No. 410 (AGC 410), is intended to be used as a follow-on document to AGC Document No. 400 (AGC 400), Preliminary Design-Build Agreement Between Owner and Design-Builder, or as a stand-alone document that addresses the entire design-build process, including the services otherwise provided under AGC 400.

This standard form agreement was developed with the advice and cooperation of the AGC Private Industry Advisory Council, a number of Fortune 500 owners' design and construction managers who have been meeting with AGC contractors to discuss issues of mutual concern. AGC gratefully acknowledges the contributions of these owners' staff who participated in this effort to produce a basic agreement for construction.

GENERAL INSTRUCTIONS

Standard Form

These instructions are for the information and convenience of the users of AGC 410, 1999 Edition. They are not part of the Agreement nor a commentary on or interpretation of the contract form. It is the intent of the parties to a particular agreement that controls its meaning and not that of the writers and publishers of the standard form. As a standard form, this agreement has been designed to establish the relationship of the parties in the standard situation. Recognizing that every project is unique, modifications may be required. See the recommendations for modifications, below.

Legal and Insurance Counsel

This Agreement has important legal and insurance consequences. Consultation with an attorney and an insurance adviser is encouraged with respect to its completion or modification.

DESIGN-BUILD FAMILY OF DOCUMENTS

In the design-build project delivery method, the owner and design-builder enter into a single contract wherein the design-builder undertakes the responsibility to provide for both the design and construction of the project in conformance with basic requirements which have been set forth by the owner. Design may be performed within the design-builder's organization, or it may be performed by design professionals under a separate contract between the design-builder and architect/engineer (AGC Document No. 420).

The AGC family of design-build standard forms has been carefully coordinated (See diagram). Use of other forms or AGC forms with different publication dates with any of this series of contract documents would require extensive modification and is not recommended.

AGC DOCUMENT NO. 410 • STANDARD FORM OF DESIGN-BUILD AGREEMENT AND GENERAL CONDITIONS BETWEEN OWNER AND DESIGN-BUILDER (Where the Basis of Payment is the Cost of the Work Plus a Fee with a Guaranteed Maximum Price)

Figure 7.1 AGC Document 410—Standard Form of Design-Build Agreement. (*By permission: Associated General Contractors of America, Alexandria, VA.*)

THE ASSOCIATED GENERAL CONTRACTORS OF AMERICA

INSTRUCTIONS FOR COMPLETION OF AGC DOCUMENT NO. 465 STANDARD FORM OF AGREEMENT BETWEEN DESIGN-BUILDER AND DESIGN-BUILD SUBCONTRACTOR

(Where the Subcontractor Provides a Guaranteed Maximum Price and Where the Design-Builder and Subcontractor Share the Risk of Owner Payment)

1999 EDITION

The Standard Form of Agreement Between Design-Builder and Design-Build Subcontractor (Where the Subcontractor Provides a Guaranteed Maximum Price and Where the Design-Builder and Subcontractor Share the Risk of Owner Payment), AGC Document No. 465 (AGC 465), is intended for use where the Subcontractor is retained by the Design-Builder early in the design phase, basically providing the same design and construction services as the Design-Builder provides to the Owner under the 1999 edition of AGC Document No. 410 or 415. Construction is performed on the basis of actual cost, plus a fee, up to a guaranteed maximum price (GMP). Furthermore, in AGC 465, payment to the Subcontractor is conditioned on the Design-Builder having received from the Owner payment for Subcontract Work satisfactorily performed. AGC Document No. 460, Standard Form of Agreement Between Design-Builder and Design-Build Subcontractor (Where the Subcontractor Provides a Guaranteed Maximum Price and Where the Design-Builder Assumes the Risk of Owner Payment), can be used when conditioned payment is not valid in the jurisdiction.

GENERAL INSTRUCTIONS

Standard Form

These instructions are for the information and convenience of the users of AGC 465, 1999 Edition. They are not part of the Agreement nor a commentary on or interpretation of the contract form. It is the intent of the parties to a particular agreement that controls its meaning and not that of the writers and publishers of the standard form. As a standard form, this Agreement has been designed to establish the relationship of the parties in the standard situation. Recognizing that every project is unique, modifications may be required. See the recommendations for modifications, below.

Legal and Insurance Counsel

This Agreement has important legal and insurance consequences. Consultation with an attorney and an insurance adviser is encouraged with respect to its completion or modification.

DESIGN-BUILD FAMILY OF DOCUMENTS

In the design-build project delivery method, the owner and design-builder enter into a single contract wherein the design-builder undertakes the responsibility to provide for both the design and construction of the project in conformance with basic requirements which have been set forth by the owner. Design may be performed within the design-builder's organization, or it may be performed by design professionals under a separate contract between the design-builder and architect/engineer (AGC Document No. 420).

The AGC family of design-build standard forms has been carefully coordinated (See diagram). Use of other forms or AGC forms with different publication dates with any of this series of contract documents would require extensive modification and is not recommended.

AGC DOCUMENT NO. 465 • STANDARD FORM OF AGREEMENT BETWEEN DESIGN-BUILDER AND DESIGN-BUILD SUBCONTRACTOR
(Where the Subcontractor Provides a Guaranteed Maximum Price and Where the Design-Builder and Subcontractor Share the Risk of Owner Payment)

Figure 7.2 AGC Document 465—Standard Form of Agreement Between Design-Builder and Design-Build Subcontractor. (*By permission: Associated General Contractors of America, Alexandria, VA.*)

THE ASSOCIATED GENERAL CONTRACTORS OF AMERICA

INSTRUCTIONS FOR COMPLETION OF
AGC DOCUMENT NO. 499
STANDARD FORM OF TEAMING AGREEMENT FOR DESIGN-BUILD PROJECT

2001 EDITION

The Standard Form of Teaming Agreement for Design-Build Project, AGC Document No. 499, has been premised in substantial part on language and concepts found in the 1999 Editions of the AGC 400 Series Documents.

AGC 499, 2001 edition, benefited from an inclusive development process. It was developed with the advice and cooperation of the AGC Private Industry Advisory Council, consisting of design and construction professionals within Fortune 500 companies representing many sectors of the U.S. economy, such as automobile manufacturing, entertainment, banking, insurance, retailing, energy generation and distribution, and health care. PIAC members meet regularly with AGC contractors to discuss construction contracting issues of mutual concern and to participate in the development and revision of AGC standard form contract documents.

AGC 499 is intended as a convenient form for agreement among team members pursuing award of a contract for a single design-build project. AGC 499 is not recommended if the award of the design-build contract already has been made. This document, also, is not intended for use as a long-term agreement among the parties for multiple projects.

GENERAL INSTRUCTIONS

Standard Form

These instructions are for the information and convenience of the users of AGC 499, 2001 Edition. They are not part of the Agreement nor a commentary on or interpretation of the contract form. It is the intent of the parties to a particular agreement that controls its meaning and not that of the writers and publishers of the standard form. As a standard form, this Agreement has been designed to establish the relationship of the parties in the standard situation. Recognizing that every project is unique, modifications may be required. See the following recommendations for modifications.

Related AGC Documents

AGC 499 is part of the AGC 400 series of contract documents. Consider also using these AGC documents.

AGC Document No. 400, *Preliminary Design-Build Agreement Between Owner and Design-Builder* Order No. 1300

AGC Document No. 410, *Standard Form of Design-Build Agreement Between Owner and Design-Builder (Where the Basis of Payment Is The Cost of the Work Plus a Fee with a Guaranteed Maximum Price)* Order No. 1302

AGC Document No. 415, *Standard Form of Design-Build Agreement Between Owner and Design-Builder (Where the Basis of Payment Is the Lump Sum Based on an Owner's Program Including Schematic Design)* Order No. 1303

AGC Document No. 420, *Standard Form of Agreement Between Design-Builder and Architect/Engineer for Design-Build Projects* Order No. 1304

AGC Document No. 450, *Standard Form of Agreement Between Design-Build Contractor and Design-Build Subcontractor* Order No. 1306

AGC Document No. 455, *Standard Form of Agreement Between Design-Build Contractor and Subcontractor (Where the Design-Builder and the Subcontractor Share the Risk of Owner Payment)* Order No. 1307

AGC DOCUMENT NO. 499 • STANDARD FORM OF TEAMING AGREEMENT FOR DESIGN-BUILD PROJECT

Figure 7.3 AGC Document 499—Teaming Agreement. (*By permission: Associated General Contractors of America, Alexandria, VA.*)

Other related ACG forms include those for performance bonds, payment bonds, payment applications, and change orders.

The CM Program Manager

As the owner's projects become more complex, issues other than pure design and construction enter the equation. In the case of a design-build project, the CM can commence work with the owner in order to assist in developing the owner's program even before any design considerations are addressed. This is a relatively new field known as program management, which asks such questions such as:

- What existing operations and what future operations are to take place in the new structure?
- How many current and future employees will there be, and what will their roles and functions be?
- What plans to expand the facility are contemplated, and will energy demands increase accordingly or exponentially?
- What is the capital budget, and what does it include? What should be added to the budget?
- If the site is a new one, has an adequate geotechnical evaluation been made?
- Are there any environmental issues that need to be addressed?
- Will the owner need assistance in permitting, licensing, and obtaining government approvals, and therefore require additional consultants?
- What are the various project delivery options open to the owner, and what are their advantages and disadvantages?
- Has the owner considered a sustainable structure—its value in both cost and public relations?

Although fairly common in petrochemical and power industries, this system of managing an owner's entire new commercial or institutional project is relatively new.

In the June 16, 2003 issue of *Engineering News-Record* magazine, writer Gary Tulacz reported that the concept of program management is now becoming more widespread and worthy of tracking as a discipline.

When looking at the myriad tasks facing some owners, the design and construction phases are only a part of the puzzle. Environmental issues, life cycle analysis, lessor/lessee considerations, insurance and liability concerns, even the review and determination of the most cost-effective project delivery system is no longer a simple matter.

Figure 7.4 is a simple chart that shows some of the components included in a program manager's responsibilities—design and construction may be the end result but the path to those activities is wide and varied.

The Role of a Program Manager during the Life of a Capital Project

Project and Site Development	
Planning	**Design phase**
Feasibility studies	Design consultant selection/award
Financing planning and structuring	Permitting
Cost time analysis and project delivery assessment	Community meetings/hearings
Cost estimating & financing review	Risk assessment
Project program assembly	Budget establishment
Land acquisition	Cost estimate vs. design analysis
Permitting	Cost and scope optimization
Project team assignment	Project delivery considerations
Geotechnical investigations	Geotechnical considerations
	Design document preparation

Project Execution		
Construction Bidding Phase	**Construction Phase**	**Project closeout phase**
Proposal and evaluation stage	Schedule control.	Completion coordination and projected move-in considerations
Schedule finalization	Cost forecasting	FF&E management
Bid analysis and negotiation	Change-order control	Owner training
Construction contract finalization	Dispute avoidance/resolution	Building commissioning
Establishing owner representative roles	Payment monitoring	Final inspections and closeouts
	FF&E procurement	Payment reconciliations
	Project cost segregation for financialanalysis and tax liability issues	Release of liens
		Consent of surety
		Convert from construction loan to long-term financing

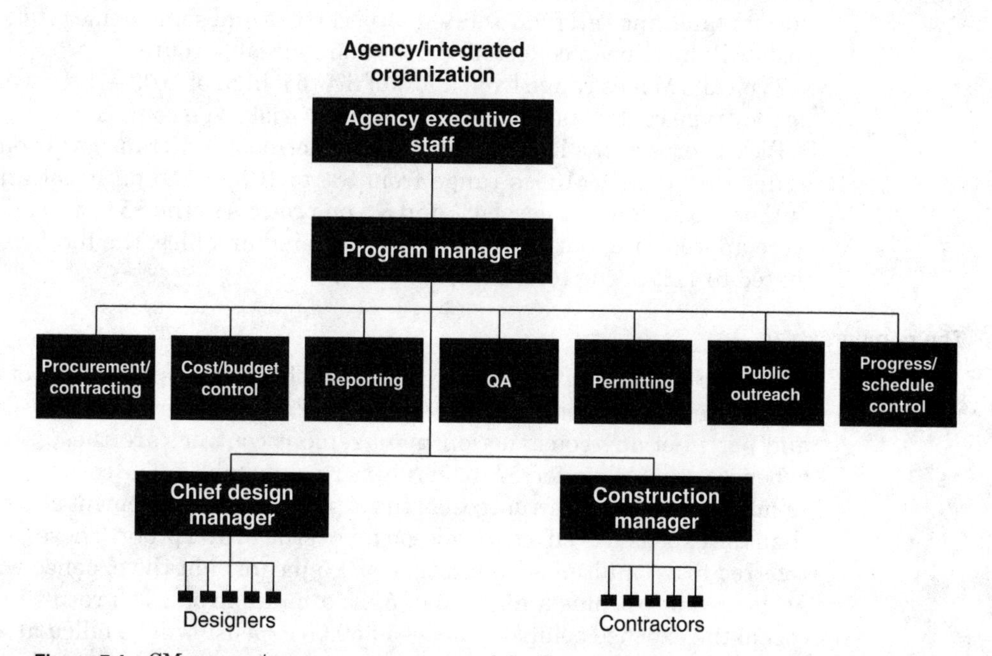

Figure 7.4 CM agency/program manager model. (*By permission: Construction Management Association of America, McLean, VA.*)

In the case of a commercial office building, a CM's responsibilities may extend to tenant fit-up issues. Given the tenant work letter, a CM may assist an owner in analyzing the tenant's design requirements, commenting on those items that exceed the work letter and those that don't, and quantifying the added costs, if the case may be, of the added costs to be borne by that tenant.

In a building where a tenant is either retail space or a restaurant, considerable interface between the base building's mechanical and electrical systems and the tenant's electrical and HVAC loads may also fall to the CM involved in program management. And in those structures where subsidies are awarded or outright annual payments are made in return for advertising space such as building signage, the program manager may also become involved in negotiating the contract based on the cost to install and remove the signage when the contract expires.

Construction Management Fees

The CM operating as an agency CM may offer the client several different fee arrangements. If the contract is a two-part affair, Part A—Design Services, Part B—Construction Services, the fee may be broken down into two parts as well. Since many two-part CM contracts can be terminated after design services are completed, a two-part fee structure is necessary. Construction management fees are lower than conventional lump sum or GMP contracts with a general contractor because the risks are less (except the CM–at risk) and costs normally associated with a general contractor's general conditions are all reimbursable to the CM. In effect, the CM's fee will travel quicker to their bottom line, since all field-related project costs and some home office personnel costs will have been collected via the reimbursable route.

Typical CM fees range from a low of 3% to a high of 10%—the lower range for agency type contracts and the higher for at-risk type contracts.

Rick Thorpe at the Los Angeles County Metropolitan Transportation Authority states that their CM fees range from 8% to 10% based on negotiation. In the author's experience, a fee of 4% and 5% on projects in the $5 to $10 million range were appropriate, but after including all reimbursables, the total cost approximated to 11% of the total contract price.

The reimbursables

The CM contract will include not only a fee, based on a percentage of total costs, but also reimbursement for expenses incurred while performing their services. Part and parcel of any construction management contract are these reimbursable expenses generally referred to merely as reimbursables. The contract may call for reimbursement of the actual cost of the expense or reimbursement at a sum greater than the expense, in effect, an add-on for overhead and profit. These "add-ons" are referred to as "multiples," a multiple of 1.5 means that the expense will be billed at its actual cost plus a markup of 50%; a multiple of 2 will result in the actual cost of the expense being doubled—a $50.00 expense will be billed at $100.00.

Some of the disagreements in an owner-CM contract frequently arise when these reimbursables are not clearly defined and are not listed in an exhibit.

It is typical to exclude any home office expenses other than accounting and estimating. Corporate overheads such as rent, light, heat, and power would be included in the CM fee, but some clients would balk at monthly visits from the company's VP billing $200.00 per hour to merely stop by and check things out.

A schedule of reimbursable expenses in the contract ought to include an hourly rate structure for those CM employees who will be actively involved in the project. Some CM contracts include a list of reimbursables with a cap on them, often included in the contract as an exhibit. For example:

EXHIBIT - Costs of Supervision and Management

Title	Quantity/weeks	Cost per week	Total cost
Project manager	34 weeks	$2,800	$ 95,200
Assistant PM	30 weeks	$2,150	$ 64,500
MEP coordinator	6 weeks	$2,500	$ 15,000
Project superintendent	30 weeks	$3,400	$102,000
Assistant super	30 weeks	$2,750	$ 82,500
Estimating	8 weeks	$2,750	$ 22,000

Note: Although these "total" costs are listed, they are based on the "contract" scope. If the contract scope increases and costs increase or if owner related delays cause an extension of time, the CM will request additional monies for those managers who are affected by the increased costs or delays.

An Owner Exercises a CM Option

Many private owners will build one or possibly two construction projects in the course of their business life, probably sprinkled with several renovation and addition projects—but they are basically businessmen and not builders, so they need some assistance when they decide on a capital project. The CM fulfills this need.

An owner deciding to explore a design-build construction project will first need to conduct some in-house exploratory work that is most likely totally alien to the main business. The owner will need to consider:

- What are the salient points in my program for this construction project I am considering?
- What do I need to investigate, and what facts do I need to assemble before I can begin to put together a program that completely defines my needs?
- How do I begin to select an architect and a contractor?
- How do I know I will accurately convey my program to an architect and engineer, since I don't know their jargon?
- When I select a designer, how will I know that my program is being developed properly?
- How can I be assured that I am being treated fairly by the design consultant or design-build team?
- Who can I turn to as my advocate if I need help at any stage of the game?

There are several answers to these questions:

1. Hire permanent staff to address these questions. But if only one or two programs are being considered, how can I attract a capable person knowing that their tenure will be relatively brief? Can I keep them on in another capacity?
2. Hire a consultant experienced in both design and construction to act as your representative for the life of the project, or to be on call when needed. Professionals will base their fee on an hourly rate plus reimbursable expenses, but they may not be available at a moment's notice for an emergency meeting or such.
3. Hire a construction management firm.

Questions a CM might wish to ask an owner:

1. Do you need assistance in analyzing and developing your construction program?
2. What is the background of the executives or managers you plan to assign to the project?
3. Will the program require expertise in more than design and construction, i.e., installation of a manufacturing plant within the structure?
4. What prior construction experience do you have?
5. Have you had a positive or negative experience in prior dealings with architects, engineers, or contractors?
6. Are you familiar with various types of project delivery systems? Are you familiar with the design-build concept?
7. Are you familiar with the construction management concept?
8. Are you considering competitive bidding to select a CM or design-build firm or do you plan to negotiate those contracts?
9. Has your lending institution expressed any desire for a particular project delivery system or, conversely, indicated a dislike for one type?
10. How will you delegate authority to the CM and how much authority do you wish to delegate?

The United States Postal Service—A Long Time CM/Design-Build Advocate

The United States Postal Service (USPS), a quasi-government agency, became the first public agency to use construction management since they were obligated to comply with the government mandated Federal Acquisition Regulations (FAR) act. The USPS was one of the first to recognize the value of design-build as a vehicle to deliver their capital projects, which now number 34,640 facilities comprising more than 300 million square feet of space.

The postal service provides a window through which to observe the practice of construction management as it relates to design-build. Many of the policies and practices developed by the USPS have applicability in both public and private sectors.

How CM delivers the mail

Mr. Robert Fraga, manager, Supply Management Facilities Portfolio at the USPS office in Arlington, Virginia, has long been an advocate of both design-build and construction management and outlined how they employ both in developing and constructing major building operations throughout the United States.

The postal service began using design-build in the 1970s and traditionally 80% of their major projects were accomplished via design-build. Although their capital facilities program slowed down in 2001 and 2002 because of budget cuts, they are now poised to start another aggressive building program and design-build will be the project delivery system of choice. However design-build is not used with any frequency for projects under $10 million.

Bob Fraga maintains a cadre of staff, officially called contracting officers to manage their construction projects. These managers operate much like project executives in the private sector overseeing several projects. They, in turn, can appoint contracting officer's representatives (CORs) for individual projects and these CORs have specific responsibilities:

- Process progress payment requests and make approval recommendations.
- Provide on-site surveillance of construction activities and routine contract administration and coordination.
- Interpret plans and specifications as required and issue clarifying information to the contractor.
- Direct the contractor to correct or remove defective work or work not in compliance with the plans and specifications.
- Review the contractor's construction schedules and make recommendations.
- Review the contractor's compliance with safety regulations.
- Prepare plans and specifications for contemplated contract modifications.
- Solicit contractor cost proposals for contemplated contract modifications, review, and make recommendations.
- Review and approve the contractor's submittals, shop drawings, catalog cuts, coordination drawings, samples, and the like for conformance with requirements.
- Review and approve operating instruction and maintenance manuals.
- Review contractor compliance with labor standards provisions and minority subcontracting programs.
- Forward copies of the following documents on a monthly basis to the major facilities purchasing officer:
  - Financial documents
  - Monthly progress reports

- RFI log and backup
- Modifications and related backup information
- Claims and correspondence
- Direct the contractor to make changes not to exceed $50,000.

The CMs engaged by the postal service act as the contracting officer's representatives and basically have the same duties and responsibilities as those listed.

Each year the USPS will award CM contracts to a group of construction managers, whom they will call when required. They may be employed for a full-scale project commitment beginning with involvement in design and development and continuing through construction. They may be called upon to review an estimate on a proposed project and nothing more.

The Individual Purchasing Plan

When a new project is planned, the USPS conducts what they call an individual purchasing plan (IPP). The purpose of the IPP is to discuss the forthcoming project: how it will be developed, i.e., design-build, design-bid-build, and the part to be assigned to each participant. If a CM will be involved in the project they will attend this meeting to learn about the extent of their involvement. At the IPP, members of the group will have their responsibilities spelled out. For example the duties of the CM will be clearly defined. The USPS designates their project managers as CORs and they imbue their CMs with much the same duties they would assign to their own COR. These duties and responsibilities can serve as guidelines for CMs employed on projects in the private sector—just remove the USPS jargon.

1. Process progress payment requests and make approval recommendations.
2. Provide on-site surveillance of construction activities and routine contract administration and coordination.
3. Interpret plans and specifications as required and issue clarifying information to the contractor.
4. Direct the contractor to correct or remove defective work or work not in compliance with the plans or specifications.
5. Review the contractor's construction schedules and make recommendations.
6. Review the contractor's compliance with safety regulations.
7. Prepare plans and specifications for contemplated contract modifications.
8. Solicit contractor cost proposals for contemplated contract modifications, review, and make recommendations.
9. Review and approve contractor submittals, shop drawings, catalog cuts, coordination drawings, samples, and the like for conformance with requirements.
10. Review and approve operating instructions and maintenance manuals.
11. Review contractor compliance with labor standards provisions and minority business subcontracting programs.

12. Forward copies of the following documents on a monthly basis to the major facilities purchasing office:
    - Financial documents
    - Monthly progress reports
    - RFI log and backup
    - Modifications and related backup information
    - Claims and related correspondence

GMP Contracts Add to the CM Responsibility

In the case of a cost-plus-a-fee GMP contract, the CM may be required to provide some audit duties. These are as follows:

1. Review the contractor GMP contract.
2. Establish a meeting with the contractor's accounting department (personnel) to review and establish the following:
   - Review the contractor's billing format (schedule of values/actual costs).
   - Review the contractor's reports for audit to establish actual costs per month.
   - Establish information flow for documents.
   - Establish audit fringe rate (use actual); establish percent factor for audited rates for field office. Provide hard copy of auditable rates.
   - Establish ground rules for contract compliance.
     - Copies of paid invoices
     - Copies of checks
     - Release of liens for subcontractors
     - Certified payroll verified by field office
     - Backup for field and office payroll for period/month
     - Format of backup of all costs for GMP
   - Establish procedures for monthly requisition adjustments after audit.
     - Adjustment of payment request to reflect latest actual costs
   - Establish procedures for disputed items not in GMP.
     - Adjustment to monthly requisition for disputed items
     - Procedure for resolution of disputed items
   - Establish monthly report format and distribution of contract status and disputed items.
3. Establish approved format of auditors monthly report to USPS.
4. Establish audit completion time (prior to next payment request), which will be based on receiving the audit package from the contractor.
5. Completion report includes savings/loss for final modification and contract closeout.
6. Audit files will be maintained with Construction Manager Support Services Contract (CMSCC) at the job site.
7. Audit files will be shipped along with other CMSSC files to the USPS for storage after project completion.

The CM's Participation in Evaluating Design-Build Proposals

When bids are received by the postal service, quite often the CM will be requested to participate in the bidder's oral presentation and also review their written submittals and assist in grading both portions of the interview process.

Robert M. "Mike" Miller, contracting officer in Bob Fraga's department, ran through this evaluation process. They receive the bids in two packets, one contains the written response to the RFP and the second is a sealed envelope containing the bid price. Mike said that they do not open the envelope with the price until the combined grades of each bidder are tallied. Opening the price first may taint an otherwise objective review of the written response, since the price factor would creep into the evaluation process. So the USPS personnel on the evaluation team along with their CM listen to the oral presentations after scrutinizing the written response, complete their numerical grade evaluation, and then open the sealed envelope with the price. Mike said that there is even an occassional surprise when the design-build team with high marks in both oral and written presentations also happens to be the low bidder.

Can there truly be an objective evaluation system?

With design-build, the design and construction work is generally evaluated on the basis of the bidder's experience, qualifications, and "best value." Both private and public agencies go to great lengths to prepare an evaluation procedure to mitigate the risk of subjective judgments.

The USPS has had excellent experience in design-build, says Bob Fraga, because of their extensive and intensive effort to requalify bidders. Fig. 7.5 contains a three-page evaluation form that the USPS evaluation team uses to prequalify bidders and prepare a short list. There are four categories with a total of 100 possible points and two pass/fail categories dealing with financial data and claims. Objectivity in rating is a nebulous thing.

Who is to say that bidder "A" has an experience rating of 32 points while bidder "B" is rated at only 29?

When evaluating contractors based on qualifications and experience, one is reminded of the architect who was asked to review a contractor's submission for an "or equal" product. The architect responding via transmittal said, "No product is equal to another, they may be similar but not equal." Especially in design-build, one design-build team may have lots of experience in the types of projects at hand, but another design-builder may have only one such experience, but it was in a project that could be the twin to the one being considered? Now who has the better experience?

It is difficult to make a comparison of alternative added value proposals in a design-build submittal. How does one compare a high-quality lighting system to a better system of on-site disposal of storm water? Some benefits require a long period of time before they prove their worth or, conversely, show that they

Date: **December 15-17, 1998**
Subject: **Prequalification Evaluation Evaluations**
Columbus, OH Processing & Distribution Center
Solicitation Number: 512582-99-A-0002

Evaluation Team: (circled member evaluating)
Mike R. Miller, Chairman
Marc Wiese, MFO
Wayne C. Perlenfein, MFO

Offerors: (circled company evaluated)
1. Axor Group Inc.
2. Baker Buildings
3. P. J. Dick, Incorporated
4. James N. Gray Const. Co.
5. H&M Construction Company
6. Hensel Phelps Constr Co.
7. Korte Construction Company
8. Park Tower Development Corporation
9. The Austin Company
10. The Clark Construction Group, Inc.
11. The Haskell Company
12. The Lathrop Company
13. The Morganti Group, Inc.
14. Walsh Group

Evaluation criteria:

Description	**Points Available**	**Points Given**	**Comments**
1. Experience: List two (2) comparable design/build projects completed or in progress during the past five years, listing a maximum of ten (10) projects. Submit the information using the format shown in Section B.2, Page 9. Other comparable design/build projects that demonstrates the design/build entity's experience, listing no more than a maximum of ten (10) projects. Submit the information using the format shown in Section B.2, Page 10. Primary Evaluation Criteria: Primary consideration will be given to comparable projects that the entire design/build entity has successfully completed as a team. Other projects which would receive favorable consideration are: design/build projects that establish contractor's past relationships; comparable design/build projects as part of another design/build entity; and separate design or construction contracts for comparable postal facilities.	**35 points**		

Figure 7.5 Form used by evaluation team for prequalification. (*By permission: United States Postal Service, Arlington, VA.*)

Date: **December 15-17, 1998**
Subject: **Prequalification Evaluation Evaluations**
Columbus, OH Processing & Distribution Center
Solicitation Number: 512582-99-A-0002

Evaluation criteria:

Description	Points Available	Points Given	Comments
2. **Past Performance**: Primary Evaluation Criteria: Past performance will be based on evaluation of responses to reference checks for listed projects. For design/construction stage, primary areas of concern will be performance of design/build entity with regard to cost, schedule and quality during design and construction stages. Nature and extent of participation in listed projects will also be validated. Operations/maintenance checks will be primarily concerned with owning and operating costs, maintenance experience, systems' reliability, energy efficiency, and repair requirements for listed projects.	**30 points**		
3. **Organization:** Primary Evaluation Criteria - Attach a statement describing the design/build entity's organization, with an organization chart. The statement and organization chart must clearly identify specific organizational elements and/or member firms that will be participating in the project, location of units; reporting relationships and functions to be performed by each unit; successful past working relationships; key executive who will have overall responsibility; key management personnel to be assigned to the project; resumes of key personnel; and professional staffing levels. In addition, describe expertise, capability, structure, and resources to adequately handle building construction and mechanization installation where required. Describe company resources that would be available for the project. Provide an employment profile for the design/build entity: total employees, total permanent employees, total professionals in each major category. Brief resumes of these individuals shall be attached. Current Form 254, as appropriate for each A/E consultant including anticipated material handling consultant must be included. (Note: Form 255 should be submitted with Part B - Management Plan.)	**25 points**		

Figure 7.5 *(Continued)*

Date: **December 15-17, 1998**

Subject: **Prequalification Evaluation Evaluations**
Columbus, OH Processing & Distribution Center
Solicitation Number: 512582-99-A-0002

Evaluation criteria:

Description	**Points Available**	**Points Given**	**Comments**
4. Backlog: Primary Evaluation Criteria - Demonstration that resources necessary to maintain backlog at a satisfactory level are available.	**10 points**		
5. Financial Capability: Primary Evaluation Criteria - Inclusion of audited and/or interim financial statement (see Page 11); completion of USPS Income Statement Evaluation Form; total bonding capacity (see Page 12); written and/or verbal bank references (see Page 12); and determination that design/build entity possesses sufficient resources to successfully complete project.	**Pass/Fail**		
6. Responsibility Verification: Primary Evaluation Criteria - No pending claims or lawsuits that could prevent successful completion of project. No debarment or other adverse legal action.	**Pass/Fail**		

Total Accumulated Points:

Other Comments:

Signature: ______________________ ______________________
Evaluation Team Member Date

Figure 7.5 *(Continued)*

were not worth much. So some value-added proposals may lack the time frame in which to back up their claim.

The tendency to award design-build projects to those companies with a long track record of successful projects may prevent a new, small firm with great ideas and innovative managers from capturing a project as a new entrant to the field. Again like the old conundrum, "How can I get the experience you require if you won't let me work for you to get it?"

The points type evaluation system is a very good approach to selecting a design-build team, but the evaluation team ought not to forget some of its limitations.

Construction Management Contracts Used by the USPS

The three basic types of contracts employed by the postal system for design-build projects are fixed price, competitively bid GMP, and a two-phase contract similar to AIA and AGC two-part forms, one for the design concept and the other for the complete contract documents and construction. Each one has its place.

Design-build fixed price

This type of contract is used for those neighborhood type, post office buildings, kind of rubber stamp affairs but with differing site, electrical, mechanical, and plumbing requirements. The design-build portion of the project will relate to site work, foundations that may vary according to soils and bearing capacities, and MEP requirements that will vary due to geographic considerations.

Design-build competitively bid GMP type contract

The USPS will provide anywhere from 10% to 30% of the design development drawings, depending on the nature of the project and the site considerations. They might provide elevations, floor plans, and detailed design criteria. This type of contract is used when "typical building designs" used in past projects, will be used again.

Design-build two-phase proposal

Phase 1 is where the USPS provides the scope of the work and will evaluate responses based on the technical qualifications of the bidder and their fee structure. The bidders are provided with a construction cost limitation (CCL) which sets the upper limits of the project's cost. The bidders are requested to submit 30% design development and the GMP price. A detailed schedule of values submitted with the bid will be used in the bid evaluation process. Phase 2 calls for complete drawings, contract award, and construction.

The postal service invites *value engineering* (VE) suggestions in their Phase 1 proposal and points will be awarded based on acceptance of valuable VE suggestions and included in the overall evaluation of the bid.

Bob Fraga mentioned one project in California, where the USPS geotechnical consultants included information in the bid instructions that there were some contaminated soils on the site. A sharp contractor asked for permission to take some soil samples and when they did, they found that the level of contamination was so slight that it would be acceptable in the local landfill. The contractor so stated, was awarded the contract based on their acceptance of the VE proposal, which saved the postal service about $400,000. This type of contract is frequently used when one-off or special projects are being considered.

In both the competitively bid GMP contract and the two-part contract, the construction management company is allowed to perform work with their own forces; however, all costs for such work must be audited. For any work exceeding $50,000, the USPS must be provided with competitive bids.

The Design-Builder Prequalification Process

Selecting the right design-builder now becomes the critical path and prequalification becomes the operative word. A careful review of each prospective bidder's background, experience, and reputation is one of the cornerstones of a successful project.

Mr. Bob Fraga of the postal service said that their prequalification process has resulted in projects with fewer Requests For Information (RFIs), fewer change orders, and more on-time completions. The USPS Design-Build Qualification Form (App. 7.2) is divided into three parts in much the same manner as the construction manager qualification statement.

Part A—General information

Part B—Past performance and capability (Total scoring limits: 100 points)

Part C—Project management (Total scoring limits: 220 points)

Emphasis is placed on the project management plan as witness the 220 point maximum award versus Part B's 100 point maximum.

Bob Fraga of the postal service said they generally have their CM participate in the design-builder review and evaluated process along with other designated postal service managers. They prepare their evaluations in a unique way. Bids are received with cost information in a separate sealed envelope. The reviewers prepare their evaluations by reviewing each component of the submission; some require only a "pass" or "fail," others require a point rating. When this evaluation process has been completed, only then do they open the sealed envelope with the costs. Their rationale is simple, if they knew that bidder A was the low bidder, it might affect their evaluation of the objective arts of the submission.

The USPS success in design-build comes from years of experience, discarding what doesn't work, and enhancing the criteria that do work. They provide guidelines and procedures that can serve as a model for both the design-build procurement and the employment of the construction management concept.

Appendix 7.1: CMAA Document A-1 Owner & Construction Manager Contract

CMAA Document A-1 (2005 Edition)

THE CONSTRUCTION MANAGEMENT ASSOCIATION OF AMERICA, INC.

CMAA Document A-1 (2005 Edition)

Standard Form of Agreement Between
OWNER AND CONSTRUCTION MANAGER
(Construction Manager as Owner's Agent)

This document is to be used in connection with the Standard Form of Contract Between Owner and Contractor (CMAA Document A-2), the General Conditions of the Construction Contract (CMAA Document A-3), and the Standard Form of Agreement Between Owner and Designer (CMAA Document A-4), all being 2005 editions.

CONSULTATION WITH AN ATTORNEY IS RECOMMENDED WHENEVER THIS DOCUMENT IS USED.

AGREEMENT
Made this __________ day of ____________________________ in the year of Two Thousand and
__________.

BETWEEN The Owner:

and the Construction Manager, (hereinafter, referred to as the "CM"):

For services in connection with the Project known as:

hereinafter called the "Project," as further described in Article 2:

The Owner and CM, in consideration of their mutual covenants herein agree as set forth below:

CMAA Document A-1 (2005 Edition)

TABLE OF CONTENTS

Article:

CMAA Document A-1 (2005 Edition)

ARTICLE 1
RELATIONSHIP OF THE PARTIES

1.1 Owner and Construction Manager

1.1.1 Relationship: The CM shall be the Owner's principal agent in providing the CM's services described in this Agreement. The CM and the Owner shall perform as stated in this Agreement. Nothing in this Agreement shall be construed to mean that the CM is a fiduciary of the Owner.

1.1.2 Standard of Care: The CM covenants with the Owner to furnish its services hereunder properly, in accordance with the standards of its profession, and in accordance with federal, state and local laws and regulations specifically applicable to the performance of the services hereunder which are in effect on the date of this Agreement first written above.

1.2 Owner and Designer

1.2.1 Owner-Designer Agreement: The Owner shall enter into a separate agreement, the "Owner-Designer Agreement", with one or more Designers to provide for the design of the Project and certain design-related services during the Construction Phase of the Project. The Project is defined in Article 2 of this Agreement.

1.2.2 Changes: The Owner shall not modify the Agreement between the Owner and Designer in any way that is prejudicial to the CM. If the Owner terminates the Designer's services, a substitute acceptable to the CM shall be appointed.

1.3 Owner and Contractors

1.3.1 Construction Contract: The Owner shall enter into a separate contract with one or more Contractors for the construction of the Project (hereinafter referred to as the "Contract"). The Contractor shall perform the Work, which shall consist of furnishing all labor, materials, tools, equipment, supplies, services, supervision, and perform all operations as required by the Contract Documents.

1.3.2 Form of Contract: Unless otherwise specified, the form of Contract between the Owner and Contractor shall be the CMAA Standard Form of Contract Between Owner and Contractor, CMAA Document A-2 (2005 Edition). The General Conditions for the Project shall be the CMAA General Conditions of the Construction Contract Between Owner and Contractor, CMAA Document A-3 (2005 Edition).

1.4 Relationship of the CM to Other Project Participants

1.4.1 Working Relationship: In providing the CM's services described in this Agreement, the CM shall endeavor to maintain, on behalf of the Owner, a working relationship with the Contractor and Designer.

1.4.2 Limitations: Nothing in this Agreement shall be construed to mean that the CM assumes any of the responsibilities or duties of the Contractor or the Designer. The Contractor will be solely responsible for construction means, methods, techniques, sequences and procedures used in the construction of the Project and for the safety of its personnel, property, and its operations and for performing in accordance with the contract between the Owner and Contractor. The Designer is solely responsible for the design requirements and design criteria of the Project and shall perform in accordance with the Agreement between the Designer and the Owner. The CM's services shall be rendered compatibly and in cooperation with the services provided by the Designer under the Agreement between the Owner and Designer. It is not intended that the services of the Designer and the CM be competitive or duplicative, but rather complementary. The CM will be entitled to rely upon the Designer for the proper performance of services undertaken by the Designer pursuant to the Agreement between Owner and the Designer.

ARTICLE 2
PROJECT DEFINITION

2.1 The term "Project", when used in this Agreement, shall be defined as all work to be furnished or provided in accordance with the Contract Documents prepared by the Designer.

CMAA Document A-1 (2005 Edition)

2.2 The Project name and location is as follows:

____
____
____
____
____
____
____

2.3 The Project is intended for use as:

____
____
____
____
____
____
____
____
____
____
____
____
____
____

2.4 The term "Contract Documents" means the Instruction to Bidders, the Contract, the General Conditions and any Supplemental Conditions furnished to the Contractor, the drawings and specifications furnished to the Contractor and all exhibits thereto, addenda, bulletins and change orders issued in accordance with the General Conditions to any of the above, and all other documents specified in Exhibit B of the Standard Form of Contract Between Owner and Contractor, CMAA Document A-2, 2005 edition.

ARTICLE 3
BASIC SERVICES

3.1 CM's Basic Services

3.1.1 Basic Services: The CM shall perform the Basic Services described in this Article. It is not required that the services be performed in the order in which they are described.

3.2 Pre-Design Phase

3.2.1 Project Management

3.2.1.1 Construction Management Plan: The CM shall prepare a Construction Management Plan for the Project and shall make recommendations to the plan throughout the duration of the Project, as may be appropriate. In preparing the Construction Management Plan, the CM shall consider the Owner's schedule, budget and general design requirements for the Project. The CM shall then develop various alternatives for the scheduling and management of the Project and shall make recommendations to the Owner. The Construction Management Plan shall be presented to the Owner for acceptance.

3.2.1.2 Designer Selection: The CM shall assist the Owner in the selection of a Designer by developing lists of potential firms, developing criteria for selection, preparing and transmitting the requests for proposal, assisting in conducting interviews, evaluating candidates and making recommendations.

3.2.1.3 Designer Contract Preparation: The CM shall assist the Owner in review and preparation of the Agreement between the Owner and Designer.

3.2.1.4 Designer Orientation: The CM shall conduct, or assist the Owner in conducting, a Designer orientation session during which the Designer shall receive information regarding the Project scope, schedule, budget, and administrative requirements.

3.2.2 Time Management

3.2.2.1 Master Schedule: In accordance with the Construction Management Plan, the CM shall prepare a Master Schedule for the Project. The Master Schedule shall specify the proposed starting and finishing dates for each major project activity. The CM shall submit the Master Schedule to the Owner for acceptance.

3.2.2.2 Design Phase Milestone Schedule: After the Owner accepts the Master Schedule the CM shall prepare the Milestone Schedule for the Design Phase, which shall be used for judging progress during the Design Phase.

CMAA Document A-1 (2005 Edition)

3.2.3 Cost Management

3.2.3.1 Construction Market Survey: The CM shall conduct a Construction Market Survey to provide current information regarding the general availability of local construction services, labor, material and equipment costs and the economic factors related to the construction of the Project. A report of the Construction Market Survey shall be provided to the Owner and Designer.

3.2.3.2 Project and Construction Budget: Based on the Construction Management Plan and the Construction Market Survey, the CM shall prepare a Project and Construction Budget based on the separate divisions of the Work required for the Project and shall identify contingencies for design and construction. The CM shall review the budget with the Owner and Designer and the CM shall submit the Project and Construction Budget to the Owner for acceptance. The Project and Construction Budget shall be revised by the CM as directed by the Owner.

3.2.3.3 Preliminary Estimate and Budget Analysis: The CM shall analyze and report to the Owner and the Designer the estimated cost of various design and construction alternatives, including CM's assumptions in preparing its analysis, a variance analysis between budget and preliminary estimate, and recommendations for any adjustments to the budget. As a part of the cost analysis, the CM shall consider costs related to efficiency, usable life, maintenance, energy and operation.

3.2.4 Management Information System (MIS)

3.2.4.1 Establishing the Project MIS: The CM shall develop a MIS in order to establish communication between the Owner, CM, Designer, Contractor and other parties involved with the Project. In developing the MIS, the CM shall interview the Owner's key personnel, the Designer and others in order to determine the type of information for reporting, the reporting format and the desired frequency for distribution of the various reports.

3.2.4.2 Design Phase Procedure: The MIS shall include procedures for reporting, communications and administration during the Design Phase.

3.3 Design Phase

3.3.1 Project Management

3.3.1.1 Revisions to the Construction Management Plan: During the Design Phase the CM shall make recommendations to the Owner regarding revisions to the Construction Management Plan. The Construction Management Plan shall include a description of the various bid packages recommended for the Project. Revisions approved by the Owner shall be incorporated into the Construction Management Plan.

3.3.1.2 Project Conference: At the start of the Design Phase, the CM shall conduct a Project Conference attended by the Designer, the Owner and others as necessary. During the Project Conference the CM shall review the Construction Management Plan, the Master Schedule, Design Phase Milestone Schedule, the Project and Construction Budget and the MIS.

3.3.1.3 Design Phase Information: The CM shall monitor the Designer's compliance with the Construction Management Plan and the MIS, and the CM shall coordinate and expedite the flow of information between the Owner, Designer and others as necessary.

3.3.1.4 Progress Meetings: The CM shall conduct periodic progress meetings attended by the Owner, Designer and others. Such meetings shall serve as a forum for the exchange of information concerning the Project and the review of design progress. The CM shall prepare and distribute minutes of these meetings to the Owner, Designer and others as necessary.

3.3.1.5 Review of Design Documents: The CM shall review the design documents and make recommendations to the Owner and Designer as to constructibility, scheduling, and time of construction; as to clarity, consistency, and coordination of documentation among Contractors; and as to the separation of the Project into contracts for various categories of the Work. In addition, the CM shall give to the Designer all data of which it or the Owner is aware concerning patents or copyrights for inclusion in Contract Documents. The recommendations resulting from such review shall be provided to the Owner and Designer in writing or as notations

CMAA Document A-1 (2005 Edition)

on the design documents. In making reviews and recommendations as to design documentation or design matters the CM shall not be responsible for providing nor will the CM have control over the Project design, design requirements, design criteria or the substance of contents of the design documents. By performing the reviews and making recommendations described herein, the CM shall not be deemed to be acting in a manner so as to assume responsibility or liability, in whole or in part, for any aspect of the project design, design requirements, design criteria or the substance or contents of the design documents. The CM's actions in making such reviews and recommendations as provided herein are to be advisory only to the Owner and to the Designer.

3.3.1.6 Owner's Design Reviews: The CM shall expedite the Owner's design reviews by compiling and conveying the Owner's review comments to the Designer.

3.3.1.7 Approvals by Regulatory Agencies: The CM shall coordinate transmittal of documents to regulatory agencies for review and shall advise the Owner of potential problems resulting from such reviews and suggested solutions regarding completion of such reviews.

3.3.1.8 Other Contract Conditions: The CM shall assist the Owner to prepare the Supplemental Conditions of the Construction Contract and separate General Conditions for materials or equipment procurement contracts to meet the specific requirements of the Project, and shall provide these to the Designer for inclusion in the Contract Documents.

3.3.1.9 Project Funding: The CM shall assist the Owner in preparing documents concerning the Project and Construction Budget for use in obtaining or reporting on Project funding. The documents shall be prepared in a format approved the Owner.

3.3.2 Time Management

3.3.2.1 Revisions to the Master Schedule: While performing the services provided in Paragraphs 3.3.1.1, 3.3.1.2 and as necessary during the Design Phase, the CM shall recommend revisions to the Master Schedule. The Owner shall issue, as needed, change orders to the appropriate parties to implement the Master Schedule revisions.

3.3.2.2 Monitoring the Design Phase Milestone Schedule: While performing the services provided in Paragraphs 3.3.1.3 and 3.3.1.4, the CM shall monitor compliance with the Design Phase Milestone Schedule.

3.3.2.3 Pre-Bid Construction Schedules: Prior to transmitting Contract Documents to bidders, the CM shall prepare a Pre-Bid Construction Schedule for each part of the Project and make the schedule available to the bidders during the Procurement Phase.

3.3.3 Cost Management

3.3.3.1 Cost Control: The CM shall prepare an estimate of the construction cost for each submittal of design drawings and specifications from the Designer. This estimate shall include a contingency acceptable to the Owner, CM and the Designer for construction costs appropriate for the type and location of the Project and the extent to which the design has progressed. The Owner recognizes that the CM will perform in accordance with the standard of care established in this Agreement and that the CM has no control over the costs of labor, materials, equipment or services furnished by others, or over the Contractor's methods of determining prices, or over competitive bidding or market prices. Accordingly, the CM does not represent or guarantee that proposals, bids or actual construction costs will not vary from budget figures included in the Construction Management Plan as amended from time to time. If the budget figure is exceeded, the Owner will give written consent to increasing the budget, or authorize negotiations or rebidding of the Project within a reasonable time, or cooperate with the CM and Designer to revise the Project's general scope, extent or character in keeping with the Project's design requirements and sound design practices, or modify the design requirements appropriately. Instead of the foregoing, the Owner may abandon the Project and terminate this Agreement in accordance with Article 10. The estimate for each submittal shall be accompanied by a report to the Owner and Designer identifying variances from the Project and Construction Budget. The CM shall facilitate decisions by the Owner and Designer when changes to the design are required to remain within the Project and Construction Budget.

CMAA Document A-1 (2005 Edition)

3.3.3.2 Project and Construction Budget Revision: The CM shall make recommendations to the Owner concerning revisions to the Project and Construction Budget that may result from design changes.

3.3.3.3 Value Engineering Studies: The CM shall provide value engineering recommendations to the Owner and Designer on major construction components, including cost evaluations of alternative materials and systems.

3.3.4 Management Information Systems (MIS)

3.3.4.1 Schedule Reports: In conjunction with the services provided by Paragraph 3.3.2.2, the CM shall prepare and distribute schedule maintenance reports that shall compare actual progress with scheduled progress for the Design Phase and the overall Project and shall make recommendations to the Owner for corrective action

3.3.4.2 Project Cost Reports: The CM shall prepare and distribute Project cost reports that shall indicate actual or estimated costs compared to the Project and Construction Budget and shall make recommendations to the Owner for corrective action.

3.3.4.3 Cash Flow Report: The CM shall periodically prepare and distribute a cash flow report.

3.3.4.4 Design Phase Change Report: The CM shall prepare and distribute Design Phase change reports that shall list all Owner-approved changes as of the date of the report and shall state the effect of the changes on the Project and Construction Budget and the Master Schedule.

3.4 Procurement Phase

3.4.1 Project Management

3.4.1.1 Prequalifying Bidders: The CM shall assist the Owner in developing lists of possible bidders and in prequalifying bidders. This service shall include preparation and distribution of questionnaires; receiving and analyzing completed questionnaires; interviewing possible bidders, bonding agents and financial institutions; and preparing recommendations for the Owner. The CM shall prepare a list of bidders for each bid package and transmit to the Owner for approval.

3.4.1.2 Bidder's Interest Campaign: The CM shall conduct a telephone and correspondence campaign to attempt to increase interest among qualified bidders.

3.4.1.3 Notices and Advertisements: The CM shall assist the Owner in preparing and placing notices and advertisements to solicit bids for the Project.

3.4.1.4 Delivery of Bid Documents: The CM shall expedite the delivery of Bid Documents to the bidders. The CM shall obtain the documents from the Designer and arrange for printing, binding, wrapping and delivery to the bidders. The CM shall maintain a list of bidders receiving Bid Documents.

3.4.1.5 Pre-Bid Conference: In conjunction with the Owner and Designer, the CM shall conduct pre-bid conferences. These conferences shall be forums for the Owner, CM and Designer to explain the Project requirements to the bidders, including information concerning schedule requirements, time and cost control requirements, access requirements, contractor interfaces, the Owner's administrative requirements and technical information.

3.4.1.6 Information to Bidders: The CM shall develop and coordinate procedures to provide answers to bidder's questions. All answers shall be in the form of addenda.

3.4.1.7 Addenda: The CM shall receive from the Designer a copy of all addenda. The CM shall review addenda for constructibility, for effect on the Project and Construction Budget, scheduling and time of construction, and for consistency with the related provisions as documented in the Bid Documents. The CM shall distribute a copy of all addenda to each bidder receiving Bid Documents.

CMAA Document A-1 (2005 Edition)

3.4.1.8 Bid Opening and Recommendations: The CM shall assist the Owner in the bid opening and shall evaluate the bids for responsiveness and price. The CM shall make recommendations to the Owner concerning the acceptance or rejection of bids.

3.4.1.9 Post-Bid Conference: The CM shall conduct a post-bid conference to review Contract award procedures, schedules, Project staffing and other pertinent issues.

3.4.1.10 Construction Contracts: The CM shall assist the Owner in the assembly, delivery and execution of the Contract Documents. The CM shall issue to the Contractor on behalf of the Owner the Notice of Award and the Notice to Proceed.

3.4.2 Time Management

3.4.2.1 Pre-Bid Construction Schedule: The CM shall emphasize to the bidders their responsibilities regarding the Pre-Bid Construction Schedule specified in the Instructions to Bidders or the Contract Documents.

3.4.2.2 Master Schedule: The CM shall recommend to the Owner any appropriate revisions to the Master Schedule. Following acceptance by the Owner of such revisions, the CM shall provide a copy of the Master Schedule to the Designer and to the bidders.

3.4.3 Cost Management

3.4.3.1 Estimates for Addenda: The CM shall prepare an estimate of costs for all Addenda and shall submit a copy of the estimate to the Designer and to the Owner for approval.

3.4.3.2 Analyzing Bids: Upon receipt of the bids, the CM shall evaluate the bids, including alternate bid prices and unit prices, and shall make a recommendation to the Owner regarding the award of the Construction Contract.

3.4.4 Management Information System (MIS)

3.4.4.1 Schedule Maintenance Reports: The CM shall prepare and distribute schedule maintenance reports during the Procurement Phase. The reports shall compare the actual bid and award dates to scheduled bid and award dates and shall summarize the progress of the Project.

3.4.4.2 Project Cost Reports: The CM shall prepare and distribute project cost reports during the Procurement Phase. The reports shall compare actual contract award prices for the Project with those contemplated by the Project and Construction Budget.

3.4.4.3 Cash Flow Reports: The CM shall prepare and distribute cash flow reports during the Procurement Phase. The reports shall be based on actual contract award prices and estimated other construction costs for the duration of the Project.

3.5 Construction Phase

3.5.1 Project Management

3.5.1.1 Pre-Construction Conference: In consultation with the Owner and Designer, the CM shall conduct a Pre-Construction Conference during which the CM shall review the Project reporting procedures and other requirements for performance of the Work..

3.5.1.2 Permits, Bonds and Insurance: The CM shall verify that the Contractor has provided evidence that required permits, bonds, and insurance have been obtained. Such action by the CM shall not relieve the Contractor of its responsibility to comply with the provisions of the Contract Documents.

3.5.1.3 On-Site Management and Construction Phase Communication Procedures: The CM shall provide and maintain a management team on the Project site to provide contract administration as an agent of the Owner, and the CM shall establish and implement coordination and communication procedures among the CM, Owner, Designer and Contractor.

3.5.1.4 Contract Administration Procedures: The CM shall establish and implement procedures for reviewing and processing requests for clarifications and interpretations of the Contract Documents; shop drawings, samples and other submittals; contract schedule adjustments; change order proposals; written proposals for substitutions; payment applications; and the maintenance of logs. As the Owner's representative at the construction site, the CM shall be the party to whom all such information shall be submitted.

CMAA Document A-1 (2005 Edition)

3.5.1.5 Review of Requests for Information, Shop Drawings, Samples, and Other Submittals: The CM shall examine the Contractor's requests for information, shop drawings, samples, and other submittals, and Designer's reply or other action concerning them, to determine the anticipated effect on compliance with the Project requirements, the Project and Construction Budget, and the Master Schedule. The CM shall forward to the Designer for review, approval or rejection, as appropriate, the request for clarification or interpretation, shop drawing, sample, or other submittal, along with the CM's comments. The CM's comments shall not relate to design considerations, but rather to matters of cost, scheduling and time of construction, and clarity, consistency, and coordination in documentation. The CM shall receive from the Designer and transmit to the Contractor, all information so received from the Designer.

3.5.1.6 Project Site Meetings: Periodically the CM shall conduct meetings at the Project site with each Contractor, and the CM shall conduct coordination meetings with the Contractor, the Owner and the Designer. The CM shall prepare and distribute minutes to all attendees, the Owner and Designer.

3.5.1.7 Coordination of Other Independent Consultants: Technical inspection and testing provided by others shall be coordinated by the CM. The CM shall receive a copy of all inspection and testing reports and shall provide a copy of such reports to the Designer. The CM shall not be responsible for providing, nor shall the CM control, the actual performance of technical inspection and testing. The CM is performing a coordination function only and the CM is not acting in a manner so as to assume responsibility or liability, in whole or in part, for all or any part of such inspection and testing.

3.5.1.8 Minor Variations in the Work: The CM may authorize minor variations in the Work from the requirements of the Contract Documents that do not involve an adjustment in the Contract price or time and which are consistent with the overall intent of the Contract Documents. The CM shall provide to the Designer copies of such authorizations.

3.5.1.9 Change Orders: The CM shall establish and implement a change order control system. All changes to the Contract between the Owner and Contractor shall be only by change orders executed by the Owner.

3.5.1.9.1 All proposed Owner-initiated changes shall first be described in detail by the CM in a request for a proposal issued to the Contractor. The request shall be accompanied by drawings and specifications prepared by the Designer. In response to the request for a proposal, the Contractor shall submit to the CM for evaluation detailed information concerning the price and time adjustments, if any, as may be necessary to perform the proposed change order Work. The CM shall review the Contractor's proposal, shall discuss the proposed change order with the Contractor, and endeavor to determine the Contractor's basis for the price and time proposed to perform the changed Work.

3.5.1.9.2 The CM shall review the contents of all Contractor requested changes to the Contract time or price, endeavor to determine the cause of the request, and assemble and evaluate information concerning the request. The CM shall provide to the Designer a copy of each change request, and the CM shall in its evaluations of the Contractor's request consider the Designer's comments regarding the proposed changes.

3.5.1.9.3 The CM shall make recommendations to the Owner regarding all proposed change orders. At the Owner's direction, the CM shall prepare and issue to the Contractor appropriate change order documents. The CM shall provide to the Designer copies of all approved change orders.

3.5.1.10 Subsurface and Physical Conditions: Whenever the Contractor notifies the CM that a surface or subsurface condition at or contiguous to the site is encountered that differs from what the Contractor is entitled to rely upon or from what is indicated or referred to in the Contract Documents, or that may require a change in the Contract Documents, the CM shall notify the Designer. The CM shall receive from the Designer and transmit to the Contractor all information necessary to specify any design changes required to be responsive to the differing or changed condition and, if necessary, shall prepare a change order as indicated in Paragraph 3.5.1.9.

CMAA Document A-1 (2005 Edition)

3.5.1.11 Quality Review: The CM shall establish and implement a program to monitor the quality of the Work. The purpose of the program shall be to assist in guarding the Owner against Work by the Contractor that does not conform to the requirements of the Contract Documents. The CM shall reject any portion of the Work and transmit to the Owner and Contractor a notice of nonconforming Work when it is the opinion of the CM, Owner, or Designer that such Work does not conform to the requirements of the Contract Documents. Except for minor variations as described in Paragraph 3.5.1.8, the CM is not authorized to change, revoke, alter, enlarge, relax or release any requirements of the Contract Documents or to approve or accept any portion of the Work not conforming with the requirements of the Contract Documents. Communication between the CM and Contractor with regard to quality review shall not in any way be construed as binding the CM or Owner or releasing the Contractor from performing in accordance with the terms of the Contract Documents. The CM will not be responsible for, nor does the CM control, the means, methods, techniques, sequences and procedures of construction for the Project. It is understood that the CM's action in providing quality review under this Agreement is a service of the CM for the sole benefit of the Owner and by performing as provided herein, the CM is not acting in a manner so as to assume responsibility of liability, in whole or in part, for all or any part of the construction for the Project. No action taken by the CM shall relieve the Contractor from its obligation to perform the Work in strict conformity with the requirements of the Contract Documents, and in strict conformity with all other applicable laws, rules and regulations.

3.5.1.12 Contractor's Safety Program: The CM shall require each Contractor that will perform Work at the site to prepare and submit to the CM for general review a safety program, as required by the Contract Documents. The CM shall review each safety program to determine that the programs of the various Contractors performing Work at the site, as submitted, provide for coordination among the Contractors of their respective programs. The CM shall not be responsible for any Contractor's implementation of or compliance with its safety programs, or for initiating, maintaining, monitoring or supervising the implementation of such programs or the procedures and precautions associated therewith, or for the coordination of any of the above with the other Contractors performing the Work at the site. The CM shall not be responsible for the adequacy or completeness of any Contractor's safety programs, procedures or precautions.

3.5.1.13 Disputes Between Contractor and Owner: The CM shall render to the Owner in writing within a reasonable time decisions concerning disputes between the Contractor and the Owner relating to acceptability of the Work, or the interpretation of the requirements of the Contract Documents pertaining to the furnishing and performing of the Work.

3.5.1.14 Operation and Maintenance Materials: The CM shall receive from the Contractor operation and maintenance manuals, warranties and guarantees for materials and equipment installed in the Project. The CM shall deliver this information to the Owner and shall provide a copy of the information to the Designer.

3.5.1.15 Substantial Completion: The CM shall determine when the Project and the Contractor's Work is substantially complete. In consultation with the Designer, the CM shall, prior to issuing a certificate of substantial completion, prepare a list of incomplete Work or Work which does not conform to the requirements of the Contract Documents. This list shall be attached to the certificate of substantial completion.

3.5.1.16 Final Completion: In consultation with the Designer, the CM shall determine when the Project and the Contractor's Work is finally completed, shall issue a certificate of final completion and shall provide to the Owner a written recommendation regarding payment to the Contractor.

3.5.2 Time Management

3.5.2.1 Master Schedule: The CM shall adjust and update the Master Schedule and distribute copies to the Owner and Designer. All adjustments to the Master Schedule shall be made for the benefit of the Project.

3.5.2.2 Contractor's Construction Schedule: The CM shall review the Contractor's Construction Schedule and shall verify that the schedule is prepared in accordance with the requirements of the Contract Documents and that it establishes completion dates that comply with the requirements of the Master Schedule.

CMAA Document A-1 (2005 Edition)

3.5.2.3 Construction Schedule Report: The CM shall, on a monthly basis, review the progress of construction of the Contractor, shall evaluate the percentage complete of each construction activity as indicated in the Contractor's Construction Schedule and shall review such percentages with the Contractor. This evaluation shall serve as data for input to the periodic Construction Schedule report that shall be prepared and distributed to the Contractor, Owner and Designer by the CM. The report shall indicate the actual progress compared to scheduled progress and shall serve as the basis for the progress payments to the Contractor. The CM shall advise and make recommendations to the Owner concerning the alternative courses of action that the Owner may take in its efforts to achieve Contract compliance by the Contractor.

3.5.2.4 Effect of Change Orders on the Schedule: Prior to the issuance of a change order, the CM shall determine and advise the Owner as to the effect on the Master Schedule of the change. The CM shall verify that activities and adjustments of time, if any, required by approved change orders have been incorporated into the Contractor's Construction Schedule.

3.5.2.5 Recovery Schedules: The CM may require the Contractor to prepare and submit a recovery schedule as specified in the Contract Documents.

3.5.3 Cost Management

3.5.3.1 Schedule of Values (Each Contract): The CM shall, in participation with the Contractor, determine a schedule of values for the construction Contract. The schedule of values shall be the basis for the allocation of the Contract price to the activities shown on the Contractor's Construction Schedule.

3.5.3.2 Allocation of Cost to the Contractor's Construction Schedule: The Contractor's Construction Schedule shall have the total Contract price allocated by the Contractor among the Contractor's scheduled activities so that each of the Contractor's activities shall be allocated a price and the sum of the prices of the activities shall equal the total Contract price. The CM shall review the Contract price allocations and verify that such allocations are made in accordance with the requirements of the Contract Documents. Progress payments to the Contractor shall be based on the Contractor's percentage of completion of the scheduled activities as set out in the Construction Schedule reports and the Contractor's compliance with the requirements of the Contract Documents.

3.5.3.3 Effect of Change Orders on Cost: The CM shall advise the Owner as to the effect on the Project and Construction Budget of all proposed and approved change orders.

3.5.3.4 Cost Records: In instances when a lump sum or unit price is not determined prior to the Owner's authorization to the Contractor to perform change order Work, the CM shall request from the Contractor records of the cost of payroll, materials and equipment incurred and the amount of payments to each subcontractor by the Contractor in performing the Work.

3.5.3.5 Trade-off Studies: The CM shall provide trade-off studies for various minor construction components. The results of these studies shall be in report form and distributed to the Owner and Designer.

3.5.3.6 Progress Payments: The CM shall review the payment applications submitted by the Contractor and determine whether the amount requested reflects the progress of the Contractor's Work. The CM shall make appropriate adjustments to each payment application and shall prepare and forward to the Owner a progress payment report. The report shall state the total Contract price, payments to date, current payment requested, retainage and actual amounts owed for the current period. Included in this report shall be a Certificate of Payment that shall be signed by the CM and delivered to the Owner.

3.5.4 Management Information System (MIS)

3.5.4.1 Schedule Maintenance Reports: The CM shall prepare and distribute schedule maintenance reports during the Construction Phase. The reports shall compare the projected completion dates to scheduled completion dates of each separate contract and to the Master Schedule for the Project.

CMAA Document A-1 (2005 Edition)

3.5.4.2 Project Cost Reports: The CM shall prepare and distribute Project cost reports during the Construction Phase. The reports shall compare actual Project costs to the Project and Construction Budget.

3.5.4.3 Project and Construction Budget Revisions: The CM shall make recommendations to the Owner concerning changes that may result in revisions to the Project and Construction Budget. Copies of the recommendations shall be provided to the Designer.

3.5.4.4 Cash Flow Reports: The CM shall periodically prepare and distribute cash flow reports during the construction phase. The reports shall compare actual cash flow to planned cash flow.

3.5.4.5 Progress Payment Reports (Each Contract): The CM shall prepare and distribute the Progress Payment reports. The reports shall state the total Contract price, payment to date, current payment requested, retainage, and amounts owed for the period. A portion of this report shall be a recommendation of payment that shall be signed by the CM and delivered to the Owner for use by the Owner in making payments to the Contractor.

3.5.4.6 Change Order Reports: The CM shall periodically during the construction phase prepare and distribute change order reports. The report shall list all Owner-approved change orders by number, a brief description of the change order work, the cost established in the change order and percent of completion of the change order work. The report shall also include similar information for potential change orders of which the CM may be aware.

3.6 Post-Construction Phase

3.6.1 Project Management

3.6.1.1 Record Documents: The CM shall coordinate and expedite submittals of information from the Contractor to the Designer for preparation of record drawings and specifications, and shall coordinate and expedite the transmittal of such record documents to the Owner.

3.6.1.2 Operation and Maintenance Materials and Certificates: Prior to the final completion of the Project, the CM shall compile manufacturers' operations and maintenance manuals, warranties and guarantees, and certificates, and index and bind such documents in an organized manner. This information shall then be provided to the Owner.

3.6.1.3 Occupancy Permit: The CM shall assist the Owner in obtaining an occupancy permit by coordinating final testing, preparing and submitting documentation to governmental agencies, and accompanying governmental officials during inspections of the Project.

3.6.2 Time Management

3.6.2.1 Occupancy Plan: The CM shall prepare an occupancy plan that shall include a schedule for location for furniture, equipment and the Owner's personnel. This schedule shall be provided to the Owner.

3.6.3 Cost Management

3.6.3.1 Change Orders: The CM shall continue during the post-construction phase to provide services related to change orders as specified in Paragraph 3.5.3.3.

3.6.4 Management Information Systems (MIS)

3.6.4.1 Close Out Reports: At the conclusion of the Project, the CM shall prepare and deliver to the Owner final Project accounting and close out reports.

3.6.4.2 MIS Reports for Occupancy: The CM shall prepare and distribute reports associated with the occupancy plan.

ARTICLE 4

ADDITIONAL SERVICES

4.1 At the request of the Owner, the CM shall perform Additional Services and the CM shall be compensated for same as provided in Article 8 of this Agreement. The CM shall be obligated to perform Additional Services only after the Owner and CM have executed a written amendment to this Agreement providing for performance of such services. Additional Services may include, but are not limited to:

Appendix 7.2: Design-Build Qualification Statement Package

Postal Service Qualification Statement *Solicitation Number 512582-99-A-0002*

UNITED STATES POSTAL SERVICE
MAJOR FACILITIES PURCHASING
DESIGN/BUILD QUALIFICATION STATEMENT PACKAGE

Completing the Qualification Statements

The Postal Service is pre-qualifying offerors for the attached referenced project. The information you provide in the Qualification Statement is the basis for the evaluation of your design/build entity. To receive favorable consideration you must do the following:

- **Qualification Statement**. Read the Qualification Statement carefully. Make sure you understand the submission requirements. Complete the entire Qualification Statement. Provide all of the information requested. Do not leave anything portion of the form blank. If the information requested is not applicable to your company, write N/A or not applicable. If your Qualification Statement is not complete, it may not receive further consideration.
- **Evaluation Process**. Your Qualification Statement will be reviewed by the Evaluation Team (ET) appointed by the Contracting Officer. The ET will evaluate your Qualification Statements against the project's evaluation criteria. The information you provide in your Qualification Statement Part II, reference checks, and financial evaluation of your company will be the basis for the evaluation of your submission. Your submission must:
  - Meet the minimum requirements established in the evaluation criteria. If you do not meet the minimum requirements your Qualification Statement will not be reviewed further nor considered for prequalification.
  - Projects you identify as comparable projects must meet or exceed all of the criteria identified in the Qualification Statement including: time frame, size, cost, building type, type of construction, material handling requirements, etc. Any project listed that fails to meet any portion of the established criteria will not be considered as comparable.
  - You must provide current references (name, telephone number and address) for your comparable projects listed. If the USPS is unable to locate the references listed to verify experience and performance that project will not be considered.
  - Projects done for the direct benefit or use of the company submitting the Qualification Statement (own corporate headquarters, self-financed projects, projects done as part of speculative or other type of development by a branch, subsidiary or affiliate of the company) will not be considered as comparable projects.
  - Submitting photographs of the projects listed is desirable but not required.
  - Your submission should be written clearly and responses should be complete, concise and relevant to the submission requirements. Present the information and example projects that best illustrate your Design/Build entity's capability, experience and performance. Whenever possible or appropriate customize your response to the specific project.
  - Your Qualification Statement must be submitted in three ring binders, four (4) copies, tabbed in accordance with the List Of Attachments.
  - This Qualification Statement Package consists of three (3) Parts:
    Part A - General Information
    Part B - Past Performance and Capability
    Part C - Project Management.

The Postal Service appreciates your submission. We encourage you to ask for a debriefing whether or not your are successfully prequalified. This is an opportunity to discuss the strength and weaknesses of your submission. You must ask for a debriefing in writing no later than three (3) calendar days after receiving a notice of your status regarding the prequalification of this project.

Postal Service Qualification Statement *Solicitation Number 512582-99-A-0002*

PART A - GENERAL INFORMATION

1. GENERAL

Your response to the Qualification Statement Package will be used to assist the Contracting Officer in determining the technical and financial capability of the Design/Build entity completing the package forms.

a. The evaluation process will be accomplished in two (2) phases.

- Phase I will require the Design/Build entities to complete Part B of the Qualification Statement Package. Part B - Past Performance and Capability will be evaluated to determine whether the entity is one of the most qualified to be placed on the Pre-qualified List for this specific project.

- Phase II will require those Design/Build entities placed on the Pre-qualified list to complete Part C of the Qualification Statement Package. Part C - Project Management will include information that is more project specific and will require the Design/Build entity to include a separate "sealed" price proposal.

b. Contract award is dependent on the Postal Service (USPS) Board of Governors approval.

c. Firms will not be reimbursed for any expense incurred in developing the prequalification package.

d. Entities not selected as most highly qualified for placement on the Pre-qualified List as a result of this effort will be notified within sixty (60) days of the response date for this statement.

e. Qualification Statements will not be returned.

2. PROJECT DESCRIPTION

a. The U.S. Postal Service intends to issue a cost reimbursement contract with a guaranteed maximum price for the design and construction of a new Processing and Distribution Center (P&DC), Columbus, Ohio. The new P&DC will contain approximately 849,315 gross square foot; a separate storage building of 30,885 gross square foot ; and a Vehicle Maintenance Facility of 11,323 gross square foot on a 64.6 acre tract of land. All areas described are approximate; exact areas will be denoted on plans and specifications to be issued with the request for proposal. The facility will include administrative office areas, associated parking and maneuvering areas, Material handling: loose mail & bulk mail system, an coordination of a Tray Management System (TMS) into the building by separate contract as part of the scope of work. Scope of work will include design and construction of the facilities described above based upon preliminary plans, standards, specifications and criteria provided by USPS.

b. Total period of design and construction for this project will be 594 calendar days.

3. MINIMUM DESIGN/BUILD ENTITY PERFORMANCE REQUIREMENTS

a. Five (5) years experience as a design/build entity, in design/build, design, or construction.

b. Two (2) comparable projects completed or ongoing within the past five (5) years one of which must be new construction. For this project "Comparable projects" are defined as design/build projects, new construction or expansion, of light industrial or similar complexity incorporating material handling systems with a minimum of 300,000 square feet in size and a construction cost of $25 million or more.

c. Demonstrate experience performing cost reimbursable guaranteed maximum price (GMP) contracts.

4. METHOD AND DATE OF SUBMITTAL

Qualification Statements must be mailed or hand delivered in a sealed envelope addressed to:

Major Facilities Purchasing
Attn: Mike Miller
U.S. Postal Service
4301 Wilson Boulevard, Suite 300
Arlington, VA 22203-1840

The Qualification Statement (four copies) must be received at the above listed address no later than December 3, 1998 by 3:00 PM EST. The outside of the envelope shall be clearly marked with following identification:

Qualification Statement of (*Entity's name*)
Project Location: Columbus, Ohio
Solicitation Number: 512582-99-A-0002

5. EVALUATION PROCEDURES

The fully completed Qualification Statement Package (Parts B and C) will be considered as the entity's Technical and Management Proposal. Part B will be first evaluated to determine the entity's overall performance and capabilities. Only those entities that demonstrate that they not only satisfy the stated minimum requirements, but through past performance possess adequate experience and capability to successfully accomplish the proposed project, will be requested to submit Part C.

Separate evaluations will be made of Part B and Part C. The rankings will be combined to determine which design/build entities are most qualified. Ranking of Part B will constitute sixty percent (60%) and Part C forty percent (40%) of the final over-all rankings.

The primary areas to be used in determining separate rankings of Part B and Part C, Project Management Plan , are listed in descending order of importance. Evaluation values have been assigned to each of the elements in lieu of assigning values to each item within an element.

Part B - PERFORMANCE and CAPABILITIES (Total Scoring - 100 points)

Part B.1 - PERFORMANCE

1. Experience

* List two (2) comparable design/build projects completed or in progress during the past five years, listing a maximum of ten (10) projects. Submit the information using the format shown in Section B.2, Page 9.

* Other comparable design/build projects that demonstrates the design/build entity's experience, listing no more than a maximum of ten (10) projects. Submit the information using the format shown in Section B.2, Page 10.

Primary Evaluation Criteria: Primary consideration will be given to comparable projects that the entire design/build entity has successfully completed as a team. Other projects which would receive favorable consideration are: design/build projects that establish contractor's past relationships; comparable design/build projects as part of another design/build entity; and separate design or construction contracts for comparable postal facilities. **(Scoring - 35 points maximum.)**

2. Past Performance:

* List references using the format shown in Section B, Pages 9 and 10.

Primary Evaluation Criteria: Past performance will be based on evaluation of responses to reference checks for listed projects. For design/construction stage, primary areas of concern will be performance of design/build entity with regard to cost, schedule and quality during design and construction stages. Nature and extent of participation in listed projects will also be validated. Operations/maintenance checks will be primarily concerned with owning and operating costs, maintenance experience, systems' reliability, energy efficiency, and repair requirements for listed projects. **(Scoring - 30 points maximum.)**

Part B.2 - CAPABILITIES

1. Organization:

* Organizational Statement (Include in your submittal as Attachment B.4)

Primary Evaluation Criteria - Attach a statement describing the design/build entity's organization, with an organization chart. The statement and organization chart must clearly identify specific organizational elements and/or member firms that will be participating in the project, location of units; reporting relationships and functions to be performed by each unit; successful past working relationships; key executive who will have overall responsibility; key management personnel to be assigned to the project; resumes of key personnel; and professional staffing levels. In addition, describe expertise, capability, structure, and resources to adequately handle building construction and mechanization installation where required. Describe company resources that would be available for the project. Provide an employment profile for the design/build entity: total employees, total permanent employees, total professionals in each major category. Brief resumes of these individuals shall be attached. Current Form 254, as appropriate for each A/E consultant including anticipated material handling consultant must be included. (Note: Form 255 should be submitted with Part B - Management Plan.) **(Scoring - 25 points maximum.)**

2. Backlog:

* Backlog - Submit the information using the format shown in Section B.4, Page 13.

Primary Evaluation Criteria - Demonstration that resources necessary to maintain backlog at a satisfactory level are available. **(Scoring - 10 points maximum)**

3. Financial Capability:

* Audited Financial Statements - (Include in your submittal as Attachment B.5)
* Bonding Capacity (Include in your submittal as Attachment B.6)
* Bank References (Include in your submittal as Attachment B.7)

Primary Evaluation Criteria - Inclusion of audited and/or interim financial statement (see Page 11); completion of USPS Income Statement Evaluation Form; total bonding capacity (see Page 12); written and/or verbal bank references (see Page 12); and determination that design/build entity possesses sufficient resources to successfully complete project. **(Pass/Fail)**

4. Responsibility Verification:

* Judgments, claims and law suits (Include in your submittal as Attachment B.1)
* Suspension or debarment (Include in your submittal as Attachment B.2)
* Affirmation (Include in your submittal as Attachment B.3)

Primary Evaluation Criteria - No pending claims or lawsuits that could prevent successful completion of project. No debarment or other adverse legal action. **(Pass/Fail)**

Postal Service Qualification Statement Solicitation Number 512582-99-A-0002

PART B: PERFORMANCE

It is required that qualification data be presented on the forms, or in the format, provided below. Failure to comply with this requirement is grounds for a determination that the submittal is unacceptable. In the case of a joint venture or formal teaming arrangement, each joint venture/team member must submit all required information

1. ENTITY BACKGROUND

a. Firm Name: ____________________

Street Address: ____________________

____

Mailing Address: ____________________

____

b. Identification of two (2) contact people within the firm:

Name Title Telephone Number

____

____

c. Entity making this submittal:

__ Parent Company __ Subsidiary __ Division __ Branch Office __ Other _______

d. Type of firm:

__ Corporation __ Partnership __ Sole Proprietorship __ Teaming Agreement

__ Joint Venture __ Other ____________________

e. Year entity was established ______________

f. Name, address, and telephone number of parent company (enter N/A if not applicable).

____

____

____

g. All former firm names (enter N/A if not applicable).

____

____

h. Is your entity recognized as an MBE/WBE or Small Disadvantaged Business?

__ No __ Yes (category: ______________)

i. Of your total subcontracting volume, what is the actual average percentage awarded to MBE/WBE/SDB over the last five (5) years? _____%

j. Joint Venture/Teaming Agreement: If this Qualification Statement is being presented by a Joint Venture or an entity formed by a formal Teaming Agreement, please indicate the participation of each Joint Venture partner/Team member. If not a Joint Venture/Teaming Agreement entity, indicate Not Applicable (N/A).

NAME OF JOINT VENTURE PARTNER/TEAM MEMBERS	TYPE OF PARTICIPATION	PERCENTAGE OF FINANCIAL PARTICIPATION	PERCENTAGE OF OPERATIONAL PARTICIPATION

k. Judgments, Claims, and Lawsuits
Are there any judgments, claims, and/or lawsuits pending or outstanding against or involving your firm? ____ No ____ Yes ***If "Yes," submit details of all judgments or claims against either parent office or division/branch that will be responsible for the accomplishment of this project on a separate sheet as Attachment B.1.***

l. Key Personnel: List Officers, Partners, and/or Owners

NAME	POSITION OR TITLE IN THE FIRM	NAME OF/YEARS WITH THE FIRM	YEARS OF EXPERIENCE

m. Is the entity, venture, or any firm comprising it, under suspension or debarment by any federal, state, or local agency?
____ No ____ Yes If "Yes," submit details on a separate sheet labeled as Attachment A.2.

n. Affirmation: I ______________________________ , hereby certify that I am the authorized representative of the entity submitting this Qualification Statement, and that the following statements are true to the best of my knowledge, information, and belief:

I affirm that neither the entity, nor any officer, controlling shareholder, partner, or principal, nor any other person substantially involved in the contracting activities of the entity has in the past five (5) years:

(1) Been convicted under state or federal statutes of a criminal offense incident to obtaining or attempting to obtain or performing a public or private contract.
(2) Been convicted under state of federal statutes of fraud, embezzlement, theft, forgery, falsification or destruction of records, or receiving stolen property.
(3) Been found civil liable under state or federal antitrust or other statutes for acts or omissions in connection with submission of bids or proposals for or performance of a public or private contract.
(4) Been criminally convicted of any violation of a state or federal antitrust statute.

Postal Service Qualification Statement *Solicitation Number 512582-99-A-0002*

(5) Been convicted under then provisions of Title 18 of the United States Code for violation of the Racketeer Influence and Corrupt Organizations Act, 18 USC, Section 1961 et seq. or the Mail Fraud Act, 18 USC, Section 1341 et seq., for acts arising out of the submission of Bids or Proposals for a public or private contract.

(6) Been criminally convicted of conspiracy to commit any act or omission that would constitute grounds for conviction or liability under any statute described in paragraphs (10), (2), (4), or (5) above; or

(7) Admitted in writing or under oath, during the course of an official investigation, or other proceeding, acts or omissions that would constitute grounds for conviction of liability under any statute described above.

____________________	____________________	__________
Signature	*Typed Name*	*Date*

If unable to make the above Affirmation, please explain why not on a separate sheet labeled as Attachment B.3

Postal Service Qualification Statement *Solicitation Number 512582-99-A-0002*

2. ENTITY EXPERIENCE

a. **Comparable Design/Build Projects**: *List projects, two (2) minimum and no more than ten (10) maximum, involving entire design/build team completed within the past five (5) years, or on-going, that meet the criteria given in General Information 3.b and or listed below.*

Attach more sheets as needed, using this format: Sheet ____ of ____.

Project Name and Location: ______________________________

____

Project Size: ____________ Building Type: ______________ Percent Complete: __________

Contract Type: GMP _____ Fixed Price:, Other: ______________ Date Awarded: __________

Orig. Contract Amount: $____________ Final Contract Amount: $____________ % Change: _____

Orig. Contract Duration: ____________ Actual Contract Duration: ____________ % Change: _____

% of Subcontracts Awarded To: WBE ______ MBE ______ SDB ______

Project contained Mechanized Conveying Systems: _____ Yes _____ No

Project was completed within Client's Schedule and Budget: _____ Yes _____No

Client Reference for Design/Construction:

Name:______________________________

Address: ______________________________

Current Telephone Number: ______________________________

Client Reference for Operations/Maintenance:

Name:______________________________

Address: ______________________________

Current Telephone Number: ______________________________

Project Description: ______________________________

____

____

____

____

Postal Service Qualification Statement *Solicitation Number 512582-99-A-0002*

b. **Other Experience:** *List up to a **maximum of ten (10)** other projects, either on-going or completed within the past five (5) years, which best demonstrate the design/build entity's and/or any member firm's qualifications and capabilities to successfully complete the design and/or construction of this project.*

Attach more sheets as needed, using this format: Sheet ____ of ____.

Project Name and Location: ____

____

Design/Build Entity Members who participated in this Project: ____________________

____

Type of Participation: ______Full D/B ______Design Only ______Construction Only

Project Size: ____________ Building Type: ____________ Percent Complete: ________

Contract Type: GMP ______ Fixed Price:, Other: ____________ Date Awarded: ________

Orig. Design Fee: $____________ Final Design Fee: $____________ % Change: ______

Orig. Constr. Amt.: ____________ Final Contract Amt.: ____________ % Change: ______

Orig. Contract Duration: ____________ Final Contract Duration: ____________ % Change: ______

% of Subcontracts Awarded To: WBE ______ MBE ______ SDB ______

Project contained Material Handling Systems: _____ Yes _____ No

Project was completed within Client's Schedule and Budget: _____ Yes _____No

Client Reference for Design/Construction:

Name:____

Address: ____

Current Telephone Number: ____________________________

Client Reference for Operations/Maintenance:

Name:____

Address: ____

Current Telephone Number: ____________________________

Project Description: ____

____

____

____

____

Postal Service Qualification Statement *Solicitation Number 512582-99-A-0002*

3. FINANCIAL CAPABILITY

a. Please provide your Dun and Bradstreet Number (DUNS #): ____________________

b. Please attach the Design/Build entity's most recent financial statements including the Balance Sheet, the Statement of Income, the Statement of Cash Flows, and the notes to the financial statements. These statements must be AUDITED by an independent, licensed CPA or CPA firm for the offeror's previous two (2) fiscal years. (This data will NOT count against the 50 page package limit.) Financial statements must be for the entity making the submittal, not the parent company, unless a guarantee of the subsidiary's obligations is provided. Label the audited financial statements as ATTACHMENT B.5. **Failure to submit audited financial statements may result in the disqualification of your submittal because of determination of financial responsibility can NOT be made without this information.** In addition, if the entity's last fiscal year-end precedes the date of this submittal by more than six (6) months, PLEASE COMPLETE the following "Offeror's Interim Financial Data" form for the entity's most recent quarterly fiscal period. If the entity's most recent fiscal year-end fell within the last six months, completion of this form is not required.

Offeror's Interim Financial Data

For the ________ month period ending ______________, 199___

ASSETS		LIABILITIES & NET WORTH	
Current Assets		*Current Liabilities*	
CASH		NOTES PAYABLE	
ACCOUNTS RECEIVABLE		ACCOUNTS PAYABLE	
CONTRACTS (COMPLETED)		ACCRUED EXPENSES	
CONTRACTS (IN PROGRESS)		BILLINGS IN EXCESS OF COST	
OTHER RECEIVABLES		DEFERRED TAXES	
LESS: RESERVE FOR UNCOL.		OTHER CURRENT LIABILITIES	
NOTES RECEIVABLE		**TOTAL CURR. LIABILITIES**	
COSTS IN EXCESS OF BILLING			
INVENTORIES		*Long Term Liabilities*	
MARKETABLE SECURITIES		NOTES PAYABLE	
PREPAID EXPENSES		*DEFERRED TAXES*	
OTHER CURRENT ASSETS		OTHER L/T LIABILITIES	
TOTAL CURRENT ASSETS		**TOTAL L/T LIABILITIES**	
Fixed Assets		*Net Worth*	
LAND		CAPITAL STOCK	
BUILDINGS			
EQUIPMENT		ADDITIONAL PAID-IN	
FURNITURE & FIXTURES		RETAINED EARNINGS	
LESS: ACCUM. DEPREC.		TREASURY STOCK	
OTHER FIXED ASSETS		OTHER ADJUSTMENTS	
TOTAL FIXED ASSETS		**TOTAL NET WORTH**	
Other Assets		**TOTAL LIAB + NET WORTH**	
LIFE INSURANCE (CASH VALUE)			
LONG TERM INVESTMENTS		EARNED REVENUES + INCOME	
OTHER ASSETS		COST OF REVENUES EARNED	
TOTAL FIXED ASSETS		GROSS INCOME	
		GENERAL & ADMIN. EXPENSE	
TOTAL ASSETS		**NET INCOME**	

USPS INCOME STATEMENT EVALUATION FORM

COMPANY: ______________________________

Date of Income Statement: ______________________________

Net Profit for the Year: ______________________________

Balance of Accumulated Retained Earnings: ______________

OPERATING MARGIN OF PROFIT:

Net Sales ______________

- Cost of Sales + ______________

Operating Expense ______________

Equals: Operating Income ______________

Divided by Net Sales ______________

Equals: Op Marg. Profit ______________

NET PROFIT RATIO:

Net Profit ______________

Divided by Net Sales ______________

Equals: Net Profit Ratio ______________

NET PROFIT TO NET WORTH RATIO:

Net Profit ______________

Divided by Stockholder Equity ______________

Equals: NP/NW Ratio ______________

c. Bonding: Please attach a letter from one or more bonding companies giving your bonding capacity with them and the amount of bonding outstanding, and stating how long they have been providing bonds to your entity. Label the bonding company's letter as Attachment B.6.

d. Banking: Please attach a letter from a bank stating the following:

- How long the entity has banked with it.
- Average balance (in general terms).
- Extent of credit available and terms of availability.
- The bank's rating of the entity as a customer.
- Name and telephone number of person(s) at bank who can be contacted by USPS evaluators.

Label the bank reference as Attachment B.7

4. BACKLOG

Please provide a statement of total entity backlog, currently and for the past two years. Include only those contracts for which the entity has responsibilities and liabilities equivalent to those of a general contractor and/or design/build entity. Exclude construction management contracts.

		A	B	A less B
	NUMBER OF ACTIVE CONTRACTS	TOTAL ORIGINAL VALUE OF ACTIVE CONTRACTS	TOTAL VALUE COMPLETED FOR ACTIVE CONTRACTS	BALANCE TO COMPLETE ($) [i.e. BACKLOG]
CURRENTLY				
ONE YEAR AGO				
TWO YEARS AGO				

6. SIGNATURE AND CERTIFICATION

Under penalty of perjury, the undersigned declares, certifies, verifies, and states to the best of his/her knowledge and belief, that the foregoing and attached information is true, correct, and complete.

____
(Typed Name of Authorized Officer)

______________________________ ______________________________ ____________
(Signature of Authorized Officer) (Title) (Date)

____
(Typed Name of Witness)

______________________________ ______________________________ ____________
(Signature of Witness) (Title) (Date)

7. LIST OF ATTACHMENTS

ATTACHMENT	DESCRIPTION	SUBMITTAL REQUIREMENT
B.1	Judgments, Claims, and Lawsuits	Only if Applicable
B.2	Details of Suspension or Debarment	Only if Applicable
B.3	Inability to make Affirmation	Only if Applicable
B.4	Company Organization	REQUIRED
B.5	Audited Financial Statements	REQUIRED
B.6	Bonding Capacity	REQUIRED
B.7	Banking Reference	REQUIRED

PART C - PROJECT MANAGEMENT (Total Scoring - 220 points)

Prequalified firms will be requested to provide a brief management plan specifically detailing how the project will be accomplished. The plan shall be separately bound and be limited to thirty five (35) pages or less, exclusive of resumes, Forms 255, and other attachments. The following elements shall be confirmed in the plan:

1. Narrative Description Of Major Building Systems:

* Outline for Narrative Building Systems

Provide a brief narrative of all major building systems using the outline as indicated in C.1: Outline for Narrative Building Systems. If a section or a heading is not applicable to your proposal please state so by indicating NA or not applicable. This narrative will be used by the Postal Service during the evaluation of your proposal to gain a better understanding of your submission. The narrative description should not exceed 10 single side, type written pages. It is important that your description of building systems and components be as concise and quantitative as possible **(Scoring - 50 points maximum)**

2. Construction Phase:

Provide organizational chart for on-site staff. Identify and provide resume for each person, above the Foreman level, who is to be assigned to the project, including the duration of each assignment. Briefly discuss systems and procedures that will be used for quality management (including prevention and quality control/assurance procedures) and cost control. **(Scoring - 40 points maximum)**

3. Design Phase:

Briefly discuss overall design approach. Describe specific systems and procedures to be used for overall design coordination, cost control, value engineering, life cycle cost analysis, constructibility reviews, coordination with field organizations, and client interface. Current Form 255 for the design portion of the design/build entity shall be attached to the plan. Form 255 should include only resumes of key staff members to be assigned to this project. **(Scoring - 30 points maximum)**

4. Project Management:

Identify specific organizational elements that will be actively participating in the project. Indicate location, reporting relationships, and functions to be performed by each unit. Provide a listing of key staff/supervisory personnel who will form the nucleus of the project team and furnish brief resumes. Give a narrative description of the management approach to the project. Describe systems and procedures to be used to provide continuity and overall control of quality, costs and schedule throughout the life of the project. **(Scoring - 25 points maximum)**

5. Material Handling:

Primary Evaluation Criteria - Past experience of management staff in supervising design and/or installation of material handling systems on comparable projects, and other material handling experience; qualifications of material handling design team; past working relationships; experience designing comparable systems; other design experience; qualifications of key staff professionals; systems and procedures to be used for overall control of cost, quality, and schedule; and qualifications and experience of proposed material handling subcontractors. **(Scoring - 25 points maximum)**

Postal Service Qualification Statement *Solicitation Number 512582-99-A-0002*

6. **Schedule:**

Provide a tentative schedule in the form of a computer-generated logic diagram or similar graphic representation indicating how the project will be accomplished within the specific overall duration. Particular emphasis shall be placed on the interface between design and construction activities, including advanced procurement and phasing or fast-tracking of construction. Indicate duration required to complete various design and construction phases, as well as the overall project. **(Scoring - 20 points maximum)**

7. **Safety Program:**

Attach a statement describing the entities' safety Program. Include an organizational chart that indicates the various levels of supervision within the program. Include a description of the company procedures that will be implemented on this project: construction site safety meetings, first-aid treatment, reporting procedures, company required safety apparel, construction site inspections, subcontractor safety program compliance requirements, and company enforcement procedures (i.e., disciplinary actions implemented after violations, etc.).

	A	B	C	D	(A or B or C) x (200,000) / D		
	# of Work Related Injuries	# of Work Related Illnesses	Lost Days of Work (in Hours)	Tot. Hrs. Worked by Co. Employees	Incidence Rates A/D	B/D	C/D
Project Name							
Project Name							
Co. Total Yr. Ago							
Co. Total Yrs. Ago							

Note: The method of calculation for the table above shall be per OSHA guidelines.
(Scoring - 15 points maximum)

8. **Accounting System:**

Include statement describing the entity's accounting system. Attach sample that demonstrates that the system is adequate for determining construction costs applicable to the contract. **(Scoring - 10 points maximum)**

9. **Procurement:**

Identify the work which will be performed by entities' own forces. Describe the systems and procedures to be followed for procurement activities. Provide subcontracting plan with emphasis on how Small, Minority-owned, and Woman-owned subcontracting goals will be met. **(Scoring - 5 points maximum)**

C.1: Outline for Narrative Building Systems

1. GENERAL:
   - Permit/Fees
   - Codes
   - Survey/Inspection/Testing
   - Temporary Utilities

2. SITE:
   - Site Utilities
   - Demolition
   - Earthwork
   - Drainage
   - Paving
   - Traffic/Traffic Controls
   - Landscaping

3. ARCHITECTURAL
   - Exterior Walls
   - Roof and Roof Drainage
   - Daylighting
   - Interior Walls
   - Mezzanines
   - Finishes
   - Special Construction
   - Equipment

4. STRUCTURAL
   - Foundation
   - Framing
   - Deck
   - Seismic

5. PLUMBING
   - Drainage Systems
   - Compressed Air
   - Plumbing Fixtures

6. FIRE PROTECTION
   - Interior Area Sprinkler
   - Exterior Area Sprinkler

7. HVAC
   - General
   - Chilled Water
   - Hot Water
   - Heating and Air Conditioning
   - DDC Controls System

8. ELECTRICAL
   - Primary Electric Service
   - Secondary Electric Service and Distribution
   - Interior Lighting
   - Exterior Lighting
   - Special Outlets
   - Fire Alarm Systems
   - Security System
   - Universal Wiring
   - Lightning Protection System

9. SPECIALTY CONSTRUCTION

10. FIXED MECHANIZATION

11. OTHER

Chapter

8

Design-Build and Sustainability

The President's Council on Sustainable Development (PCSD), established in June 1993 by President Bill Clinton, was given the mission to develop and implement bold new approaches for integrating economic, social, and environmental policies to guide the United States to a more environment-friendly approach in the coming new century. In 1996, the council issued their report *Sustainable America* that essentially started the country down the road to a new way of looking at the impact we all have on nature's fragile and intricate framework. The word *sustainability* entered the lexicon of architectural, engineering, and construction communities and the green building movement received national recognition.

The design-build delivery system appears to be a perfect vehicle by which to pursue sustainable or green building construction. The process whereby an owner in a design-build situation, assembles a team of contractors, architects, and engineers who bring vendors and subcontractors onboard, seems to be the ideal setting in which to strategize and formulate a game plan for a building project. The experience of all parties and some brainstorming can work together to develop the most effective approach for reaching the owner's program goal. The back-and-forth of capital versus operating expense, initial cost versus long-term and life-cycle costs must invariably touch on the same topics that environmentalists have been harping on for years; how can we design our buildings to be more environmentally sensitive and preserve our physical resources? This subject of sustainability—green buildings, is now in the mainstream and both public and private owners recognize the savings that can accrue from incorporating many of these environmentally friendly schemes into their current building program and save some money as well.

The advocates of green buildings can no longer be viewed as tree huggers as more communities and corporations view new opportunities to effect savings, protect the environment, and create more public awareness of the growing need to preserve our planet. The process of building factories, office buildings, and homes has had a major impact on our ecosystem in past years, but it is a process that can be mitigated and turned around without too much difficulty.

The Impact of Construction on the Environment

Commercial and institutional buildings have a dramatic impact on the environment:

- Buildings in the United States consume 36% of the total energy use and 65% of all electrical consumption.
- Buildings are responsible for 30% of all greenhouse gas emissions.
- Buildings consume 30% of the raw materials.
- Buildings produce 30% of the total waste output, approximately, 136 million tons annually.
- Buildings consume 12% of all potable water.

The U.S. Department of Energy reports that there are 4.6 million commercial buildings in the United States, occupying more than 67 billion square feet of space, and these buildings consume one-sixth of the world's fresh water supply, one-half of the virgin wood harvested, and two-fifths of materials and energy reserves.

We all have the responsibility to ourselves and the rest of the world to control our voracious appetite for global harvests and preserve as much of our renewable resources as possible—this is the essence of this sustainable movement.

What do we mean by sustainability?

Sustainability is the term applied to the quest to *sustain* economic growth while maintaining long-term environmental health. When applied to construction, sustainability means creating designs that seek to balance the short-term goals of a project with the long-term goals of efficient operating systems that protect the environment and nature's resources. Sustainable buildings represent a holistic approach to construction that combine the advantages of modern technology with proven construction practices, using nature to enhance the building's efficiency rather than fighting it. Using fenestration to let natural light into the building while employing the latest technology of inert gas-filled insulated glass panels, low-emission coatings, and thermal break frames helps not only to reduce interior space lighting, but also to reduce building heating and cooling loads. Oriented strand board (OSB) and medium density fiberboard (MDF) are two perfect examples of sustainability, using waste and recycled wood products to create new products that, in some cases, are more durable and more maintenance free than the virgin wood from which they are made.

Whole Building Design

The process of design-build lends itself to the whole building design process, or possibly vice versa, the whole building design lends itself to the design-build delivery system. Whole building design is a process where the building's structure, envelope, interior components, mechanical and electrical systems, and site orientation are viewed holistically. Each party to the design-construction process will be called upon for their input, ideas, and solutions. The whole building concept considers site, energy, materials, indoor air quality, acoustics, natural resources, and their interrelationship with each other.

This whole building approach allows a design-build firm to really show its stuff. Bringing the experience of the total group, including vendors and subcontractors, to the table with a desire to present not only a functional design, but also one that affords the owner the most cost-effective initial cost and the lowest life-cycle costs is the goal of the team. New and proven technologies can be discussed, weighed, debated, and incorporated or discarded.

The benefits of the whole building design should be directed toward the following goals:

Reduce energy costs

Reduce both capital and maintenance costs

Reduce the environmental impact of the building to the site and environs

Increase occupant comfort, health, and safety

Increase employee productivity

The history of green building construction in this country is proof that all these requirements can be met at little or no initial cost to the project. The cost-effectiveness of these green buildings, over the somewhat long term, is just beginning to be documented and it validates green buildings' reason for being. But let's discuss the term *sustainability* in today's vernacular a little closer.

LEED Is Not Sustainability

Sustainability is the process involved in designing and building an environmentally friendly structure. LEED (Leadership in Energy and Environmental Design) is a trademark-protected rating system developed by the United States Green Building Council (USGBC), a program of standards and certification for accreditation purposes. LEED addresses a variety of types of construction, but all with one purpose—to define high-performance green buildings that are environmentally responsible, healthy, and profitable. The LEED program encompasses:

LEED-NC—New construction

LEED-EB—Existing buildings

LEED-CI—Commercial interiors

LEED-C&S—Core and shell

LEED-H—Homes

LEED-ND—Neighborhood development

The rating systems were developed by the USGBC committees and allow for four progressive levels of certification:

Certified—the lowest level

Silver

Gold

Platinum—the highest level

There are six credit areas in each category with points awarded for degree of compliance:

1. Sustainable sites
2. Energy and atmosphere
3. Water efficiency
4. Indoor environmental quality
5. Materials and resources
6. Innovation in design

Within each credit area there are a number of points available, and the number of points a building earns will determine the level of certification achieved.

For example, the total number of points awarded is 69.

Basic certification requires 26 to 32 points.

Silver certification requires 33 to 38 points.

Gold certification requires 39 to 51 points.

Platinum certification requires 52 points or more.

The basic certification level must meet 40% of the LEED system; silver must meet 50%, gold 60%, and platinum must meet 80% of the rating system.

A rating system checklist for new construction is shown in Fig. 8.1. More detailed information about each item in the checklist can be found in the full LEED program. Figure 8.2 contains details relating to site selection; Fig. 8.3 pertains to the energy and atmosphere portion of the checklist, called optimize energy performance.

Rating System

Version 2.0

Including the Project Checklist

June 2001

Figure 8.1 LEED rating system for new construction. (*Courtesy: U.S. Green Building Council.*)

LEED

Materials & Resources

13 Possible Points

Y			Prereq 1	**Storage & Collection of Recyclables**	Required
Y	?	N	Credit 1.1	**Building Reuse**, Maintain 75% of Existing Shell	1
Y	?	N	Credit 1.2	**Building Reuse**, Maintain 100% of Shell	1
Y	?	N	Credit 1.3	**Building Reuse**, Maintain 100% Shell & 50% Nonshell	1
Y	?	N	Credit 2.1	**Construction Waste Management**, Divert 50%	1
Y	?	N	Credit 2.2	**Construction Waste Management**, Divert 75%	1
Y	?	N	Credit 3.1	**Resource Reuse**, Specify 5%	1
Y	?	N	Credit 3.2	**Resource Reuse**, Specify 10%	1
Y	?	N	Credit 4.1	**Recycled Content**, Specify 25%	1
Y	?	N	Credit 4.2	**Recycled Content**, Specify 50%	1
Y	?	N	Credit 5.1	**Local/Regional Materials**, 20% Manufactured Locally	1
Y	?	N	Credit 5.2	**Local/Regional Materials**, of 20% Above, 50% Harvested Locally	1
Y	?	N	Credit 6	**Rapidly Renewable Materials**	1
Y	?	N	Credit 7	**Certified Wood**	1

Indoor Environmental Quality

15 Possible Points

Y			Prereq 1	**Minimum IAQ Performance**	Required
Y			Prereq 2	**Environmental Tobacco Smoke** (ETS) **Control**	Required
Y	?	N	Credit 1	**Carbon Dioxide** (CO_2) **Monitoring**	1
Y	?	N	Credit 2	**Increase Ventilation Effectiveness**	1
Y	?	N	Credit 3.1	**Construction IAQ Management Plan**, During Construction	1
Y	?	N	Credit 3.2	**Construction IAQ Management Plan**, Before Occupancy	1
Y	?	N	Credit 4.1	**Low-Emitting Materials**, Adhesives & Sealants	1
Y	?	N	Credit 4.2	**Low-Emitting Materials**, Paints	1
Y	?	N	Credit 4.3	**Low-Emitting Materials**, Carpet	1
Y	?	N	Credit 4.4	**Low-Emitting Materials**, Composite Wood	1
Y	?	N	Credit 5	**Indoor Chemical & Pollutant Source Control**	1
Y	?	N	Credit 6.1	**Controllability of Systems**, Perimeter	1
Y	?	N	Credit 6.2	**Controllability of Systems**, Non-Perimeter	1
Y	?	N	Credit 7.1	**Thermal Comfort**, Comply with ASHRAE 55-1992	1
Y	?	N	Credit 7.2	**Thermal Comfort**, Permanent Monitoring System	1
Y	?	N	Credit 8.1	**Daylight & Views**, Daylight 75% of Spaces	1
Y	?	N	Credit 8.2	**Daylight & Views**, Views for 90% of Spaces	1

Innovation & Design Process

5 Possible Points

Y	?	N	Credit 1.1	**Innovation in Design**: Specific Title	1
Y	?	N	Credit 1.2	**Innovation in Design**: Specific Title	1
Y	?	N	Credit 1.3	**Innovation in Design**: Specific Title	1
Y	?	N	Credit 1.4	**Innovation in Design**: Specific Title	1
Y	?	N	Credit 2	**LEED™ Accredited Professional**	1

Project Totals

69 Possible Points

Certified 26-32 points **Silver** 33-38 points **Gold** 39-51 points **Platinum** 52-69 points

U S Green Building Council

Figure 8.1 *(Continued)*

Project Checklist

LEED

Sustainable Sites

14 Possible Points

Y	Prereq 1	**Erosion & Sedimentation Control**	Required
Y ? N	Credit 1	**Site Selection**	1
Y ? N	Credit 2	**Urban Redevelopment**	1
Y ? N	Credit 3	**Brownfield Redevelopment**	1
Y ? N	Credit 4.1	**Alternative Transportation**, Public Transportation Access	1
Y ? N	Credit 4.2	**Alternative Transportation**, Bicycle Storage & Changing Rooms	1
Y ? N	Credit 4.3	**Alternative Transportation**, Alternative Fuel Refueling Stations	1
Y ? N	Credit 4.4	**Alternative Transportation**, Parking Capacity	1
Y ? N	Credit 5.1	**Reduced Site Disturbance**, Protect or Restore Open Space	1
Y ? N	Credit 5.2	**Reduced Site Disturbance**, Development Footprint	1
Y ? N	Credit 6.1	**Stormwater Management**, Rate or Quantity	1
Y ? N	Credit 6.2	**Stormwater Management**, Treatment	1
Y ? N	Credit 7.1	**Landscape & Exterior Design to Reduce Heat Islands**, NonRoof	1
Y ? N	Credit 7.2	**Landscape & Exterior Design to Reduce Heat Islands**, Roof	1
Y ? N	Credit 8	**Light Pollution Reduction**	1

Water Efficiency

5 Possible Points

Y ? N	Credit 1.1	**Water Efficient Landscaping**, Reduce by 50%	1
Y ? N	Credit 1.2	**Water Efficient Landscaping**, No Potable Use or No Irrigation	1
Y ? N	Credit 2	**Innovative Wastewater Technologies**	1
Y ? N	Credit 3.1	**Water Use Reduction**, 20% Reduction	1
Y ? N	Credit 3.2	**Water Use Reduction**, 30% Reduction	1

Energy & Atmosphere

17 Possible Points

Y	Prereq 1	**Fundamental Building Systems Commissioning**	Required
Y	Prereq 2	**Minimum Energy Performance**	Required
Y	Prereq 3	**CFC Reduction in HVAC&R Equipment**	Required
Y ? N	Credit 1.1	**Optimize Energy Performance**, 20% New / 10% Existing	2
Y ? N	Credit 1.2	**Optimize Energy Performance**, 30% New / 20% Existing	2
Y ? N	Credit 1.3	**Optimize Energy Performance**, 40% New / 30% Existing	2
Y ? N	Credit 1.4	**Optimize Energy Performance**, 50% New / 40% Existing	2
Y ? N	Credit 1.5	**Optimize Energy Performance**, 60% New / 50% Existing	2
Y ? N	Credit 2.1	**Renewable Energy**, 5%	1
Y ? N	Credit 2.2	**Renewable Energy**, 10%	1
Y ? N	Credit 2.3	**Renewable Energy**, 20%	1
Y ? N	Credit 3	**Additional Commissioning**	1
Y ? N	Credit 4	**Ozone Depletion**	1
Y ? N	Credit 5	**Measurement & Verification**	1
Y ? N	Credit 6	**Green Power**	1

LEED™ Rating System 2.0

Figure 8.1 *(Continued)*

1 Point

Credit 1 **Site Selection**

Intent

Avoid development of inappropriate sites and reduce the environmental impact from the location of a building on a site.

Requirement

Credit 1.0 (1 point) Do not develop buildings on portions of sites that meet any one of the following criteria:

-Prime farmland as defined by the American Farmland Trust

-Land whose elevation is lower than **5 feet above** the elevation of the 100-year flood as defined by FEMA

-Land which provides habitat for any species on the Federal or State threatened or endangered list

-Within **100 feet** of any wetland as defined by 40 CFR, Parts 230-233 and Part 22, OR as defined by local or state rule or law, whichever is more stringent

-Land which prior to acquisition for the project was public parkland, unless land of equal or greater value as parkland is accepted in trade by the public landowner (Park Authority projects are exempt)

Technologies & Strategies

During the site selection process, give preference to those sites that do not include sensitive site elements and restricted land types. Select a suitable building location and design the building with the minimal footprint to minimize site disruption. Strategies include stacking the building program, tuck under parking, and sharing facilities with neighbors.

U S Green Building Council

Figure 8.2 Rating system details—site selection. (*Courtesy: U.S. Green Building Council.*)

SS	WE	EA	MR	EQ	ID

Credit 1

2-10 Points

Credit 1 **Optimize Energy Performance**

Intent

Achieve increasing levels of energy performance above the prerequisite standard to reduce environmental impacts associated with excessive energy use.

Requirements

Reduce design energy cost compared to the energy cost budget for regulated energy components described in the requirements of ASHRAE/IESNA Standard 90.1-1999, as demonstrated by a whole building simulation using the Energy Cost Budget Method described in Section 11:

New Buildings	Existing Buildings	Points
20%	**10%**	2
30%	**20%**	4
40%	**30%**	6
50%	**40%**	8
60%	**50%**	10

Regulated energy components include HVAC systems, building envelope, service hot water systems, lighting and other regulated systems as defined by ASHRAE.

Credit 1.1 (2 points) Reduce design energy cost by **20% / 10%.**
Credit 1.2 (4 points) Reduce design energy cost by **30% / 20%.**
Credit 1.3 (6 points) Reduce design energy cost by **40% / 30%.**
Credit 1.4 (8 points) Reduce design energy cost by **50% / 40%.**
Credit 1.5 (10 points) Reduce design energy cost by **60% / 50%.**

Technologies & Strategies

Design the building envelope and building systems to maximize energy performance. Use a computer simulation model to assess the energy performance and identify the most cost effective energy efficiency measures. Quantify energy performance as compared to a baseline building.

U S Green Building Council

Figure 8.3 Rating system details—optimize energy performance. (*Courtesy: U.S. Green Building Council.*)

Government takes the LEED

According to a study released by USGBC in February 2005, 41 cities in the United States have adopted some type of LEED certification program for construction or major renovation work in their public facilities. Bidders on these designated projects will have to show proficiency in delivering LEED certified buildings in order to be qualified.

Of the 41 nationwide municipal participants, here are some specifics:

Atlanta, Georgia. All city-funded projects larger than 5,000 square feet (465 square meters) or costing at least $2 million must meet a LEED silver rating level.

Austin, Texas. LEED certification required on all public projects larger than 5,000 gross square feet (4,000 square meters).

Berkeley, CA. Municipal buildings greater than 5,000 square feet (465 square meters) were required to be LEED certified in 2004; in 2006, buildings of this size must achieve silver certification.

Dallas, Texas. All city buildings larger than 10,000 square feet (929 square meters) are required to have at least LEED silver certification.

Boston, MA. This city established LEED silver as the goal for all city-owned buildings.

Chicago, Illinois. All new city-funded construction and major renovation projects will require LEED silver certification at minimum.

Kansas City, MO. All new city buildings must be designed to meet LEED silver at minimum. The city is participating in a LEED-EB (existing buildings) pilot program for their city hall.

San Francisco, CA. All new municipal construction, additions, and major renovation projects larger than 5,000 square feet (465 square meters) must achieve LEED silver certification.

Scottsdale, Arizona. In March 2005, the city passed Resolution 6644, requiring all new public buildings to be certified as LEED gold.

In Canada the number of sustainable buildings are growing:

Calgary. The city's sustainable building policy requires all new or significant renovations larger than 500 square meters (5,380 square feet) to achieve LEED silver certification as a minimum.

Vancouver. All new civic buildings larger than 500 square meters (5,380 square feet) have adopted green building standards LEED-British Columbia (LEED-BC).

New public buildings must achieve LEED gold certification as a minimum.

Green Buildings in the Private Sector

Private developers have recognized the value of green buildings both in terms of costs and public relations. The Swiss Reinsurance Tower in London reported

50% less energy consumption than a conventional building. Closer to home, the Conde Nast Building in Manhattan uses 35 to 40% less energy than standard construction design requires, and the Solaire, a 27 story, 293 unit apartment building further downtown in Battery Park City, is 35% more energy efficient than required by code, resulting in 67% lower power demands. During construction 93% of recoverable materials were diverted from the local landfills.

Out West, the Robert Redford Building in Santa Monica, California, reported using 60% less water than a conventional building because of its green water management system. In that same general area, Toyota embraced green buildings with its new $87 million sales campus in Torrance. This 624,000 square foot facility has 53,000 square feet of solar panels that generate 536 kilowatts and is projected to pay for itself in seven years. Motion sensors control the building's lighting, and ceramic floor tiles are made from recycled glass and recycled concrete.

Pennsylvania in the LEED

Pennsylvania's Department of Environmental Protection (DEP) has been at the forefront of green construction with five LEED registered projects on stream as of 2005. The state's first LEED gold-level green building was built in Cambria, and this 40,000 square foot project came in at $90.00 per square foot, slightly under comparable costs for conventional buildings. This building had triple pane high-performance windows installed that ultimately reduced their heating and cooling loads savings by $20,000 in initial costs and continue to reduce operating costs. The DEP reports that their LEED silver-level buildings cost virtually the same as conventional construction.

Even the Pentagon is interested in savings. Hensel-Phelps Construction Company, while working on a Pentagon renovation project, discovered a wheat strawboard product that was suitable to use as backer boards in electrical closets. This simple substitution of product saved the government $30,000.

Some design-build/sustainable building guidelines. There are eight simple principles of sustainable design that Tony Loyd and Donald Caskey, senior vice presidents and principals of Orange County, California-based Carter & Burgess set as guidelines to design, construction, and operation:

- A multidiscipline, integrated approach is the key to success.
- Simple is better than complex.
- The overriding framework in these types of projects reflects a respect for nature so that it is not depleted or harmed.
- Life-cycle costs are more significant than first costs—the age-old battle of capital versus expenses.

- Minimize the energy uses in the selection of building materials, mechanical systems, and appliances.
- Since maintenance of the structure is important, plan accordingly.
- Build with local materials whenever possible to reduce transportation costs. Local materials may be better suited to that environment.
- Consider passive strategies whenever possible—building orientation, overhangs and sunshades, thermal mass, and natural lighting.

Are Green Buildings More Expensive than Conventional Construction?

A study of the cost and benefits of green buildings was conducted by the State of California after Governor Gray Davis issued Executive Order D-16-00 in August 2000 that funded the research. The complete study, *A Report to California's Sustainable Building Task Force-October 2003*, is available on the Internet at *http://www.usgbc.org/Docs/News/News477.pdf*.

This detailed study showed that while green buildings may cost more than conventionally designed buildings, the premium for sustainability is much lower than generally perceived. And green costs are coming down every year as more architects and engineers, equipment manufacturers, and builders become more familiar with the concept and gain more experience in its development.

This California study indicated that minimal increases in upfront costs of about 2% would, on an average, result in life-cycle cost savings of about 20% of total construction costs. For example, an initial investment of $100,000 to incorporate green building features into a $5 million project would result in a savings of $1 million in today's dollars over the life of the building, according to the findings in this report.

The financial benefits of green buildings, as pointed out in the survey, include lower energy costs, lower waste disposal costs, lower water costs, lower environmental and emissions costs, lower operating and maintenance costs, and increased productivity and health of the workers occupying these types of buildings.

The energy costs and water savings were rather easy to predict but the productivity and health gains were much less precise and much harder to predict.

The report recognizes the difference between present value (the value of a future stream of benefits) and net present value (the present value of the long-term benefits minus the initial investment)—Fig. 8.4.

The average green cost premium varies with the level of LEED certification; certified being the least demanding certification level and platinum the most demanding. Figures 8.5*a* and *b* compare green building premiums for various levels of certification. While the total number of buildings surveyed is not large, only 33, it does reveal that, in general, green buildings have an average premium cost of just about 1.84%.

The growth in interest in green buildings is evidenced in the growth in membership in USGBC (Fig. 8.5*c*).

Use of Present Value (PV) and Net Present Value (NPV)

The overarching purpose of this report is to answer the following question: Does it make financial and economic sense to build a green building? Green buildings may cost more to build than conventional buildings, especially when incorporating more advanced technologies and higher levels of LEED, or sustainability. However, they also offer significant cost savings over time.

This report will seek to calculate the current value of green buildings and components on a present value (PV) or net present value (NPV) basis. PV is the present value of a future stream of financial benefits. NPV reflects a stream of current and future benefits and costs, and results in a value in today's dollars that represents the present value of an investment's future financial benefits minus any initial investment. If positive, the investment should be made (unless an even better investment exists), otherwise it should not.[69] This report assumes a suitable discount rate over an appropriate term to derive an informed rationale for making sustainable building funding decisions. Typically, financial benefits for individual elements are calculated on a present value basis and then combined in the conclusion with net costs to arrive at a net present value estimate.

Net present value can be calculated using Microsoft's standard Excel formula:

$$NPV = \sum_{i=1}^{n} \frac{values_i}{(1+rate)^i}$$

The formula requires the following:

- **Rate**: Interest Rate per time period (5% real)
- **Nper (n)**: The number of time periods (20 years)
- **Pmt (values):** The constant-sized payment made each time period (annual financial benefit)

This provides a calculation of the value in today's dollars for the stream of 20 years of financial benefits discounted by the 5% real interest rate. It is possible to calculate the net present value of the entire investment—both initial green cost premium and the stream of future discounted financial benefits—by subtracting the former from the latter.

Discount Rate

To arrive at present value and net present value estimates, projected future costs and benefits must be discounted to give a fair value in today's dollars. The discount rate used in this report is 5% real. This rate is stipulated for use by the California Energy Commission[70] and is somewhat higher than the rate at which the state of California borrows money through bond issuance.[71] It is also representative of discount rates used by other public sector entities.[72]

Figure 8.4 Use of present value and net present value. (*Source: State of California, Sustainable Building Task Force, October 2003.*)

Term
California's Executive Order D-16-00, committing California to provide energy efficiency and environmental leadership in its building design and operation, stipulates that "a building's energy, water, and waste disposal costs are computed over a twenty-five year period, or for the life of the building."[73] Buildings typically operate for over 25 years. A recent report for the Packard Foundation shows building life increasing with increasing levels of greenness. According to the Packard study, a conventional building is expected to last 40 years, a LEED Silver level building for 60 years and Gold or Platinum level buildings even longer.[74] In buildings, different energy systems and technologies last for different lengths of time – some energy equipment is upgraded every 8 to 15 years while some building energy systems may last the life of a building. This analysis conservatively assumes that the benefits of more efficient/sustainable energy, water, and waste components in green buildings will last 20 years, or roughly the average between envelope and equipment expected life.

Inflation

This report assumes an inflation rate of 2% per year, in line with most conventional inflation projections.[75] Unless otherwise indicated, this report makes a conventional assumption that costs (including energy and labor) as well as benefits rise at the rate of inflation – and so present value calculations are made on the basis of a conservative real 5% discount rate absent any inflation effects. In reality, this is quite an oversimplification and a more detailed analysis might attempt to make more accurate but complicated predictions of future costs. In particular, energy costs are relatively volatile, although electricity prices are less volatile than primary fuels, especially gas.

Figure 8.4 *(Continued)*

Level of Green Standard and Average Green Cost Premium

Level of Green Standard	Average Green Cost Premium
Level 1 – Certified	0.66%
Level 2 – Silver	2.11%
Level 3 – Gold	1.82%
Level 4 – Platinum	6.50%
Average of 33 Buildings	1.84%

Source: USGBC, Capital E Analysis

Figure 8.5a Levels of average green cost premiums. (*Source: State of California, Sustainable Building Task Force, October 2003.*)

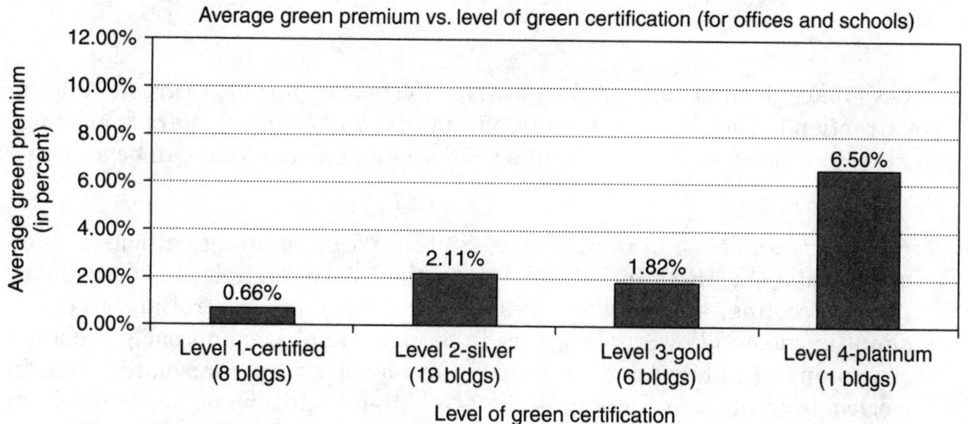

Source: USGBC, capital E analysis.

Year of completion	Average green cost premium
1997–1998	2.20%
1999–2000	2.49%
2001–2002	1.40%
2003–2004	2.21%
Avg. of 18 silver buildings	2.11%

Figure 8.5*b* Average green building premium for offices and schools. (*Source: State of California, Sustainable Building Task Force, October 2003.*)

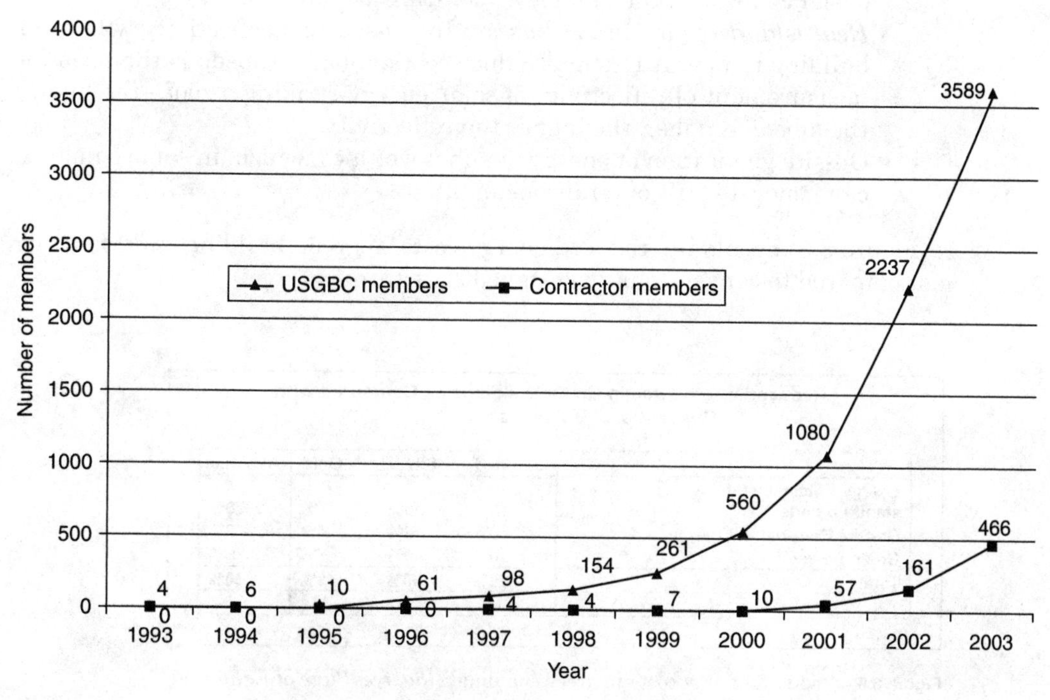

Figure 8.5*c* Growth in membership in USGBC. (*By permission: U.S. Green Building Council.*)

Let's take a look at some of the positive effects attributable to green building construction in the California study, effects that will obviously vary from state-to-state, but nonetheless represent an order of magnitude that can be adjusted accordingly.

Energy use. These buildings were 25 to 30% more energy efficient when compared to ASHRAE 90.1-1999. Interactions between lighting versus heating and cooling, and between fresh air and humidity control are analyzed simultaneously allowing designers to prepare a holistic approach to energy consuming equipment and building performance. Except for isolated areas in this country, air conditioning is the overriding requirement, particularly in buildings with high-occupancy rates and therefore particular attention needs to be paid to this building component. Innovative approaches to satisfying cooling loads are:

- Incorporation of more efficient lights, task lighting, sensors to cut unnecessary lighting, and use of daylight that not only reduces power consumption but also reduces cooling loads.
- Increase ventilation effectiveness that will help to cut cooling loads during peak periods through improved system optimization.
- *Underfloor air distribution systems.* The use of an underfloor plenum to deliver space conditioning typically cuts fan and cooling loads.
- Commissioning in a systematic approach to ensure that systems as designed are installed and are operating as planned.
- *Heat island reduction measures.* Increased roof reflectivity will lower building temperatures and reduce cooling loads. Albedo is the term for measurement of reflectivity of solar energy striking a roof—the higher the albedo number, the higher the reflectivity.
- On-site generation of energy via photovoltaics, which, in some climates, can generate 20% of total consumption.

Figure 8.6 reveals the reduced energy costs in green buildings in this study as compared to conventionally designed structures.

Reduced Energy Use in Green Buildings as Compared with Conventional Buildings

	Certified	Silver	Gold	Average
Energy Efficiency (above standard code)	18%	30%	37%	28%
On-Site Renewable Energy	0%	0%	4%	2%
Green Power	10%	0%	7%	6%
Total	**28%**	**30%**	**48%**	**36%**

Source: USGBC, Capital E Analysis

Figure 8.6 Reduced energy costs in green buildings. (*Source: State of California, Sustainable Building Task Force, October 2003.*)

Projected savings

The California study showed that the reduction in energy costs will provide the following energy savings over 20 years using the present value cost analysis:

30% reduced consumption at an electricity price of $0.11/kWh is about ($0.44/sf)/yr × 20 yr = $5.48/sf.

The additional value of peak demand reduction from green buildings was estimated at ($0.025/sf)/yr × 20 years = $0.31/sf.

Together, the total 20 yr present value of energy savings from a typical green building is $5.79/sf.

Water conservation—Green building water conservation is divided into four sectors:

1. Efficient use of potable water through use of better design and new technologies.
2. Capturing gray water—nonfecal wastewater from bathroom sinks, tubs, showers, washing machines, and drinking fountains—to be used for lawn and planting irrigation.
3. On-site storm water capture for use on site or to recharge groundwater tables.
4. Recycled or reclaimed water for other uses.

The information provided by California showed that, taken all together, these measures can reduce water consumption in the building to levels 30% lower than code requirements and can reduce exterior water demands by as much as 50%. In areas where water supplies are being overloaded, reclaimed water projects are taking on added importance. The Bay Area of California expects 50% of their new water supply to come from reclaiming. These reclaiming projects typically cost about $600 to $1100 per acre/foot, based on estimates from the East Bay Municipal Utility District.

Waste reduction. We are known as the disposable generation—use a couple of times and discard; the packaging costs often exceed the value of the item being packaged and are always 500% larger than the product itself. Reducing waste is a national concern and a nationwide problem. Not only are trucking and removal costs higher due in no small part to increases in gasoline and diesel fuel, but many states are also simply running out of room and have no place to dump their waste. California estimates that their total annual waste, as of 1998, amounted to 33 million tons, 21 million of which is generated by nonresidential buildings. An updated study would most likely show a much higher figure. Green building attempts to reduce waste focus on recycling and reuse and one or both can begin during the construction process and continue during the lifetime of the building.

During construction

- Reuse and minimize construction and demolition debris, and divert some of this debris from landfills to recycling facilities. Good examples are recycling cast-in-place concrete to remove rebars and convert the concrete to aggregate. Recycling of masonry materials for use as a base course under paving has proven to be an effective use of construction debris.
- Use materials that are more durable and easier to repair/maintain.
- Use of reclaimed materials, as indicated above, aggregate for the base-course underpaving and ground glass as a reflective material in asphalt paving.
- Use of materials that can function in a dual role, i.e. exposed structural systems, exposed ductwork, etc., staining concrete floor slabs.
- Incorporate an existing structure into a new building program where it can be updated and renovated in lieu of demolishing it.

During the life of the building

- Develop an indoor recycling program
- Design for deconstruction
- Design for flexibility via use of movable walls, modular furniture, movable task lighting, and other reusable building components

Construction and demolition diversion rates reached as high as 97% on some California projects and are typically 50 to 75% in green buildings.

Recycling creates jobs

An interesting sidebar to this question of disposal or recycle is how it affects employment. The total impact from diversion is nearly twice as much as the impact from disposal.

A study conducted by University of California, Berkeley, revealed that one additional ton of waste disposed of in a landfill generated $289 of total output in the state economy. One additional ton of waste diverted as recyclable, generated an average of $564. Only 2.46 jobs were created for every 1,000 tons of waste disposed, but 4.73 jobs were created for waste diverted as recyclable.

The Sustainable Approach to Design

Sustainable structures begin life during design and continue their objective through the construction cycle. These design-construction options can be incorporated into the project's drawings and/or included as requirements in the specifications. Some of these design phase considerations are:

- Simplify construction details—consider "constructability issues."
- Standardize design components.

- Attempt to utilize repeatable details and components.
- Verify all materials and equipment dimensions. (When wood framing member sizes changed years ago, some structural steel rolled sections changed dimensionally.)
- Consider alternative ways to bring in utilities to disrupt existing terrain and consider ways to dispose of site drainage by finding other solutions to disposal methods.
- Simplify building systems and components with an eye to future expansion or alteration projects at this structure.
- Take into consideration design elements that may affect safety and worker productivity.
- Whenever possible incorporate structural elements that require no finish materials.
- Optimize dimensions to utilize the entire product to reduce waste.
- Minimize piping and ductwork bends.
- Select fittings and fasteners and sealants that permit quicker assembly.
- Use local materials.
- Contact manufacturers to determine how to reduce their packaging waste.
- Investigate sources to accept salvage materials that can be recycled.
- Consider donating excess materials to nonprofit organizations such as Habitat for Humanity.

The Sustainable Approach to Construction

The process of designing and constructing a structure adhering to green standards involves not only the building itself but also the site on which it will be located and access to that site.

The following goals and objectives can be viewed as a primer that forms sustainability:

The site

Site-work goal—meet or exceed standards for erosion and sedimentation control.

- Prevent loss of soil during construction by storm water runoff and wind erosion.
- Prevent siltation of existing storm sewers and streams.
- Protect topsoil stockpiles for reuse, or modify soils to meet topsoil acceptable standards.

Site utilities goal—reduce storm water runoff, and reuse.

- Minimize or totally eliminate storm water runoff by carefully planning infiltration swales and basins to reduce impermeable surfaces instead of installing detention ponds.
- Retain or recharge existing water tables by minimizing disturbances, saving trees and natural vegetation, support and enhancing natural landforms and drainages.
- Store roof runoff for future use as gray water or reclaimed water.
- Install wastewater on-site, small footprint, and state-of-the-art treatment plant to recycle water for irrigation purposes.

Open space and landscaping goal—protect and restore existing vegetation.

- Protection of trees enhances value of the site and lowers cooling loads. Indigenous landscaping supports wildlife and biodiversity and does not require the level of irrigation required for new ground cover, it also eliminates need for chemical treatment.
- Minimize pesticide use by installing weed cloth, mulches, and dense plantings.

Circulation and transportation goal—improve circulation and decrease need for private transportation. Tie development or building to transit nodes and emphasize alternatives such as organized carpooling, water taxies (if available), buses, and car sharing.

The building

- During construction goal—reduce waste, and divert at least 75% of construction, demolition, and land clearing from disposal as landfill.
- Deconstruct all existing structures with substantial recoverable materials and dispose them off to recyclers.
- Adjust new site contours to provide for a balanced site. Modify nontopsoil soils to acceptable topsoil requirements.

The Holistic Approach—Again

Energy efficient building components are all-encompassing. Energy efficient heating and cooling systems and building envelope products like double/triple glazed windows come easily to mind—so do advanced, programmable control systems. And what about foundation insulation, roof insulation, and albedo values? Energy efficient plumbing fixtures and lighting fixtures with built-in power management systems improve every year. Office equipment that goes into a sleep mode when not used not only reduces electrical costs but also lowers the heat load.

Passive solar design, the technology of heating, cooling, and lighting by converting sunlight into a power source, can work effectively with other energy efficient materials and products. Photovoltaics can supplement or replace power from local utility companies.

NREL and Oberlin College's Pilot Program

The National Renewable Energy Laboratory (NREL) was established in 1974 and is the principal laboratory for the Department of Energy's Office of Energy Efficiency and Renewable Energy. Their mission is to develop renewable and energy efficient technologies.

Oberlin College in Oberlin, Ohio, wanted to design and construct a building to serve as a model and teaching aid for students in their environmental studies program, and to that end built the 13,600 square foot (1,260 square meters) Adam Joseph Lewis Center on campus containing classrooms, offices, and an atrium.

The goal of the project was to construct a building that was not only energy efficient, but also one that was able to export energy to the local grid system. In order to do so, they would install passive solar designs, use natural ventilation wherever possible, design an enhanced thermal building envelope, and use geothermal heat pumps for heating and cooling. The building's roof would incorporate an integrated photovoltaic (PV) system to allow for solar generation of electricity for the building.

After the building was completed in 2000, the NREL began to monitor the structure to evaluate its energy performance. Their findings would serve three purposes:

- Evaluate the performance of the building and several of its subsystems
- Provide suggestions to improve the initial performance
- Document lessons learned to improve the design of future low-energy buildings

This study, while sophisticated in its analysis of performance of the mechanical and electrical systems, graphically describes steps that can be taken to reduce energy demands.

NREL's study of the Oberlin College building ended in 2003, and it stated that more work was required to fulfill the original goal of the project as being one of a net energy exported, but the strides taken in this venture further the cause of energy self-sufficiency.

A brief description of the building and its systems is set forth in Fig. 8.7.

Some of the lessons learned by NREL are generic in nature and would apply to any sustainable building project:

- PV systems must be engineered to minimize transformer balance and system losses. These losses can represent a significant portion of the overall system production.

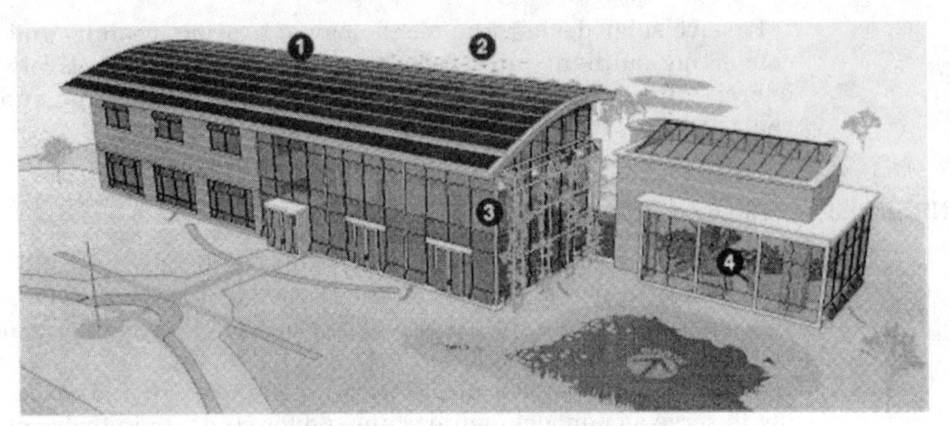

1 = PV array
2 = Location of ground wells for heat pump loop
3 = Passive solar heating and ventilation; daylighting
4 = Sunspace for ecological wastewater system

Figure 8.7 The Oberlin College Project. (*Source: National Renewable Energy Lab, U.S. Department of Energy.*)

- PV systems may not significantly reduce the building demand. In this case, any small reduction in demand due to PV is from load diversity.
- During summer months, on average, large PV systems in commercial buildings can export electricity from 8:00 a.m. to 6:00 p.m. From the utility perspective, this building was a net positive during daylight hours in the summertime and provided power when it was most needed by the grid.
- Control design must be fully integrated with the full capabilities of the equipment in the building including CO_2 sensors, motion sensors, and thermostats. A balance must be achieved between the human operations and the automation.
- Dark ceilings must be avoided to take full advantage of the daylighting and uplighting.
- Daylighting sensors are needed in all daylit areas. It is not sufficient to rely on manual controls.
- Daylighting must be designed to reach all occupied areas. The daylighting design should consider additional heating and cooling loads imposed upon the building. Overglazed areas such as the atrium in this building provided abundant daylighting but resulted in additional heating and cooling loads.
- Specifications for heat pumps must work with appropriate groundwater temperatures.

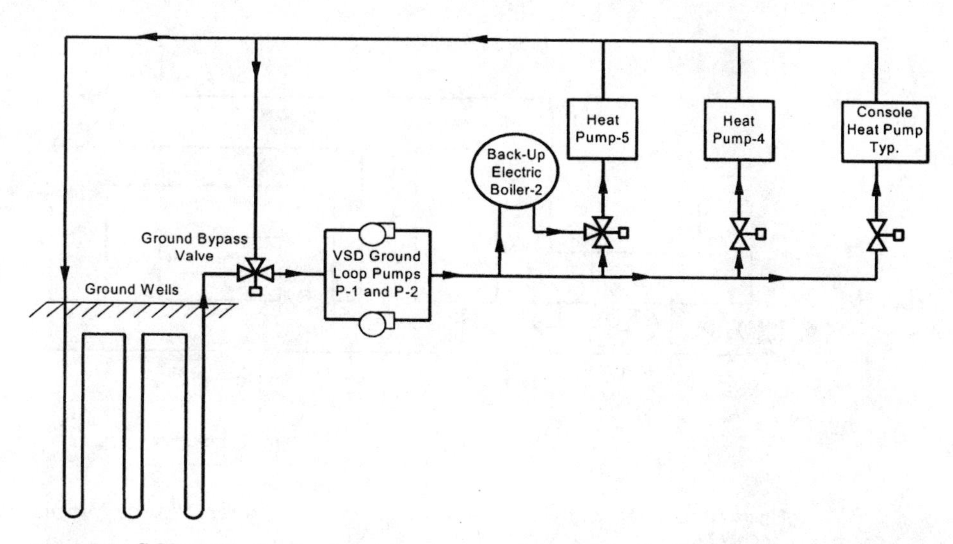

Figure 8.8 Schematic of ground loop heat pump piping—Oberlin College. (*Source: National Renewable Energy Lab, U.S. Department of Energy.*)

- Electric boilers can be used as a backup source, if they are used sparingly and do not cause excessive demand charges on the building. Controls and staging are essential for integration of limited use systems, such as these.

Figure 8.8 is a schematic of the ground loop heat pump piping installation; Fig. 8.9 is a schematic of the heat pump operation; and Fig. 8.10 is another mechanical schematic.

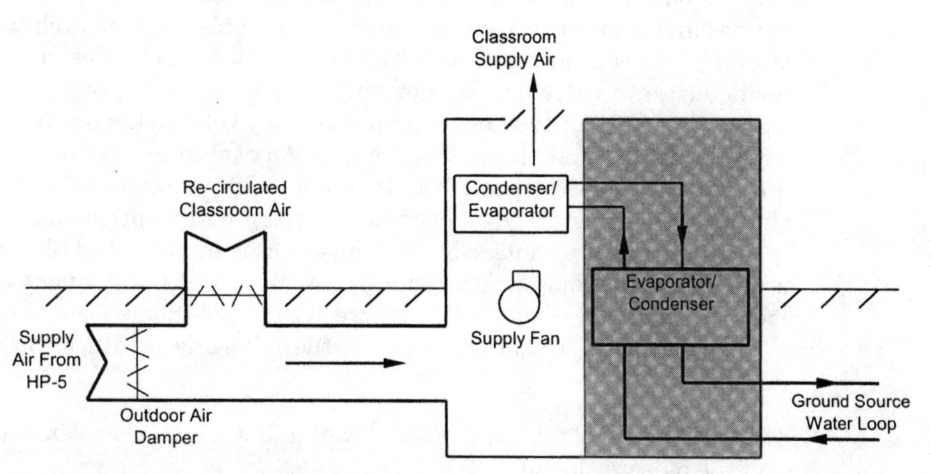

Figure 8.9 Schematic of console heat pumps and how they operate—Oberlin College. (*Source: National Renewable Energy Lab, U.S. Department of Energy.*)

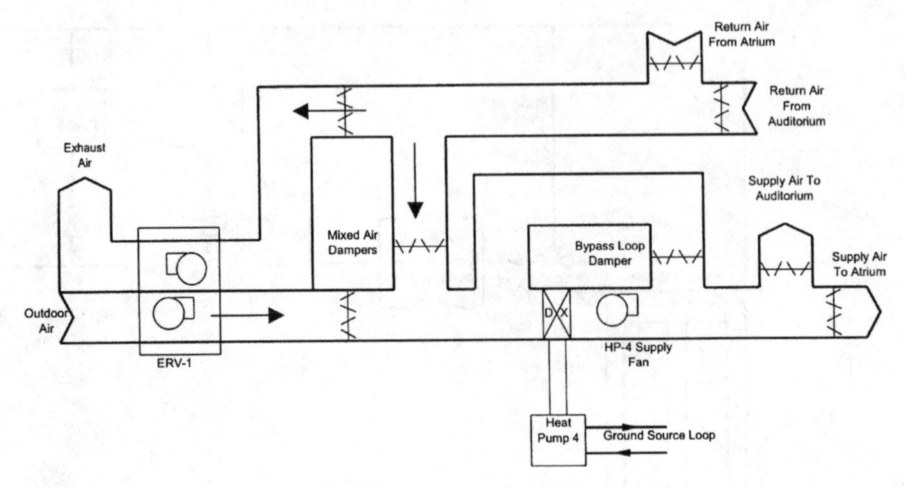

Figure 8.10 Another mechanical schematic—Oberlin College. (*Source: National Renewable Energy Lab, U.S. Department of Energy.*)

Greening of Existing Buildings

The LEED–EB Certification was established to deal with the upgrading of existing buildings to green standards.

JohnsonDiversey upgrade of an existing building

JohnsonDiversey is a manufacturer of cleaning and hygiene products located in Sturtevant, Wisconsin. It is housed in a 277,440 square foot building, built in 1997; 70% of which is office space and 30% is devoted to research laboratories. It is a breakaway company from the SC Johnson, Inc. in Racine, Wisconsin, the well-known producer of Johnson's Wax. JohnsonDiversey's legacy for innovation extends back to that parent company in Racine that was one of the first corporations in America to recognize the important effect of good architecture on the working lives of their employees. The Frank Lloyd Wright–designed SC Johnson headquarters in Racine was not only a monument to progressive corporate policy when built in 1936 but remains so today with its famous lily pad columns in the building's main room. SC Johnson Wax voluntarily eliminated CFCs, an ozone-depleting refrigerant, from their aerosol product line in the 1970s and led the development to more environmentally compatible propellant products.

In 2004, the Sturtevant facility at JohnsonDiversey, earned its LEED certification as an existing building for a structure containing 80,000 square feet (7,435 square meters), about 30% of its 278,000 square foot (25,836 square meters) building.

Their LEED certification included the following modification/remediation measures:

- Native prairie plants and restored wetlands were developed on more than half of the 57 acre site.
- Storm water collection for turf grass reduced potable water use by 2 to 4 million gallons per year.

- Low-flow fixtures reduced water use.
- More than 50% of solid waste was recycled.
- Ninety percent of interior space received reflected light.
- Personal environment controls were installed at each workstation.
- Rapidly renewable, locally available materials such as maple wood were used throughout.

There are several innovative programs at the site, some that do not involve substantial cash outlays and one as simple as encouraging alternative transportation choices. Of the 580 parking spaces provided, 10% or 58 are reserved for hybrid vehicles; 16 car/vanpool spaces are allotted to encourage carpooling.

The Personal Environment Modules (PEMs) installed in 93% of the total building office areas allow for individual control of temperature, airflow, lighting, and acoustics at each designated workstation.

JohnsonDiversey converted water usage from 2.5 to 0.5 gallons per minute (gpm) by installing aerators on all lavatory faucet fixtures and they reduced usage from 2.5 to 1.8 gpm by the installation of aerators on all shower fixtures. In combination with flush valve replacement diaphragms rated at 1.6 gallons per flush (gpf) for toilets to 0.5 gpf for urinals they have reduced water use performance to very low levels.

They have reduced waste disposal by a vigorous recycling program and employee awareness. They have distributed a recycling card to each employee providing information on what is to be recycled, where to take recyclables, and whom to contact for questions. They have 24 recycling areas for cans, plastics, and glass throughout the building, which are collected and emptied into large containers on the loading dock.

The JohnsonDiversey Annual Waste Generation Profile

Garage	208,000 lb
Waste recycled	74,800 lb
Paper	116,480 lb
Commingle (cans, glass, plastic)	5,200 lb
Total wastestream	404,480 lb
Total recycled	196,480 lb
Percent recycled	49%

For all construction projects within the building, they require that the staff or contractors recycle and/or salvage at least 30% by weight of any construction, demolition, or land clearing waste.

Items like toxic materials source reduction were addressed by inventorying items such as existing light fixtures and bulbs. They now purchase 32W T-8 Alto lamps from Phillips that have a mercury content of 18.6 parts per million (ppm), considerably under the limit of 25.0 ppm per code.

The green building rating system for existing buildings was issued in October 2004 and is referred to as LEED-EB. A project checklist for LEED-EB is shown in Fig. 8.11

The USGBC is planning to launch version 2.2 of green building rating systems in late 2005 reflecting the experience gleaned from comments to the previous iteration. A direct dialog with ASHRAE resulted in new calculations to achieve some performance goals. New application guides for health-care facilities, schools, and laboratories are also in the works.

As the green building movement spreads across the private and public sector, new opportunities await those design and construction firms that become intimate with the requirements of sustainable structures.

Green Building Rating System

For Existing Buildings

Upgrades, Operations and Maintenance

(LEED-EB)

Version 2

October 2004

Figure 8.11 A project checklist for LEED-EB existing buildings. (*Courtesy: U.S. Green Building Council.*)

LEED-EB Project Checklist

Sustainable Sites		14 Possible Points
Prereq 1	**Erosion and Sedimentation Control**	Required
Prereq 2	**Age of Building**	Required
Credit 1	**Plan for Green Site and Building Exterior Management**	2
Credit 2	**High Development Density Building and Area**	1
Credit 3.1	**Alternative Transportation:** Public Transportation Access	1
Credit 3.2	**Alternative Transportation:** Bicycle Storage & Changing Rooms	1
Credit 3.3	**Alternative Transportation**: Alternative Fuel Vehicles	1
Credit 3.4	**Alternative Transportation:** Car Pooling & Telecommuting	1
Credit 4	**Reduced Site Disturbance:** Protect or Restore Open Space	2
Credit 5	**Stormwater Management:** Rate and Quantity Reduction	2
Credit 6.1	**Heat Island Reduction:** Non-Roof	1
Credit 6.2	**Heat Island Reduction:** Roof	1
Credit 7	**Light Pollution Reduction**	1
Water Efficiency		**5 Possible Points**
Prereq 1	**Minimum Water Efficiency**	Required
Prereq 2	**Discharge Water Compliance**	Required
Credit 1	**Water Efficient Landscaping:** Reduce Water Use	2
Credit 2	**Innovative Wastewater Technologies**	1
Credit 3	**Water Use Reduction**	2
Energy & Atmosphere		**23 Possible Points**
Prereq 1	**Existing Building Commissioning**	Required
Prereq 2	**Minimum Energy Performance**	Required
Prereq 3	**Ozone Protection**	Required
Credit 1	**Optimize Energy Performance**	10
Credit 2	**On-site and Off-site Renewable Energy**	4
Credit 3.1	**Building Operations and Maintenance:** Staff Education	1
Credit 3.2	**Building Operations and Maintenance:** Building Systems Maintenance	1
Credit 3.3	**Building Operations and Maintenance:** Building Systems Monitoring	1
Credit 4	**Additional Ozone Protection**	1
Credit 5.1-5.3	**Performance Measurement:** Enhanced Metering	3
Credit 5.4	**Performance Measurement:** Emission Reduction Reporting	1
Credit 6	**Documenting Sustainable Building Cost Impacts**	1
Materials & Resources		**16 Possible Points**
Prereq 1.1	**Source Reduction and Waste Management:** Waste Stream Audit	Required

LEED for Existing Buildings

Figure 8.11 *(Continued)*

Prereq 1.2	**Source Reduction and Waste Management:** Storage & Collection of Recyclables	Required
Prereq 2	**Toxic Material Source Reduction:** Reduced Mercury in Light Bulbs	Required
Credit 1	**Construction, Demolition and Renovation Waste Management**	2
Credit 2	**Optimize Use of Alternative Materials**	5
Credit 3	**Optimize Use of IAQ Compliant Products**	2
Credit 4	**Sustainable Cleaning Products and Materials**	3
Credit 5	**Occupant Recycling**	3
Credit 6	**Additional Toxic Material Source Reduction:** Reduced Mercury in Light Bulbs	1
Indoor Environmental Quality		**22 Possible Points**
Prereq 1	**Outside Air Introduction and Exhaust Systems**	Required
Prereq 2	**Environmental Tobacco Smoke (ETS) Control**	Required
Prereq 3	**Asbestos Removal or Encapsulation**	Required
Prereq 4	**PCB Removal**	Required
Credit 1	**Outside Air Delivery Monitoring**	1
Credit 2	**Increased Ventilation**	1
Credit 3	**Construction IAQ Management Plan**	1
Credit 4.1	**Documenting Productivity Impacts:** Absenteeism and Healthcare Cost Impacts	1
Credit 4.2	**Documenting Productivity Impacts:** Other Impacts	1
Credit 5.1	**Indoor Chemical and Pollutant Source Control:** Non-Cleaning – Reduce Particulates in Air Distribution	1
Credit 5.2	**Indoor Chemical and Pollutant Source Control:** Non-Cleaning –High Volume Copying/Print Rooms/Fax Stations	1
Credit 6.1	**Controllability of Systems:** Lighting	1
Credit 6.2	**Controllability of Systems:** Temperature & Ventilation	1
Credit 7.1	**Thermal Comfort:** Compliance	1
Credit 7.2	**Thermal Comfort:** Permanent Monitoring System	1
Credit 8.1	**Daylighting and Views:** Daylighting for 50% of Spaces	1
Credit 8.2	**Daylighting and Views:** Daylighting for 75% of Spaces	1
Credit 8.3	**Daylighting and Views:** Views for 40% of Spaces	1
Credit 8.4	**Daylighting and Views:** Views for 80% of Spaces	1
Credit 9	**Contemporary IAQ Practice**	1
Credit 10.1	**Green Cleaning:** Entryway systems	1
Credit 10.2	**Green Cleaning:** Isolation of Janitorial Closets	1
Credit 10.3	**Green Cleaning:** Low Environmental Impact Cleaning Policy	1
Credit 10.4-5	**Green Cleaning:** Low Environmental Impact Pest Management Policy	2
Credit 10.6	**Green Cleaning:** Low Environmental Impact Cleaning Equipment Policy	1

LEED for Existing Buildings

Figure 8.11 *(Continued)*

Innovation in Operation, Upgrades and Maintenance — 5 Possible Points

Credit 1.1	**Innovation in Operation & Upgrades**	1
Credit 1.2	**Innovation in Operation & Upgrades**	1
Credit 1.3	**Innovation in Operation & Upgrades**	1
Credit 1.4	**Innovation in Operation & Upgrades**	1
Credit 2	**LEED Accredited Professional**	1

Project Totals

80 possible base points plus 5 for IOUM

Certified	32–39 points
Silver	40–47 points
Gold	48–63 points
Platinum	64–85 points

LEED for Existing Buildings

Figure 8.11 *(Continued)*

Chapter

9

Interoperability and Building Information Modeling

Interoperability is the ability to share intelligent building information seamlessly among all participants of the construction project. Building information modeling (BIM) is the process of developing three-dimensional (3D) and four-dimensional (4D) images, which will change forever the way we design and construct buildings.

Architects, engineers, and contractors move cautiously and test the water with one toe before jumping into this information mainstream. The highly competitive, fragmented nature of our industry and not overly abundant profit margins are mainly responsible for our "go-slow" approach.

Looking at the Past Three Decades

The first commercially produced personal computers to reach the marketplace were made by IBM in 1981 and coupled with the sale of Microsoft's Windows software offering in 1985 the electronic revolution being played out today got its beginning. It was the federal government's Telecommunications Act of 1996, permitting local telephone companies to compete for customers with long distance carriers, that was at the forefront of the information explosion that contributed to the dot com bubble. Each of these new telecommunication companies sought to have their own infrastructure, and with the change from electronic pulses traveling down copper wires to digital bits transmitted as coded light waves on fiber optic cables, the race was on.

One fiber optic company, Global Crossing, gambled that these local, national, and international phone companies would have a huge demand for transmission lines, and banking on an explosion of the new digital technology, began laying fiber optic cables to bind the world together. Now some eight years later, Global Crossings as a company is no more, but the residual of the fiber optic networks they installed now connect the world.

This fiber optic infrastructure was then in place to complete those connections, but it wasn't until "open protocols" were developed to allow digital devices to "talk" to each other and retrieve information that global communications became a reality. With the advent of HTML (hypertext markup language) and URL (uniform resource locator) that allowed web pages to be located and displayed universally, and HTTP (hypertext transfer protocol) to move these documents around, the quest for interoperability was a work-in-progress as many new software and program developers introduced their own proprietary language.

Contractors and Architects and the Early Electronic World

Contractors in the early 1980s began to use computers for payroll and accounting functions, and when digitized estimating software became available, they plunged wholeheartedly into that program. They signed on to the project management software and scheduling software, and the old method of hexagonal CPM modules gave way to computerized CPM scheduling programs.

Architects purchased more advanced and less costly CAD software and were able to progress from simple two-dimensional designs to layered graphics, where architectural drawings could be lifted from their structural skeleton or MEP designs could be overlaid on floor plans. The ability to electronically transmit design development and contract documents was a major step toward increasing the flow of information from architects to owners, engineers, contractors, subcontractors, and vendors.

But always, all parties were looking for better ways to do things.

By the latter part of 1990s and into the first decade of the twenty-first century, data-based CAD systems permitted the electronic storage of building components such as doors and windows concurrently with the design. That created a process whereby any change in design concurrently affected and was reflected in a change of the list of building materials. This innovation was closely followed by 3D drafting focused on creating geometry to support the visualization and a realistic rendering of the structure. The further development of 3D drafting allowed building components to be displayed in multiple views and by adding a degree of "intelligence," any change in one element in the building's design immediately affected all other related elements. This process was referred to as object-oriented computer-assisted design (OOCAD). BIM can be viewed as the latest iteration of OOCAD software, whereby all parts or pieces in the building design reside in the project's database and therefore represent all the information associated with the design software. Programs developed by several software makers provide a running total of all building components. Design an 8-foot-high, 10-foot-long wall, with 25-gauge steel studs in the center and a list of square footage of drywall, and the number and length of all 25 gauge studs will be produced. At the end of the day, a document is produced showing every item in the design.

The transfer from diskette to Internet provider

Initially, electronic information was either stored in the computer's hard drive or backed up on floppy disks. Subsequently, diskettes and compact disks were used. Transferability of information was performed by sending these disks or diskettes back and forth by hand or mail. The Internet became a main vehicle for the transportation of digital information contained in those diskettes and disks and now data transfer could be instantaneous, literally, throughout the world, by merely clicking on Send. Today 3D modeling and Web-based collaborative services can be shared among all parties of a construction project. Architects in New York City, collaborating with engineers in Bangalore, India, would have virtually extended the workday nearly twofold since information sent to India at day's end would be processed as the day begins in the East and would return to Manhattan by morning coffee break next day.

Not only was more speed achievable, but very low costs also made thoughts about outsourcing a real option for small as well as larger design firms.

Offshore engineering companies in India advertise their services all over the Internet. One company says "We offer top flight engineering service. Why pay $53,240 for a CAD drafter in Los Angeles, when we can supply fully qualified people at about $12,000 per year? They have a 35-hour workweek in Europe but here in India, we have a 35 hour workday."

The architectural and engineering professions are making giant strides in the production of design documents but builders have taken a more "wait and see" approach.

Contractors slow to embrace

Although architects and engineers immediately saw the benefits, opportunities, and savings the new hardware and software could provide, many in the construction industry were reluctant to change. Builders cited many reasons for hanging onto paper documents:

- The cost of hardware and software is still too expensive.
- They still don't have full confidence that information won't be lost through computer "crashes" or from temporary loss of power.
- The old saw, "We've always done it that way, it works, so why change"
- Contractors routinely communicate with subcontractors and vendors who don't have computers for any use other than payroll or other accounting functions.
- Local, county, and state offices frequently require some paper format and documentation for filing.
- Requirements for original seals/signatures on documents filed with various government agencies are still out there.
- The use of electronic media on the construction site by employees unaccustomed to this medium will be inefficient and therefore prone to inaccuracies.

- Paper records are more official and are legally more acceptable than stored electronic data.
- There is no real incentive to work electronically.

The Construction Finance Management Association (CFMA) reported in a recent study that Excel is the most widely used software application for all businesses and that the predominant construction industry software includes AccuBid, Bidmaster Plus, McCormick Estimating, and Precision Collection. The most common forms of project management software are Primavera Enterprise and Expedition and Prolog Manager. Only 25% of the construction firms surveyed by CFMA used collaborative software such as Buzzsaw, Constructware, or Meridian Project Talk. Scheduling software was primarily Suretrak and Primavera.

Both the design and construction people know there is a better way to do things, and finally some innovators are beginning to show the rest of us what can be done and why it is so important to improve the way we go about our work.

The Role Owners Can Play

Owners have taken the lead in several areas where change was needed and where the industry was slow to respond. One such area was safety. Owners, from a humanitarian and public relations standpoint, wanted safe working conditions on their new building project, and they did not want the adverse publicity an accident or fatality on their construction site could create. Many owners required bidders to submit their safety plan along with their proposals, and in some cases, also provide a record of past safety history. They were saying, in effect, "If you want to bid on my work, I'll need to be convinced that you will provide sufficient manpower and policies to make my site a safe place in which to work," and it worked.

In another example, owners desiring to connect all parties via the Internet for communication purposes would include in their RFPs (Requests For Proposal) a statement that the general contractor, all subcontractors, and all vendors must have email, certain types of software, and access to the Internet, and again it worked.

The green building movement received a huge boost when local and state governments entered the picture and required some of their projects to meet the U.S. Green Building Council LEED standards.

Owners know that they bear the costs for these requirements and, as the ultimate customer, they will do so as long as they receive benefits commensurate with those added costs.

Owners may now begin to demand a seamless flow of communication to take advantage of the cost savings this process has promised to provide as it relates to design, construction, maintenance, and business process systems.

Those companies that are not able to provide these services will find themselves either rushing to implement them, or having to forego some high-value projects that will be awarded to a more elite group of design consultants and contractors who recognized the many advantages of becoming interoperable-capable.

Interoperability—What Is It and Why Is It So Important?

One definition of *interoperability* is the ability to exchange and manage electronic information seamlessly and the ability to comprehend and integrate this information across multiple software systems. Another definition is—an open standard for building data exchanges, in effect, interoperability means that your system can "talk" to mine, and we can all "talk" to the designers, contractors, subcontractors, vendors, and owner's representatives in the same language. There is little interoperability in the architect, engineer, contractor, owner (AECO) community today but many organizations recognizing its importance are aggressively attacking the problem.

And this problem is present in other industries as well. One German automobile manufacturer was alerted to the problem of interoperability after receiving a fair amount of customer warranty complaints about various system component failures in the electronics installed in their high-priced models. Apparently there was no central protocol in place governing or controlling the "language" of computer chips supplied by each of those disparate component vendors, and when all these parts from a variety of suppliers were installed they could not "talk" to each other, which manifested itself, in the eyes of the customer, as a system failure. It took some time to uncover the cause and correct the problem, but in the meantime there were a lot of customers who were very unhappy.

Recently several trade and private organizations have begun to recognize the missed opportunities and tremendous costs of not fully engaging the interoperability arena and its relationship to other systems such as BIM technology. Transmitting 3D imaging to all parties of the design and construction process, if it is to be fully utilized, requires a single, seamless integration of the entire project's database—from design and construction to commissioning and continuing on through the building's life cycle and that's where interoperability plays a major role.

The NIST Report

The National Institute of Standards and Technology (NIST) concluded their 2002 study to quantify the cost for inefficient interoperability in commercial, institutional, and industrial facilities for both new and "in place" construction. According to NIST, this inability to seamlessly exchange and manage electronic information in the construction industry adds an astounding $6.18 per square foot to project costs in addition to operations and maintenance costs of $0.23 per square foot. In total, inefficient interoperability cost to the construction industry, per this report, was a whopping $15.6 billion in 2002. Figure 9.1 reveals how these costs are divided between architect/engineers, general contractors, subcontractors, and owners. Figure 9.1*a* contains the cost of inadequate interoperability for owners and operators; Fig. 9.1*b* the cost for contractors; Fig. 9.1*c* the cost for specialty fabricators; and Fig. 9.1*d* the cost for architects and engineers.

Although the manufacturing sector has dealt with this problem with considerable success, it must be kept in mind that, on the whole, they deal with a flow of similar products in a controlled environment and they also enjoy economies of scale. The construction industry deals in mainly one-off products, and even when a similar product is built, say a motel chain or fast food restaurant, varying zoning and building regulations frequently impact the structure's basic design.

In the September 2004 issue of *Architectural Record* magazine, Mr. Ken Sanders, FAIA, author of the midnineties book, *The Digital Architect*, in his article entitled *Why Building Information Isn't Working...Yet*, compared the automobile and aerospace industries with the construction industry in the use of technology. "Finally and most importantly, cars and planes are the products of an integrated *design-build* (this author's italics not Mr. Sanders) process. The

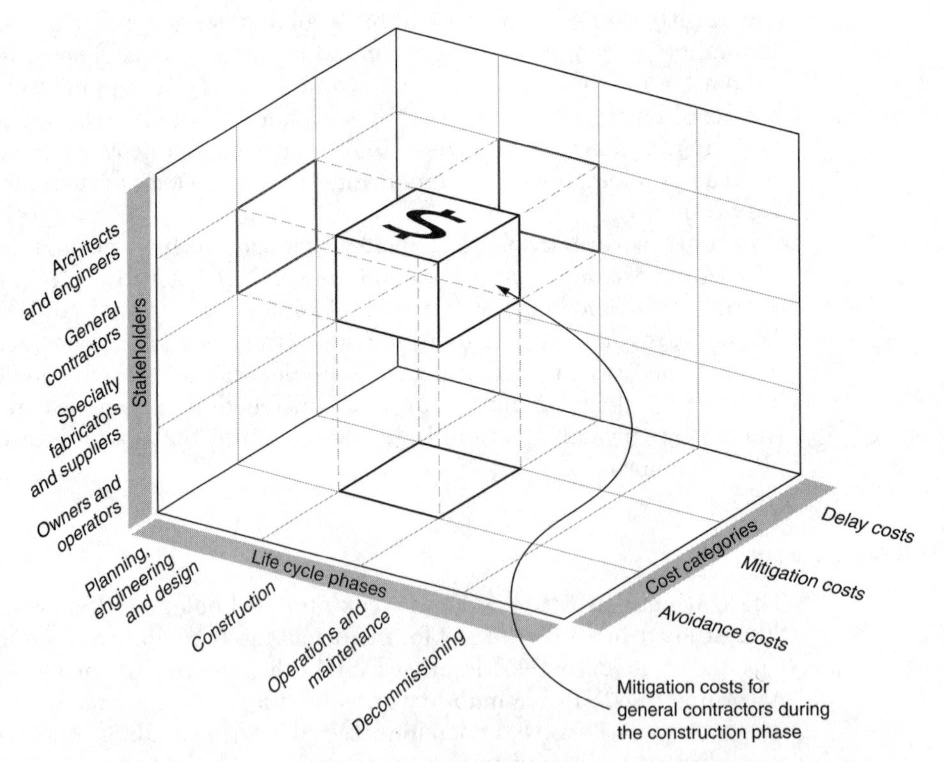

Figure 9.1 Inadequate interoperability costs for project participants. (*Source: National Institute for Standards & Technology.*)

Costs of Inadequate Interoperability for Owners and Operators

Life-Cycle Phase	Cost Category	Cost Component	Average Cost per Square Foot	Average Cost per Square Meter	Inadequate Interoperability Cost Estimate ($Thousands)
Planning, Engineering, and Design	**Avoidance Costs**	Inefficient business process management costs	0.38	4.07	430,111
		Redundant CAx systems costs	—	—	—
		Productivity losses and training costs for redundant CAx systems	—	—	—
		Redundant IT support staffing for CAx systems	—	—	—
		Data translation costs	—	—	—
		Interoperability research and development expenditures	0.0039	0.042	4,422
	Mitigation Costs	Manual reentry costs	0.16	1.67	176,882
		Design and construction information verification costs	0.0056	0.061	6,415
		RFI management costs	0.092	0.99	104,966
	Subtotal	Avoidance costs	0.38	4.07	434,533
		Mitigation costs	0.25	2.73	288,263
		Subtotal	**0.64**	**6.80**	**722,796**
Construction	**Avoidance Costs**	Inefficient business process management costs	0.49	5.32	561,926
		Redundant CAx systems costs	—	—	—
		Productivity losses and training costs for redundant CAx systems	—	—	—
		Redundant IT support staffing for CAx systems	—	—	—
		Data translation costs	—	—	—
		Interoperability research and development expenditures	0.003	0.03	3,618
	Mitigation Costs	Manual reentry costs	0.15	1.59	167,975
		Design and construction information verification costs	0.0068	0.07	7,701
		RFI management costs	0.14	1.48	156,793
	Subtotal	Avoidance costs	0.50	5.32	565,544
		Mitigation costs	0.29	3.15	332,469
		Subtotal	**0.79**	**8.47**	**898,013**
Operations and Maintenance	**Avoidance Costs**	Inefficient business process management costs	0.04	0.46	1,638,915
		Redundant CAx systems costs	—	—	—
		Productivity losses and training costs for redundant CAx systems	—	—	—
		Redundant IT support staffing for CAx systems	—	—	—

(continued)

Figure 9.1*a* Costs of inadequate interoperability for owners and operators. (*Source: National Institute for Standards and Technology.*)

Costs of Inadequate Interoperability for General Contractors

Life-Cycle Phase	Cost Category	Cost Component	Average Cost per Square Foot	Average Cost per Square Meter	Inadequate Interoperability Cost Estimate ($Thousands)
Planning, Engineering, and Design	**Avoidance Costs**	Inefficient business process management costs	0.14	1.55	163,674
		Redundant CAx systems costs	—	—	—
		Productivity losses and training costs for redundant CAx systems	—	—	—
		Redundant IT support staffing for CAx systems	—	—	—
		Data translation costs	—	—	—
		Interoperability research and development expenditures	0.0006	0.006	630
	Mitigation Costs	Manual reentry costs	0.16	1.74	184,028
		Design and construction information verification costs	0.006	0.06	6,302
		RFI management costs	0.12	1.24	131,299
	Subtotal	Avoidance costs	0.14	1.55	164,304
		Mitigation costs	0.28	3.05	321,629
		Subtotal	**0.43**	**4.59**	**485,933**
Construction	**Avoidance Costs**	Inefficient business process management costs	0.82	8.78	927,487
		Redundant CAx systems costs	—	—	—
		Productivity losses and training costs for redundant CAx systems	—	—	—
		Redundant IT support staffing for CAx systems	—	—	—
		Data translation costs	—	—	—
		Interoperability research and development expenditures	0.003	0.03	3,571
	Mitigation Costs	Manual reentry costs	0.11	1.19	126,047
		Design and construction information verification costs	—	—	—
		RFI management costs	0.16	1.74	183,818
		Construction site rework costs	0.01	0.11	11,356
	Delay Costs	Idle employees costs	0.01	0.12	12,988
	Subtotal	Avoidance costs	0.82	8.78	931,059
		Mitigation costs	0.28	3.04	321,221
		Delay costs	0.01	0.12	12,988
		Subtotal	**1.11**	**11.94**	**1,265,268**
Operations and Maintenance	**Mitigation Costs**	Post construction redundant information transfer costs	**0.04**	**0.48**	**50,419**
Total Cost					**1,801,620**

Source: RTI estimates; totals may not sum correctly due to rounding.

Figure 9.1*b* Cost of inadequate interoperability for general contractors. (*Source: National Institute for Standards and Technology.*)

Costs of Inadequate Interoperability for Specialty Fabricators and Suppliers

Life-Cycle Phase	Cost Category	Cost Component	Average Cost per Square Foot	Average Cost per Square Meter	Inadequate Interoperability Cost Estimate ($Thousands)
Planning, Engineering, and Design		Inefficient business process management costs	0.25	2.65	279,652
		Redundant CAx systems costs	0.0001	0.0007	70
		Productivity losses and training costs for redundant CAx systems	0.0002	0.002	230
		Redundant IT support staffing for CAx systems	—	0.0004	44
		Data translation costs	0.005	0.05	5,366
	Avoidance Costs	Interoperability research and development expenditures	0.0008	0.009	953
		Manual reentry costs	0.11	1.21	128,119
	Mitigation Costs	Design and construction information verification costs	0.02	0.19	20,019
		RFI management costs	0.007	0.08	7,944
	Subtotal	Avoidance costs	0.25	2.70	286,316
		Mitigation costs	0.14	1.48	156,081
		Subtotal	**0.39**	**4.18**	**442,397**
		Inefficient business process management costs	1.39	15.00	1,584,696
		Redundant CAx systems costs	0.0001	0.0007	70
		Productivity losses and training costs for redundant CAx systems	0.001	0.012	1,305
		Redundant IT support staffing for CAx systems	0.0002	0.0024	249
		Data translation costs	0.027	0.29	30,410
	Avoidance Costs	Interoperability research and development expenditures	0.005	0.05	5,402
		Manual reentry costs	0.10	1.10	115,726
		Design and construction information verification costs	0.01	0.15	16,015
	Mitigation Costs	RFI management costs	0.007	0.075	7,944
		Construction site rework costs	0.0003	0.004	374
	Delay Costs	Idle employees costs	—	—	—
	Subtotal	Avoidance costs	1.43	15.30	1,622,132
		Mitigation costs	0.12	1.33	140,059
		Delay costs	—	—	—
Construction		**Subtotal**	**1.55**	**16.63**	**1,762,190**
Total Cost					**2,204,588**

Source: RTI estimates; totals may not sum correctly due to rounding.

Figure 9.1c Cost of inadequate interoperability for specialty fabricators and suppliers. (*Source: National Institute for Standards and Technology.*)

Costs of Inadequate Interoperability for Architects and Engineers

Life-Cycle Phase	Cost Category	Cost Component	Average Cost per Square Foot	Average Cost per Square Meter	Inadequate Interoperability Cost Estimate ($Thousands)
Planning, Engineering, and Design	Avoidance Costs	Inefficient business process management costs	0.31	3.37	356,126
		Redundant CAx systems costs	0.0001	0.001	158
		Productivity losses and training costs for redundant CAx systems	0.04	0.45	47,947
		Redundant IT support staffing for CAx systems	0.0004	0.005	501
		Data translation costs	0.002	0.02	2,139
		Interoperability research and development expenditures	0.02	0.21	22,234
	Mitigation Costs	Manual reentry costs	0.41	4.38	462,734
		Design and construction information verification costs	0.10	1.08	114,342
		Reworking design files costs	0.0009	0.009	968
	Subtotal	Avoidance costs	0.38	3.85	429,106
		Mitigation costs	0.51	5.47	578,044
		Subtotal	**0.89**	**9.32**	**1,007,150**
Construction	Avoidance Costs	Inefficient business process management costs	0.04	0.41	43,290
		Redundant CAx systems costs	0.00003	0.0003	28
		Productivity losses and training costs for redundant CAx systems	0.007	0.08	8,461
		Redundant IT support staffing for CAx systems	0.00008	0.0008	88
		Data translation costs	0.0003	0.004	378
		Interoperability research and development expenditures	0.003	0.04	3,924
	Mitigation Costs	Manual reentry costs	0.024	0.26	27,750
		Design and construction information verification costs	0.006	0.07	7,377
		RFI management costs	0.05	0.53	55,656
	Subtotal	Avoidance costs	0.05	0.49	56,169
		Mitigation costs	0.08	0.86	90,783
		Subtotal	**0.13**	**1.35**	**146,952**
Operations and Maintenance	Mitigation Costs	Post-construction redundant information transfer costs	**0.01**	**0.15**	**15,660**
Total Cost					**1,169,762**

Source: RTI estimates; totals may not sum correctly due to rounding.

Figure 9.1*d* Cost of inadequate interoperability for architects and engineers. (*Source: National Institute for Standards and Technology.*)

designer and builder are one and the same. This is rarely the case with building design and construction."

Sounds like another case for design-build.

Why are contractors deficient in information management?

The NIST study uncovered many reasons why the construction industry suffers from inefficiency in information management, often operating in isolation and not effectively communicating with other internal and external partners of the design and construction process. The main reasons listed in the study are:

1. Collaboration software is not integrated with the other systems of the contractor. Some builders use collaborative software, but it is generally not integrated with other systems—it is used in a stand-alone application, defeating the purpose of the software.
2. Many parties work together on one project only so there is little incentive to invest in long-term collaborative software, each project frequently being unique, with different participants, scope, workforce, teams, and operating in a different location.
3. Life-cycle management processes are fragmented and not integrated across the project's life cycle.
4. There are inefficiencies and communication problems when participants of the project from all parts of the life cycle have various versions of the same software or different software.
5. A lack of data standards inhibits the transfer of data between different phases in the life of a project and their associated systems and applications.
6. Internal business processes are fragmented and inhibit interoperability. NIST found that in some firms, an estimated 40% of engineering time is dedicated to locating and validating information gathered from disparate sources.
7. Many firms use automated and paper-based systems to manage data and information, while hard copy construction documents are routinely used on the job site.
8. Many smaller construction firms do not employ or have only limited use of technology in managing their business processes and information.

The Federal Government Push for Interoperability and Building Information Modeling

On January 24, 2005, the General Services Administration sent out a Request For Information (RFI) to the capital facilities industry (design consultants, general contractors, subcontractors, and vendors) with the following statement:

> Interoperability problems in the capital facilities industry stem from the highly fragmented nature of the industry's continued paper-based business practices, a lack of standardization and inconsistent technology adoption among stakeholders. Based on interviews and surveys it is found that $15.8 billion in annual interoperability costs were quantified for the capital facilities industry in 2002. Of these costs, two-thirds are borne by owners and operators that incur most of these costs during on-going facility operation and maintenance (O&M).
>
> The U.S. General Services Administration (GSA)/Public Buildings Service (PBS) is seeking information from industry partners on Industry Foundation Classes (IFC)-based integrated and interoperable Building Information Model (BIM) technology as part of its effort to improve project deliveries in the capital construction program. The GSA/PBS currently has an active pipeline of more than 200 major capital construction projects conclusively exceeding a value of $11 billion.

The GSA, in this RFI, announced an opportunity for firms in the design, construction and facility management, and real property industries to submit information on the use of IFC-BIM technology. This information will be used by the government in establishing potential sources in the marketplace with knowledge and experience in the use and practice of this state-of-the-art technology.

It looks like the federal government is taking an active role in this seamless approach to integration and will provide some interesting opportunities to those design and construction firms willing to add an important tool to their sales and marketing approach.

Several industry leaders point to the success of the LEED program as an answer to those who carp that the fragmented nature of the industry will be a deterrent to the acceptance of new technologies. Many of the country's foremost trade associations are making their members aware of the opportunities that await those that begin learning about jumping on the interoperability bandwagon.

Interoperability and BIM issues

In August 2004, Mr. Norbert W. Young, Jr, board chairman of the International Alliance for Interoperability, had this to say about the NIST report that got everyone's attention and started the ball rolling:

> While to us the benefits have always been clear, we see this report as a catalyst for wide acceptance of interoperability as a practical, profit-enhancing goal, especially in terms of funding for the work still to be done.

The Industry Movement Toward Interoperability

The International Alliance for Interoperability (IAI) has made accommodations with industry leaders in 19 countries around the world to define a single building information framework. Using heating and cooling as an example, IAI asked ASHRAE in the U.S., their counterparts in CIBSE in England, and DIN in Germany, to get together and define a process for calculating a building's HVAC requirements. They wanted to develop a generic model for systems development to provide a seamless flow of information for mechanical systems across all national boundaries.

This is the process that is termed industry foundation classes (IFC) that must be repeated by other design and construction teams to develop the specific nongraphic common language required for interoperability. Each IFC thereby becomes a dictionary for project component information to be shared by owners, architects, engineers, general contractors, and specialty contractors.

Just as the HTML and HTTP protocols allowed for the transmission of web pages to become a universal event, so is there a need for technology for cross-referencing and dissemination of design and construction information.

The Current State of Affairs

FIATECH, a nonprofit research and development consortium based in Austin, Texas, focuses on developing and delivering technologies to the construction industry to improve the design, engineering, and construction of capital projects. They are currently working on several approaches to advance the interoperability of construction software.

Extensible markup language (XML) is a simple and flexible text format originally designed to meet the needs of large-scale publishing, but is now playing a major role in exchanging data over the Internet. AecXML was chartered in 1999 to promote and facilitate *interoperability* among software applications and information exchange of architecture, engineering, and construction. AgcXML, a program sponsored by the Associated General Contractors of America, planned for delivery in 2005–2006 will create an XML schema (plan) to deal with the following common construction documents:

- Requests for information
- Submittals
- Purchase orders
- Contracts—both AGC forms and other industry standard forms
- Pay applications
- Change order requests (CORs) and change order approvals
- Punch lists
- Daily reports
- Addendum notifications
- Meeting minutes
- Requests for proposals and pricing

The Open Building Information Xchange, referred to as oBIX, is a movement backed by facility managers and industry sources to use the programming of

XML for seamless Internet- and intranet-based communications between building systems and enterprise applications, for running a building on standard protocols and techniques, and for permitting a standard way for buildings, facility managers, and owners to interface with the Internet.

The civil engineering profession has developed LAND XML and the steel industry has created their interoperability protocol called CIS/2.

The Steel Industry Becomes a Leader

In 2004, the American Institute of Steel Construction, Inc. (AISC) issued a white paper entitled *Interoperability and the Construction Process* in which they put forth their efforts and the steel industry's efforts to achieve interoperability. AISC has initiated the CIM Steel Integration Standards Version 2 (CIS/2) enabling designers and specialty steel contractors to exchange data seamlessly. CIS/2 is compatible with other software programs such as Bentley, RAM, GT Strudl, Robot, and ISS drafting software, Tekla and Design Data detailing software, and Fabtrol shop fabrication software.

ASCI, in their report says, "The neutral file format allows stand-alone programs such as structural analysis and design, detailing and manufacturing information systems, as well as CNC driven fabrication equipment to communicate with each other by translating a program's native format into a neutral format to allow data interchange across multiple platforms."

A structural engineer can now design a steel structure in the BIM (3D) mode, and concurrently and instantaneously transmit the design to the architect and MEP design consultants so they can begin to incorporate their work into this "skeleton." If a general contractor is on board at the time, a copy of the 3D design can be forwarded to them, and possibly also to a steel fabricator. Each of these design subcontractors will be able to "talk" to each other and to the general contractor and subcontractors in a paperless fashion. Suggested changes offered by the general contractor or their subcontractors can be distributed, reviewed, addressed immediately, changes affected, and distributed so that steel shop drawings can be emailed to the engineer of record for approval—all these procedures dramatically speeded up.

All this can be accomplished without handling all those rolls of shop drawings, the time-consuming packaging, labeling, and the delivery charges back and forth. Just think about those savings that can accrue, maybe small on small projects but possibly thousands of dollars on larger ones.

One of the goals of 3D modeling is to create systems designs that really work—a set of drawings that are really coordinated and that eliminate the need for RFIs to resolve questions that have been addressed and resolved during design development not after the construction contract award.

The architect as captain of the team is the focus during this entire process, reviewing, commenting, changing, and approving every aspect of the design and without the lengthy time required to generate, distribute, question, and respond to the paper blizzard that is part and parcel of the process today. Just imagine the savings in costs and time when such a system is routinely up and running. How many more projects can each one of the participants manage, reducing stress and overhead and freeing up human resources to perform important functions by not getting bogged down in pushing paper and generating RFIs, RFCs, and hundreds of transmittals to send to multiple sources?

Interoperability and BIM as envisioned by the steel industry

According to Tom Faraone, Senior Regional Engineer for AISC Marketing, LLC—an organization affiliated with the American Institute of Steel Construction, the steel industry is already using bar codes to speed up product fabrication and delivery, and is working on other ways to utilize them more effectively. With the increased interest in radio frequency identification devices (RFID), a micro–radio transmitter affixed to each structural steel member as it enters the fabricator's shop could provide exact information on the time it took for fabrication and on leaving the shop could potentially convey to a computer-operated crane its precise position within the structural framework.

The goal of AISC is to develop a system in conjunction with its members that will accelerate the entire design-fabricate-deliver-erection process of a structural steel building. If time is money, it surely applies to this industry.

The *New York Times*, in an article dated April 13, 2005, on a project in Boston called the Charles Street Jail, reported the redevelopment of this historic building into a four-star hotel. The developer budgeted the project at $50 million in 2003 but was devastated by the sharp increase in structural steel that was occurring at that time. An eight-month redesign was required to reduce the updated cost of $74 million down to a more manageable $64 million. Although the consumer price index (CPI) showed an inflation rate in the 2 to 3% range, not so in the building business where some estimating services pegged inflation in the industry at 12% for the year 2004.

Developers all over the country were complaining about spiraling costs in steel and cement, where the product in this global economy goes to the highest bidder.

The final design of the Charles Street Jail required the architect to delete the basement which had been planned, reduce the floor-to-floor height, and add a 15th floor. Mr. Richard Friedman, CEO of Carpenter & Company, the developer, summed it all up in four words, "It's been a nightmare."

More rapid design and review cycles can become an effective guard against the forces of inflation.

Puma Steel, Cheyenne, Wyoming says their CSI/2 can affect a 50% savings in scheduling (Fig. 9.2). The NIST U.S. Capital Facilities Industry Final Report (Fig. 9.3) graphically shows how design changes occurring in various stages impact cost.

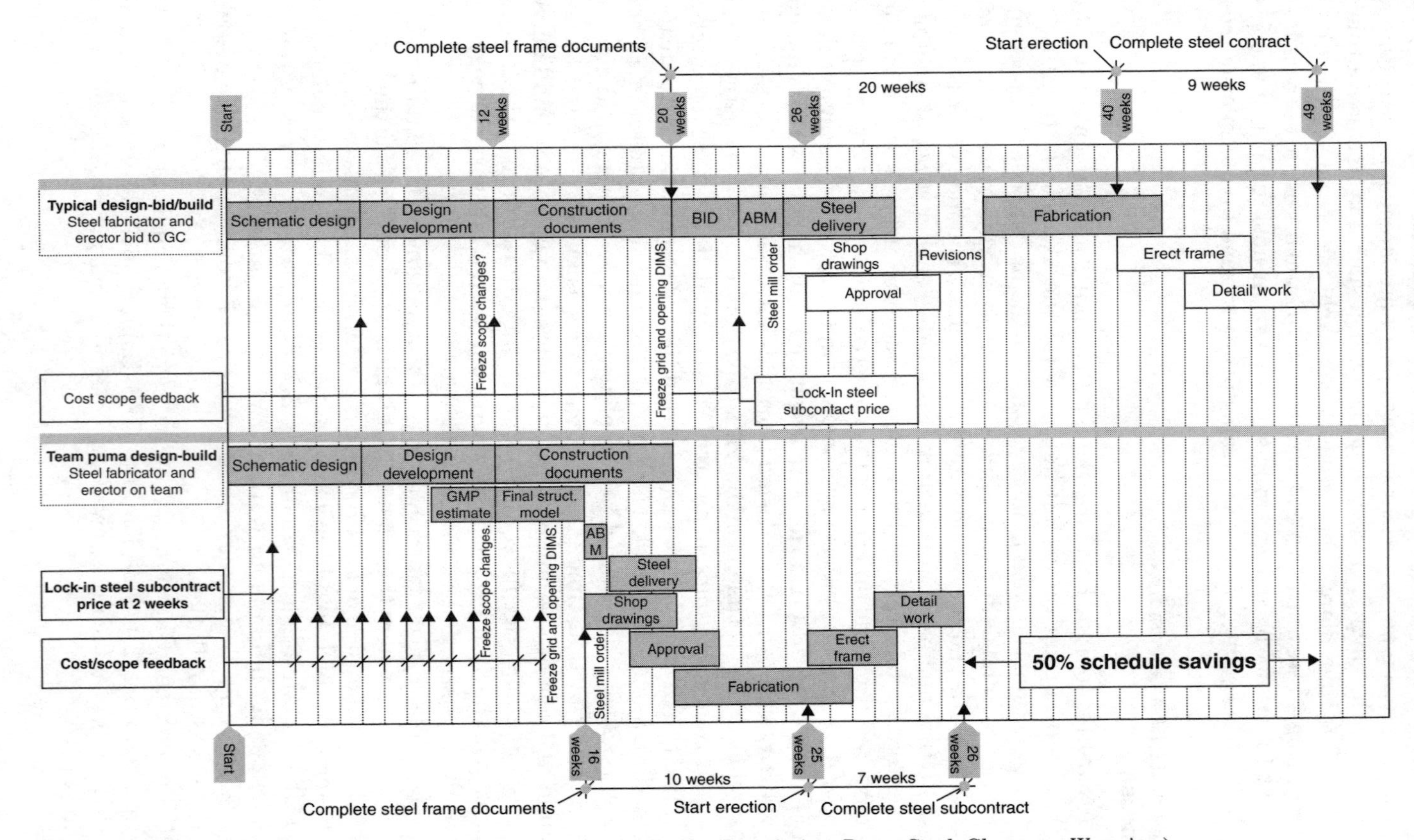

Figure 9.2 Reduction in schedule via design-build & scope feedback. (*Permission: Puma Steel, Cheyenne, Wyoming.*)

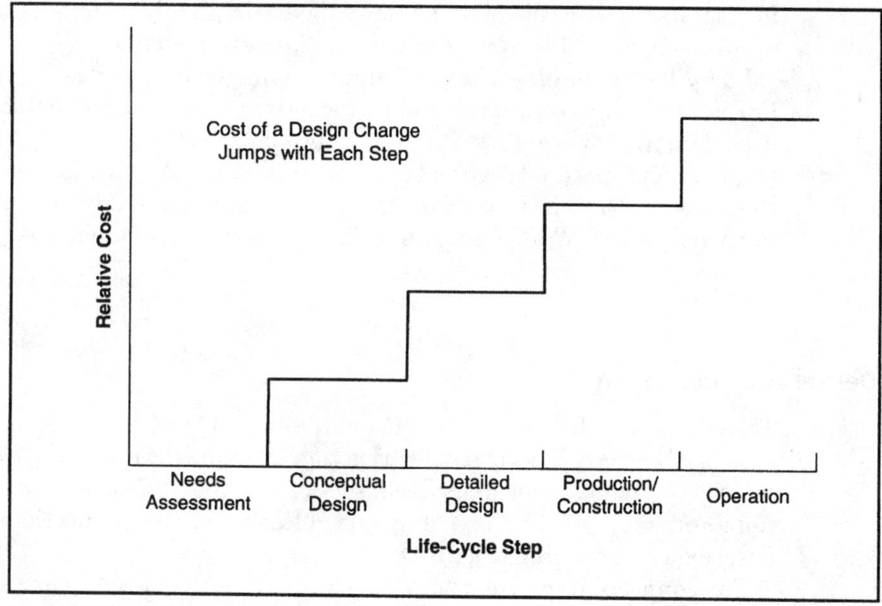

Source: LMI.

Expected Design Life by Facility Type

Facility or Infrastructure Element	Expected Design Life (in years)
Commercial buildings	30 to 50 years
Industrial buildings	50 to 60 years
Utility systems	75 to 100 years

Sources: Cotts, 1998; Hudson, Haas, and Uddin, 1997; NRC, 1998.

Figure 9.3 Cost impact of changes during various stages of project. (*Source: National Institute for Standards and Technology.*)

Case Study—The Lansing Community College Project, Lansing, Michigan

The interoperable process, by maximizing efficiencies between the designer and fabricator, allowed the Lansing Community College Health & Human Services Career and Administration Building project to lower their costs to add a 4th floor by $315,000 or $2.35/sf, according to AISC. The electronic transfer of information between the designers and fabricator allowed the building team to rapidly review alternative design schemes, make changes, and get them reviewed and approved resulting in the elimination of 700 members and a savings of 190 tons of steel. Without this interoperability process, changes of this nature would have required multiple manual reentries of data; long delays in the revision, review, and approval of shop drawings; and, almost certainly, a justifiable delay in completion, the cost of which may have completely negated a large portion, or all the savings that would accrue to the design change.

Larry Kruth, engineering and safety manager at Douglas Steel Fabricating Corporation, the contractor that fabricated and erected the project's structure, is sold on interoperability. On an unrelated project, Larry said that the design engineer had specified several large rolled sections, W44 × 265s, which were only available at a mill in Luxembourg. Larry quickly notified the engineer, suggesting a switch to a W40, available in this country. The change was made and the project's progress continued seamlessly.

The Denver Art Museum Project

The addition to the Denver Art Museum was a 147,000 square foot structure consisting of 16,500 pieces of steel with a total combined weight of 2,750 tons. There were 3,100 pieces of primary steel sections, 50,000 bolts, and 28,500 pounds of field and shop welds. The complexity of some of the connections can only be described as daunting (Fig. 9.4).

The connection information was passed from the design team to the detailer using simple sketches of each individual connection. Each sequence of steel fabrication was detailed in a 3D model and 2D shop drawing details were created. As each sequence of shop drawings was completed the detailer provided the design/construction team with 3D electronic models in addition to the hard copy drawings so that the architect could verify and check the geometric control and coordination with other architectural elements. The end result of this design/fabricate/erect process was:

- Conflict resolutions were based on the overall impact to the project where each member of the team increased their efforts to facilitate the work of others. This had a major impact on the project's schedule and costs.
- 3D graphic aids were freely shared by designer/contractor/subcontractor to facilitate each one's work and improve the overall product.
- Minimal shop issues were encountered due to the level of coordination during the 3D design.
- Minimal field issues were encountered and erection proceeded without any major field adjustments or fixes.
- The fast track approach of overlapping design, fabrication, and erection resulted in a faster start and more rapid completion.
- The preliminary interactive work by all members of the team during design smoothed out the fabrication and erection process resulting in completion of the erected steel 3 months ahead of schedule.

The AISC white paper quoted Mr. David I. Ruby, P.E., a principle in the firm of Ruby & Associates who succinctly described the current process:

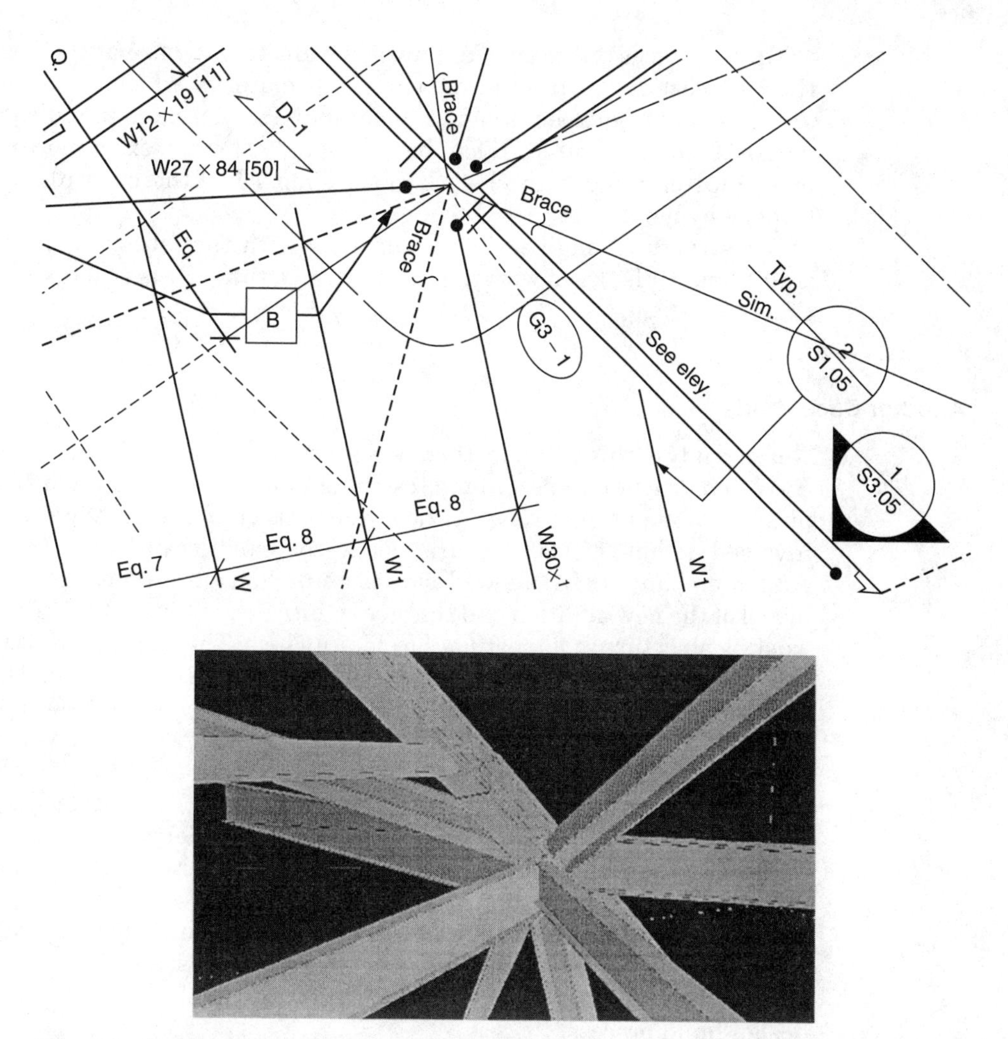

Figure 9.4 Complex steel Connections Denver Art Museum project. (*Permission: American Institute of Steel Construction.*)

The architect would present a defined building concept to the structural engineer who would design the structure utilizing a structural analysis program, prepare design drawings, and submit to the fabricator.

The fabricator would take the drawings and have a material specialist prepare a full take-off by hand to determine the material required for the structure. The fabricator would review all the material from the engineer, page by page, sheet by sheet, floor by floor. They'd take a yellow crayon to mark off every beam, and another person would recheck with a red crayon indicating it was checked again so the fabricator knew that the shop bill accounts for all the materials. Manually, this process took a week or more. And we're not talking just 40 hours of labor, but two or three people putting in 40 hours or more to pull that all together. With interoperability, this process takes just a few hours. We can now send files at noon and by 3 o'clock, the fabricator has the bill of materials to order.

Mr. Ruby goes on to say, "You always want to purchase at the best cost and the best cost comes from purchasing mill material which is normally rolled and/or stocked between 40 and 60 feet long. So you have to multiply it. That means if you need three 18-foot beams, you don't order exact pieces, you order one 55-footer and cut it to length in the shop. All of these calculations used to be done by hand."

The structural engineer works directly with the fabricator instead of the traditional RFIs. Even with the use of emails, these RFIs still take up valuable time.

Another Case Study

A case study of this CIS/2 method was presented in the AISC magazine *Modern Steel Construction* in describing the design-build three-story addition to a hospital in Albuquerque, New Mexico. Renovation of the Presbyterian Hospital involved adding 150,000 square feet to the existing building. The electronic data interchange (EDI) allowed the design-build team to prepare a 3D structural model of the new addition and the steel fabricator conducted a site survey of the existing structure, not trusting the old drawings. They now knew exactly where the old members were located so tie-ins would be accurate. Using RAM Structural System software, the engineer added this existing information to the previous 3D model and was therefore able to analyze both new and existing structures as one project depicted in the revised model. When the design was about 95% complete, via a CIS/2 software "translator," this 3D model was transmitted to the detailing software. This detailing model was sent as a computer numerical system file to the fabricator where it entered electronic instructions into the fabrication equipment.

The design-build team estimated that the structural steel design and fabrication process was more accurate and it saved at least a couple of months from start to finish and this for just the structural steel component of the entire design-build process.

The structural steel industry has taken the lead in this process of interoperability, a logical step since all designs emanate from the structural skeleton of the building. When this system is incorporated into MEP and architectural design, the entire construction cycle will have been speeded up immeasurably and at significant cost savings.

What Is Building Information Modeling All About?

The process used to conceive, design, and manage projects hasn't changed much in the past 100 years. Isolated building components are still being designed by an array of design consultants requiring a central control point to pull all these parts together. In most cases the architect, in the role of team captain assumed this responsibility. But we continue experiencing the same old problems—insufficient review to pick up all the errors and omissions and not enough time and money to complete a really thorough coordination effort.

BIM was conceived as a system to create a single building model recognized universally as a repository for all elements of a building, including its properties and interrelationships.

BIM is sometimes used synonymously with virtual building model (VBM) or virtual design and construction (VDC), referring to the ability to produce a 3D view of a construction project as building components are designed, modified, or deleted as the progression of design proceeds from design development to contract document.

The International Alliance for Interoperability (IAI) was one of the forces behind the push for this international acceptance of a single building model and their work was furthered by the International Standards Organization (ISO), which endorsed a draft of the standard.

BIM and the design-build delivery system present unique opportunities for a synergistic approach to future project delivery systems. Coupled with a seamless interoperable system tying the owner, designer, contractor, subcontractor, and vendors together, these two approaches address important concerns of the industry—time and costs, and in the process, reduce both.

Building information models permit collaboration among all parties for the construction project through the development of a digital database that can be distributed from the architect to the engineer and vice versa, to the contractor, and to the contractor's subcontractors and vendors—all through file sharing. Some recipients of this information can "read only" while others can review and recommend changes which, if implemented, show up in all parts of the design affected by the change or changes.

So there are more acoustical ceiling changes that appear on a floor plan, but are not updated on the finish schedule thereby generating one or two RFIs.

Ambiguities in the design can be highlighted by any team member and resolved before the design has been finalized. Constructability issues can be raised and debated, and changes suggested and considered, rejected or enacted with a certainty that all other affected components are adjusted accordingly and equally important, that all members of the team are instantly apprised of these changes.

This means that the time normally spent manually checking all the drawings by the design consultants, and by the contractor and their subcontractors, will be reduced considerably or totally eliminated, allowing all parties more productive time to spend on project management, quality control, and scheduling matters.

Various software vendors have seized on the opportunity to gain a commercial advantage by developing and selling BIM software informing the industry of the advantages of BIM in general, and, specifically, of their own product.

Autodesk's *Revit* program is a central database system providing the user with the ability to coordinate every building element into its database from the start of design. Any design revisions are reflected in all related drawings and schedules and coordination issues are detected immediately. The day's design production concurrently produces an adjusted quantity takeoff, a by-product of the database concept.

Practicing 4D Modeling

Joel Hardt, President of HCI CPM Consultants, Inc., headquartered in Gaithersburg, Maryland, was quick to grasp the importance of BIM, not only as an adjunct and enhancer to his CPM scheduling services, but as a contractor marketing device, a visual aid when presenting or defending a claim and a project management tool. He lists several team goals that can be achieved through 3D and 4D modeling (Fig. 9.5). Joel is a practitioner of 4D modeling, and using Bentley software, has been an early innovator in employing this process to further communication between owners, design consultants, and contractors. Owners are much taken with a 4D presentation that helps them visualize their entire construction process as each design discipline is added in

3D/4D BIM Modeling and Visualization
Team Project Goals

Goal	How	How (more detail)	Solution
Construction Costs "On-Budget"	Improve the quality of the contract documents by identifying design issues before bid and construction	Improve coordination between building systems. Reduce RFIs, ASIs, and Change Order requests.	Interference identification and management of building systems from Design Documents
Construction Costs "On-Budget"	Optimize construction by identifying issues before construction	Refine schedule, improve material management, utilities coordination, occupied space, move management	Schedule Simulator - 4D construction simulation from design model and construction schedule
Startup - Operations and Maintenance "Reduce Cost"	Reduce time required to bring on-line Initial Operations and Maintenance	Improve quality of and save time collecting data by collecting data during design and construction	Data capture and quality control from Design through Construction
Project delivery "Reduce Cost"	Reduce time to access the most current information by all project participants	Improve project documentation management through a secured central repository, document version control, access rights, document linking	Document management and publishing via desktop client and/or project extranet

Figure 9.5 3D and 4D team goals. (*By permission: HCI CPM Consultants, Gaithersburg, MD.*)

layers—first civil work, followed by the structure, MEP systems, and then architectural treatment (Fig. 9.6). Hardt says, "This is a powerful presentation tool."

One age-old problem nagging architects, engineers, and contractors is system interference management, simply put, making sure that all MEP systems fit in their allotted space and don't interfere with each other. Hardt, having been employed early in his career as a project manager with several top mechanical contractors, is particularly sensitive to these sorts of problems. Four-dimensional modeling brings all these interference concerns to light during the design stage (Fig. 9.7) and not the construction stage. Hardt says that as each building system is put in place by their respective designers, a "walk"-through the virtual building will uncover any coordination or interference conflicts that can be resolved quickly at that time.

Correcting interference problems before they occur eliminates scores of RFIs—a time-consuming and frustrating process as field problems surface and productivity dives while solutions are being sought.

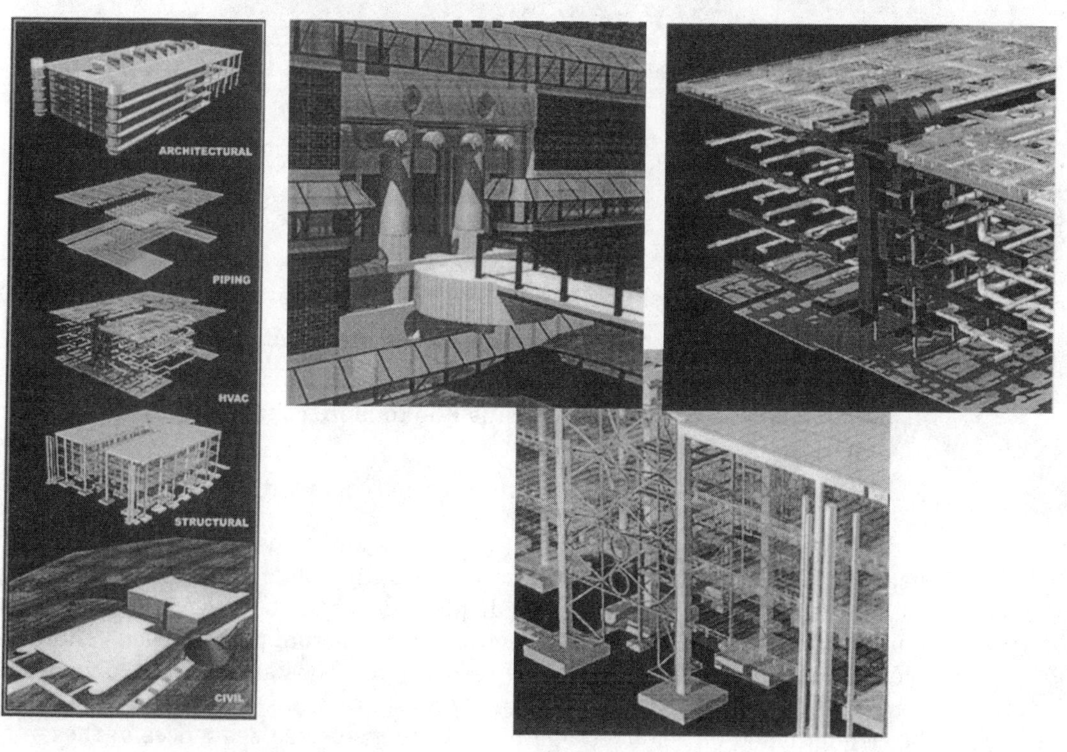

Figure 9.6 The 4D layered design process. (*By permission: HCI CPM Consultants, Gaithersburg, MD.*)

Systems Interference Management

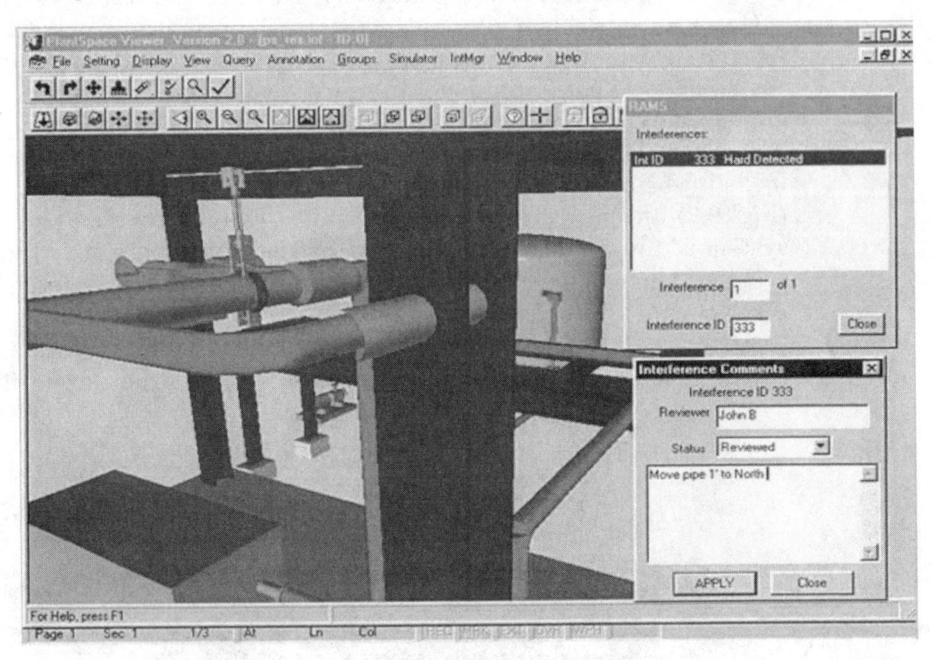

Figure 9.7 4D modeling highlights systems interference problems. (*By permission: HCI CPM Consultants, Gaithersburg, MD.*)

HCI CPM Consultants advise their clients that this 4D modeling employed during design review offers the following benefits:

1. Reduces or possibly eliminates RFIs
2. Reduces or possibly eliminates architects supplementary instructions (ASIs)
3. Drastically reduces change orders related to coordination/component conflict (interference) problems
4. Reduces potential for cost overruns due to more control over change order generation
5. Reduces delays in design and construction schedules

Schedules become more than just paper presentations when 4D modeling is used. Weekly project meetings can now visually display specific parts of the "planned schedule" and graphically display the "as built" field condition at that point in time. Joel said that his proposed versus actual presentations allow all parties to focus on cause-and-effect recovery methods and be able to view the results of their efforts in a follow-up 4D presentation. Delay claims can either be strengthened or defended against by using selected sequences of these 4D visual actual versus planned presentations.

And imagine presenting a 4D two-week look-ahead schedule at a subcontractor's meeting and viewing actual progress at the next meeting.

Joel answers the question "Why 4D modeling?" succinctly. "Where time is key in this industry, 4D models support timely and integrated decision making needed to move projects forward quickly."

These visual 3D and 4D models are attractive to all stakeholders in the construction process, helping to improve the quality of design, lessening even further the design-construction cycle—all of which translate into a more cost-effective approach to capital facility projects.

BIM—Its Promises and Its Problems

As a single source for building information, BIM presents the following advantages over the old 2D and 3D design approach:

- The plans, elevations, wall sections, and schedules are always consistent—change one, change all related work.
- The coordination across different disciplines eliminates the problems previously associated with ensuring that everything fits in its allotted space—horizontally and vertically.
- Schedules for finishes, doors, windows, and hardware are easily generated and updated as changes occur in the plan and elevation design.
- The ability to generate quantities of materials during design facilitates procurement and, particularly in the design-build mode, constantly tracks design and budget.
- The data created by BIM continues to have a useful life during commissioning and during the continuing operation and maintenance of the building.

BIM can provide higher quality

Because changes to one system or one item are reflected back through the database to related systems, we may have finally gotten rid of that typical problem where several window sizes are made smaller, but no corresponding change is made to the exterior masonry wall where the windows are to be installed. So masonry openings and window rough openings don't match and may not be discovered until window shop drawings are submitted. With BIM, these types of problems that previously created confusion and ate up man-hours will no longer exist.

And because it is a database system, one change is recognized and adjusted throughout the design instantaneously and architects and engineers may find that they have a little more time to review and tweak their designs. The contractor relieved somewhat, or completely, from the task of issuing RFIs to obtain answers to drawings lacking complete coordination or missing information can spend more time on processes, schedules, and quality.

Owners, tired of the finger pointing that happens whenever errors and omissions type-change orders occur, will have one less argument to resolve and one less cost to pay.

BIM has its caveats

Ian Howell, CEO, and Bob Batcheler, vice president of Newforma, coauthored a white paper in 2005 titled, *Building Information Modeling Two Years Later—Huge Potential, Some Success and Several Limitations.* That white paper responded to a seminal report on BIM issued in April 2003 by Jerry Laiserin. Newforma, a venture-funded development company serving the design and construction industries, is developing software to enable the seamless flow of information between all project team members pursuing the quest for interoperability.

Their report included the following comments learned from early adopters of BIM:

- The size and complexity of the files that BIM systems create on complex projects will represent a major challenge to persons charged with dealing with them.
- There will be a need for increasingly sophisticated data management at the building object level. Pioneer model server technology is now being developed to help address issues that surface when multidisciplinary design teams try to adopt a single BIM such as object versioning, object-level locking, and real-time multiuser access.
- A contradiction in work process occurs when using a single detailed BIM to try to represent a number of the alternative design schemes under consideration. While a parametrically defined building object can be quickly recreated based on the input of selected dimension and properties, the need to maintain separate BIM models for different design alternatives is prohibitive.
- Managing "what if" scenarios for engineering design using a single BIM model for building performance modeling (i.e., energy analysis, sun/shade studies, egress simulation, and the like) does not provide the flexibility needed by consulting engineers to conduct a multitude of "what if" scenarios to study alternative approaches and to optimize design alternative in order to maximize energy efficiency, ensure fire and life safety compliance, and achieve structural integrity at minimum cost.
- The expectation that everyone on the project team will adopt one BIM system (project teams comprise a collection of different companies), each of which has its own preferred and trusted software applications for design and analysis is another questionable issue. It is very rare that a single technology will be used on any one building project between different companies and/or across all phases of the project life cycle.

Design-build and BIM

It is fairly easy to see the benefits that would accrue a design-build firm employing BIM.

At the 44th Annual Meeting of Invited Attorneys' sponsored by Victor O. Shinnerer & Company in 2004, one of the participants referred to a recent survey of general contractors in the southeastern United States, 78% of whom were involved in traditional design-bid-build projects. These contractors reported the following frequency of problems:

- Problems with specifications—100%
- Unrealistic schedules—84%
- Physical interference problems—75%
- Tolerance problems—73%

This same survey revealed that 75% of responding general contractors attributed constructability problems to their inability to provide input during design. All these problems affecting contractors, not only in the Southeast, but throughout the country, could have been alleviated or possibly totally eliminated if 4D modeling had been available and if the design-build team concept had been employed.

Ken Stowe, Revit construction manager at Autodesk, lists his 10 modeling commandments as follows:

1. Win a higher percentage of proposals with compelling presentations, reliable cost, and schedule forecasts.
2. Do leaner and faster estimating during preconstruction.
3. Investigate more design options with cost and schedule ramifications analyzed.
4. Create better project plans with construction visualization.
5. Eliminate interferences via 3D viewing of ductwork, piping, structure, ceiling, lighting, and so on.
6. Prevent or considerably reduce change orders with resultant increased owner satisfaction.
7. Communicate digital documents and the 4D plan to subcontractors and foremen at the jobsite.
8. Extract *x-y-z* coordinates for field layout and control.
9. Deliver a powerful and accurate as-built to the owner.
10. Store a powerful project story as a digital archive.

The $15.8 annual cost attributed to the inadequacy of today's interoperability, as reported by NIST, is too large an amount to be ignored, and as large owners such as the federal government and those in the private sector along with design professionals' and contractors' industries become more familiar with this interoperability process, it will gain the urgency it deserves.

Even though the use of BIM and 4D modeling may become more prevalent in the industry, experts say that *real* strides will be achieved, not only by acceptance of more digital technology, but in creating true partnerships between owners, designers, and contractors. And what better way to create these partnerships than to bring the architect-engineer-contractor-owner (AECO) together working as one team on a new project. Sounds just like design-build!

Chapter

10

Bond and Insurance Considerations and Issues

The approach to insurance and bonds in design-build may be quite different from an architect and contractor's previous experience with risk management procedures. This chapter will acquaint the design-build team with some of the basic differences, but insurance and bonding professionals need to be consulted when such matters are under consideration.

Before discussing bonds and insurance, perhaps a simple definition of each is in order:

Bonds. Provide guarantees of project performance and completion to a third party, typically an owner.

Insurance. A loss-sharing mechanism guarding the policyholder against damages from potential future losses.

Bonds versus Insurance

Most contractors are familiar with bonds if they have worked in the public sector. The Miller Act of 1935, still in effect today, requires contractors working on federal projects exceeding $100,000 in value, where taxpayer money is involved, to provide surety bonds on all such projects. The states followed with passage of similar acts referred to as "Little Miller" acts.

Design consultants, by comparison, may have had little or no experience in furnishing bonds since they are generally not required when offering their professional services. However, architects and engineers carry Errors & Omissions (E&O) insurance as a normal course of business while builders may have little or no experience with this type of policy, other than making claims against an owner evoking the E&O provisions in their contract and upsetting the architect. Owners may be familiar with their architect's E&O policy and their contractor's payment and performance bonds, but may not be familiar with the different ways in which these two instruments function when applied to a design-build project. So it looks like we all may have some new things to learn regarding design-build bonds and insurance coverage.

This Risky Business

The construction business is a risky business by its very nature. Some risks are apparent: manufacturing a one-off product in an outdoor factory subjected to the vagaries of the weather, the potential of rising costs against a fixed-price contract, which often extends for two years or more, labor disputes, and material shortages—all where razor-thin profit margins that can be wiped out in a minute if control of the process is lost. Other risks are not so apparent: default of key subcontractors at crucial stages of the project, key personnel jumping ship, and a sudden and deep downturn in the market that wipes out a potential back log—all have been known to affect this risky business.

The Surety Information Office published a white paper in 2002, "Why Do Contractors Fail," in which they included a study of 823,830 building (nonsingle family), heavy/highway and specialty contractors in operation in 2000, and found that only 589,850 were still in business two years later. That equates to a 28.4% failure rate. A Dun & Bradstreet study revealed that 39% of contractors operating for more than 10 years failed; contractors 6 to 10 years old failed at a rate of 29%, and the failure rate for companies in operation for less than 5 years was 32%.

The top five factors for contractor failure

The Surety Association of America (SAA) examined 86 claims and identified the top five factors for contractor failure and the percentage of occurrence. These are as follows:

1. Unrealistic growth—37%
   Change in the type of work performed
   Change in the location of the work performed
   Significant increase in the size of projects undertaken
   Rapid expansion
2. Performance issues—36%
   Inexperience with new types of work
   Staff lacking adequate training or experience
   Lack of sufficient personnel
3. Character/personal issues—29%
   The owner retires, dies, and sells the company, thereby changing leadership and focus of the company.
   No ownership or management plans in place for transition in case of death or sale.
   Key staff members leave the company.
   Staff is inadequately trained on company policy and operations.

TABLE 10.1 Major Financial Difficulties Facing Contractors

	2005	1996
Low profit margins	64%	45%
Slow collections	57%	—
Insufficient capital/excessive debt	50%	43%
Inadequate volume	47%	20%
Inadequate controls	38%	26%
Mismanaging the business	35%	28%
High overheads	34%	23%
Imprudent risk taking	29%	24%
Weak project execution	21%	16%
Poor estimating	21%	18%
Change order volume	19%	—

4. Accounting and management issues—29%
   Inadequate cost and project management systems in place
   Estimating and procurement inadequacies
   Lack of adequate insurance
   Improper accounting practices (not adhering to AICPA Audit Guide for Construction Contractors)

Another study about contractor failures

The Grant Thornton LLP 2005 Surety Credit Survey for Construction Contractors compares the major financial difficulties facing contractors in 1996 and again in 2005. This survey shows how risks have shifted, some slightly and some rather significantly during that 9-year period (see Table 10.1).

Warning Signs That a Contractor Is in Trouble

SAA also prepared a list of warning signs of when a contractor is in trouble. These warning signs provide a "what not to do" scenario, a veritable minefield for contractors to avoid. The list of warning signs is as follows:

- Ineffective financial management system.
  Inability to forecast cash flow or present cash flow is tight.
  Receivables are turning over too slowly.
  Vendors are demanding cash on delivery for supplies and materials.
  Bills from vendors/suppliers are past due.
- Bank lines of credit constantly borrowed to their limits.
  All credit is fully secured.
  Credit lines are not being renewed.

- Poor estimating or job cost reporting.
  Revenue and margins decrease over time.
  Continued operating losses.
  Loss or reduction of bonding capacity.
  Bidding jobs too low.
- Poor project management.
  Inadequate supervision.
  Inability to administer and collect change orders.
  Project(s) not completed on time.
  One or more contracts have a claim.
  Company is constantly involved in litigation.
- No comprehensive business plan.
  Contingency plans are not developed.
  Company does not have strategy plan with goals and objectives.
- Communication problems.
  Disputes between contractor and owner.
  Poor communication from field to management and management to field.

Risk and Risk Avoidance

A design-builder provides a single source responsibility to an owner and therefore assumes risk for both design and construction. This risk exceeds the individual risk of a designer and that of a contractor, much the same way that an apple represents more than the sum of its parts. A design-builder is exposed to the liabilities of design professionals as well as the liabilities associated with the uncertainties of the building end of the business. Both these risks have to be handled successfully by any design-builder planning to remain in the business for the long haul.

A subtle but significant change in liability

In the more traditional design-bid-build system, the architect does not warrant their services as being perfect, but relies on the *standard of care* proviso that implies that they will *perform in a manner consistent with the skills of their profession.* Any errors, omissions, or inconsistencies may be picked up by the contractor and would be cause for a change order to the owner. In design-bid-build, the owner is held to a somewhat different standard in the relationship with a contractor and can't use a standard of care proviso to relieve themselves of responsibility for plan and specifications shortcomings. The owner, in effect, provides an implied warranty to the contractor that the plans and specifications are adequate, and they, the owner, assumes the risk for deficiencies in design and may, in fact, be liable to the contractor for any design errors or defects. A classic case supporting this concept was firmly established in the landmark case of Spearin v. U.S. Navy—subsequently referred to as the Spearin doctrine.

The Spearin doctrine

Spearin, a large, respected contractor in the first decade of the twentieth century, was awarded a contract on a Navy drydock project that included replacing a 6-foot section of storm sewer pipe, which they replaced. The replacement pipe proved to be inadequate to handle the volume and pressure of water flowing through it and eventually broke due to high internal pressures. The Navy held Spearin responsible and demanded that the company replace the pipe, but Spearin declined, taking the case all the way to the Supreme Court in 1918. The Court's ruling, which became known as the Spearin doctrine, stated, "If the contractor is bound to build according to the plans and specifications prepared by the owner, the contractor will not be responsible for the consequences of the defects in the plans and specifications."

Because the design-builder furnishes both the design and the plans and specifications, all defects attributable to design shift from the owner onto the design-builder's shoulders. But this shift of responsibility is also one of its virtues, and a selling point when marketing a design-build project delivery system. So it would seem that the shift in responsibility is both boon and bane for the design-builder.

It would seem that only if an owner supplied false or intentionally misleading program information to a design-builder would any transfer of risk back to the owner have any credibility. Risk management will be an important facet in the design-builder's operation, and owners will seek assurances that this risk is managed to their satisfaction. A payment and performance bond is one such assurance that owners frequently demand.

Bonds and Letters of Credit

Bonds are essentially a three-party arrangement between the surety (bonding company), the principal (contractor, or in this case the design-build firm), and the obligee (owner). The surety obligates itself to pay the obligee if the principal does not meet the performance criteria of the construction or the "build" portion of a design-build contract. The bond may also be called if the design-builder fails to pay for all labor, materials, and equipment incorporated into that structure. This is the essence of the performance bond and the payment bond.

A bid bond, the third primary type of bond, provides financial assurance that a bid has been submitted by a contractor in good faith and that the contractor intends to enter into a contract at the price of the submitted bid. If the contractor fails to accept the offer of an award, they will forfeit the value of the bond and the owner will be free to use those proceeds toward the contract award of the next lowest bidder.

The terminology of bonding

Calling the bond. Notification to the bonding company by the owner that the contractor or subcontractor has failed to live up to the terms of the contract for construction, or in the case of design-build, of the contract for design or construction. Therefore, the bonding company (surety) is being requested by the owner to provide sufficient funds to cover the unsatisfied contract commitments.

Consent of surety. On successful completion of a construction or design-build project, when all bills have been paid and all provisions of the contract (and therefore the bond) have been met, the contractor or design-builder will request the owner to "sign-off" on the bond indicating that the terms and conditions of the contract have been met. The contractor or design-builder is, in effect, asking for the consent of surety in recognizing that all bond obligations have been met so that the bond can be terminated.

Dual obligee. When two parties have a financial interest in the project, such as an owner and a lending institution, the bonding company will have a financial obligation to both parties—a dual obligation.

Guarantor. The underwriter or surety company.

Obligee. The project owner and others if there is a dual obligee.

Penal sum. The amount of the bond (generally the amount of the contract).

Premium. The cost of the bond.

Principal. The entity requesting the bond (contractor, architect/engineer, or design-builder).

Surety. The bonding company (not the insurance agency transmitting the bond).

The Letter of Credit

A Letter of Credit (LOC) is not quite a substitute for a bond. It can be used in those instances where bonding may not be available to the design-builder, or the builder has reached the limits of his bonding ability, but the owner wants some financial assurance that the designer-builder has a significant incentive to fulfill their obligation(s).

A bank LOC is a cash guarantee, whereby the bank would, in the case of a design-build project, freeze a predetermined portion of a design-builder's liquid assets in an amount equal to the value of the LOC. If the commitments under which an LOC was issued are not met, the owner would "call" the LOC and receive its proceeds. A "conditional" LOC, in the case of a design-build contract, requires some burden of proof from an owner that the design-builder has failed to perform in some aspect. A "standby" LOC deals with the payment of a specific sum within a specified period of time. A "transactional" LOC applies to one specific transaction.

Insurance, as opposed to the protection afforded by an LOC or bond, defends the insured from losses incurred due to unexpected or unusual claims related to personal liability and/or property damage.

Another way of looking at bonds, LOCs, and insurance is that the underwriting of insurance anticipates losses and protects the insured from these losses; bonds and LOCs guarantee performance.

The Bonding Process

In some cases, an owner may not require a bond, but may inquire whether the design-build team is "bondable." Because the process for qualifying for payment and performance bonds is a rather complex one and is deeply intrusive into the contractor's/designer's financial condition, merely being bondable may provide sufficient assurance to an owner rather than having to pay for the additional cost of a bond. Bonds are generally priced at a low of 1% or slightly less, to a high of 2% or slightly more, of the construction costs.

As a contractor or design-builder successfully completes larger and larger bonded projects, their rates will fall. An indicator of a financially strong company can often be determined by how low their bonding rates are.

Prequalifying for a bond

A contractor, or in this case a design-builder, will need to provide the following information to the bonding company for review and consideration:

- An organization chart of key employees, noting their responsibilities along with their resumes.
- A business plan outlining the type and size of work sought, prospects for that work (a sales development plan), the geographic area in which the entity plans to work, the company's growth, and profit goals.
- Current work in progress, a history of completed projects (with name, address, phone/email of the owner, contract sum, completion date, and gross profit earned) and current backlog.
- A continuity of business plan outlining how the business will continue on death or retirement of the present owner. Life insurance policies on key personnel will also be required.
- Evidence of a bank line of credit.
- Letters of recommendations from owners, architects/engineers, subcontractors, and major suppliers.

Financial statements must accompany the application and include:

- Fiscal year-end statements for at least the past three years along with the latest statement audited and certified by an accountant.
- Balance sheet showing assets, liabilities, and net worth.

- Income statement—gross profit on contracts, operating profits, and the net profit before and after tax provisions.
- Statement of cash flow.
- Accounts receivable and payable schedules.
- Schedule of general and administrative expenses.
- Explanatory footnotes—qualifications made by the accountant.
- Management letter conveying the certified public accountant's (CPA's) findings, observations, and recommendations.

Suggestions for the Newly Created Design-Build Teams—the Three Cs

The basics of bonding can be summed up simply by the three Cs—character, capital, and capacity. If these three Cs have been met, chances of obtaining the first bond are pretty good.

Character

- Have the two firms, designer and contractor, been involved in litigation defending themselves against questionable business practices?
- Has either company been "blackballed" from bidding on public projects?
- Has either company failed to complete a project for reasons other than an owner deciding to cancel the work for legitimate reasons?
- Do both companies enjoy a good reputation in their respective industries?

A no answer to all except the last "bullet" are what sureties are looking for.

Capital

- Financial statements that comply with good accounting practices
- Financial statements with a qualified opinion indicating that accepted accounting practices have been followed in all cases except for a few exceptions and each exception is spelled out in the auditor's letter
- No adverse opinions in the financial statement

Capacity

- A look at both companys' human resources, depth and experience of the management team, their approach, marketing, and planning procedures
- History of acceptable estimating and cost control procedures
- History of producing profitable projects

TABLE 10.2 Changing Bonding Criteria

	2005	1996
Strength of balance sheet	98%	90%
Financial statement presentation	97%	91%
Equity	92%	75%
Debt	81%	68%
History of successful projects	81%	68%
Consistent profitability	78%	63%
Experience in type of project	75%	67%
Use of CPAs with industry knowledge	75%	68%
Claims history	67%	N/A
Reputation of firm and/or principals	66%	65%
Financial statement disclosure	65%	70%
Accounting policies	63%	59%
Experience in geographic area	59%	43%
Size of over/under billings	55%	36%
Overhead expenses	47%	35%
Contract volume	36%	33%
Succession planning	30%	N/A
Safety record	10%	20%

When the design-build team is formed, it is a good idea to begin establishing a relationship with a reputable surety even though a bond is not needed immediately.

The first "C," character, can be a door opener in building the surety's confidence in the new firm. Even if not specifically required by a client, it might be wise to procure a small bond for a small project. Successful completion of a bonded project not only builds confidence with the surety but is also the beginning of a performance resume. And lastly, the design portion of the team can pursue obtaining a bond along with the contractor team member, so that the surety can look to both firms if required.

The Grant Thornton 2005 Surety Credit Survey also looked at the change in important criteria for obtaining bonding that occurred during the period between 1996 and 2005 (see Table 10.2).

Traveling in Newly Charted Waters

The bond in a conventional design-bid-build project guarantees the construction portion of the project, but for design-build projects, this distinction is not so clear-cut because a bond would apply to both design and construction. Although some bonding companies may not be reluctant to provide bonds for large, well-established design-build firms, sureties are reluctant to provide bonds that include both design and construction for smaller and newer entrants to the field. In the design-build concept, an owner buys a final product that includes design and construction and since a bond guarantees the construction project, it may be construed that the bond also covers design errors—something the surety may not have contemplated or proposed in their issuance of the bond. In a landmark decision in a lawsuit filed in Louisiana in 1992, the courts found the surety liable for design flaws under the terms of the performance bond that they issued and this raised a red flag in the industry.

Design services have probably been adequately covered when the architect/engineer was engaged in their traditional business but those methods of risk management may not be so effective when applied to this new design-build project, and a contractor having experience in obtaining bonds may have to rethink the way in which they approach requesting a bond for that new client, if they have teamed up with a design firm.

Bonds and Design-Build

When contractors and design consultants are considering joining together to form a design-build entity, the insertion of a standard "hold harmless" clause or indemnification clause may not be sufficient to protect them from design shortcomings or failures. A typical hold harmless agreement between designer and contractor may shift the responsibility for design liability from the contractor to the design consultants and it is possible that the architect's insurance policy relating to design errors and omissions may address these issues, but both aspects of this risk shift need to be fully explored by consulting surety and insurance professionals.

When requested by design-builders to issue a bond that would include design concerns, sureties can do either of the following:

1. Insert explicit language and disclaimers limiting their exposure to design.
2. Urge the design-builder to insert language in their contract with the owner stipulating that the bond does not cover the design portion or services of the contract. A typical proviso would look something like this:

> The bond does not cover any responsibility for negligence, errors, and omissions in design, or warranty of design. Coverage under the bond is limited to only the construction phase and postconstruction phase of the contract. The bond premium is based solely upon the value of construction and postconstruction portions of the contract and does not include the design aspect of the contract.

Bonds and the contractor-led design-build team

On design-bid-build projects there is a demarcation between the design and construction activities, and therefore it is rather easy for surety to underwrite the risks of one activity or another. But under the design-build concept, these distinctions become blurred because surety may be asked to guarantee performance of the design as well as the construction. Now add to the mix a new form of business entity such as a joint venture (JV) and surety is being asked to guarantee a new process to be performed by a new business entity. Sureties use two methods of dealing with situations like this as mentioned below:

1. Use explicit language limiting their exposure and insert disclaimers in the body of the surety bond.
2. Urge the design-builder and the owner to include language in their contract recognizing that surety will issue a bond that does not secure the design portion of the contract and make this a condition of issuance.

A provision such as this in the contract between the design-build (DB) entity and owner would read something like this:

> The bond does not include any responsibility for errors or omissions in design, negligence, nor does it warrant design. Coverage under the bond is limited to the construction and postconstruction phase of the contract only. The premium for the bond is based upon the value of the construction and postconstruction phase of the contract and not upon the design included in the contract.

From a practical standpoint, it is questionable whether an owner would allow language like this without some other safeguards to protect them against design problems or failures.

Even though a surety may have had positive experience with a general contractor when providing payment and performance bonds on conventional design-bid-build projects or even negotiated projects, a bond request for a new entity led by that same contractor may present problems for the bonding company.

There may be lack of past performance by the contractor in the design-build venue, thereby causing reluctance on the part of the surety to bond this unknown quantity. A proven track record of the designated A/E team, with respect to performance on design-bid-build projects backed up with satisfactory relationships with those project owners, will be welcomed by the surety.

In this new design-build business, who assumes responsibility for differing site conditions, adequacy of design, express and implied warranties and guarantees, and how payment will be received and distributed are all new issues for a general contractor and a design consultant. The surety will look at the A/E's performance record, its financial strength, claims on its E&O insurance policy, track record of paying subcontractors (outside engineers and other consultants), and other issues that will also be new to the A/E in their new design-build mode.

Bonds and the A/E led team

In this relatively new area of architect-led design-build delivery systems, sureties have had to become familiar with the increased exposures facing design consultants. No longer does an owner view the architect as their agent, providing checks and balances during the construction cycle. Under the conventional design-bid-build process, the architect can hang their "liability" on the agency concept with the owner, and should the end product fail to meet the owner's expectations, the design consultant may be able to fall back on the standard of care doctrine as a possible defense for inadequate performance and may suffer a proverbial black eye but no great financial loss to the project. But under the design-build process, where a payment and performance bond is furnished by the A/E, if the design consultants fail to meet the expectations of the owner, the bond may be called allowing surety to recover any loss it incurs to satisfy the bond and then go after the assets of the A/E to make itself whole.

When the architect is the lead member in the design-build team, surety will look at the financial strength of the design firm and how the firm will manage their financial obligations created through any subcontract agreements with the builder on their team. The surety will also look at the design firm's financial ability to perform on the project under the conditions for which a bond is issued as well as how the financial obligations of the firm's other capital requirements will be addressed and handled.

Possibly the best way for an architect-led DB project to obtain a bond is by teaming up with a contractor having positive past performance with the surety. If the structure of the A/E led team is a JV, then it could become the named principal on the bond and this will expose the A/E to any owner-imposed liability if the bond is called. Usually all parties in the JV will execute a surety agreement of indemnity that serves as the consideration for the surety's providing the payment and performance bonds. This agreement of indemnity allows the surety to attempt recovery from the venture partners, referred to as indemnitors, if there is a loss expense incurred by the surety as a result if its having to act on the bond.

Another approach to bonding in an architect-led design-build project is by assigning this responsibility to others via contract provisions.

So for all parties, design-build bond requests are, in effect, a whole new ballgame.

Bond Provisions in the Contract

The Associated General Contractors of America (AGC), like other professional design organizations, has developed specific contract language to define the nature of the bond to be submitted by the DB team. Will the bond be all inclusive and cover both design and construction performance or will it focus only on one activity or the other?

AGC Form 470—Design-Build Performance Bond, where surety is liable for design costs of the work, states:

> This bond shall cover the costs to complete the work, but shall not cover any damages of the type specified to be covered by the design-builder's liability insurance or by the professional liability insurance required pursuant to the contract, whether or not such insurance is provided or in an amount sufficient to cover such damages.

AGC Form 471—Design-Build Performance Bond, where surety is not liable for design services, says:

> Pursuant to Article 2 of the Bond, the surety shall be liable for all construction costs of the work, up to the bond sum, but shall not be liable for any costs or damages arising from any design services provided pursuant to the contract.

The surety market which, at first, was reluctant to provide performance bonds to design-build contractors, has become more receptive to issuing these instruments based on the successful completion of so many design-build projects. However, bond premiums will be priced according to the degree of risk included, particularly those covering design defects and negligent design.

The architect's insurance policy regarding design errors and omissions can address some of those types of issues, thereby avoiding the need to provide the payment and performance bonds while still accepting liability for design errors.

But when the owner, whose decision to use design-build was based, initially, on a single-source contact for design and construction, finds the DB team pointing fingers at each other rather than resolving design issues, then the DB team will have a very short shelf life.

Bonding companies acutely aware of one major problem with design-build, i.e., costs associated with design errors and omissions may look askance at a design-build team that is not adequately protected against the potential costs for design deficiencies. For this purpose, a close look at insurance requirements for the design-build team becomes essential.

Insuring Design-Build Risks

Contractors and designers need to focus on the new types of risks facing them as design-builders and the types of insurance policies that will be required to lessen the impact on those risks. No longer will E&O insurance, Commercial General Liability (CGL), and Contractors Professional Liability Insurance (CPL) fit as neatly into the project matrix as they did before. The design-build team will have to become accustomed to new risks across the entire spectrum of insurance matters.

The hold harmless clause has been previously inserted in many contracts to limit the liability of a contractor or designer; however, these types of disclaimers may not produce their intended result. Some states maintain strict provisions that limit the shifting of liability for one's negligence, and failure to recognize and become aware of these rules and regulations may invalidate the hold harmless clause. Insurance provided by an insurance specialist to fit the needs of a specific design-build project is the best protection against potential risks.

A design-builder has liability exposure relating to their design and the construction that follows; any deficiencies in design will continue through or become magnified during construction. Negligence in the preparation of the design, committed by any one of the many consultants, architect, civil or structural engineer, or MEP engineers, will ultimately arrive at the doorstep of the design-builder. Claims of economic loss, property damage, and bodily injury may be filed by participating parties or third parties.

General liability insurance

General liability insurance (GLI) protects the insured against liability claims due to direct bodily injury and damage to property to third parties caused by an accident. This form of insurance is also referred to as commercial general liability insurance (CGL) or comprehensive general liability insurance.

Commercial general liability

While most CGL policies do not include a specific exclusion for design-build work, coverage usually falls within the provisions of the bodily injury and property damage clause arising out of an occurrence as defined in the policy. Many of the claims for design errors are of a consequential damages nature such as damage from delays, inability to occupy the building, and so forth. These claims are not included in a CGL policy.

In some cases the CGL form of insurance will offer a design-builder some protection from design defects as long as those design defects result in either bodily injury or property damage. However, most general liability insurers add an endorsement to policies offered to design-builders that exclude liability occurring because of design errors.

Two endorsements were offered to insurers beginning in 1996 that specifically relate to design errors:

Insurance Services Organization (ISO) Form CG 22 79 (Fig. 10.1) permits some coverage for design services arising out of a contractor's ability to control the means and methods as described in a standard contract for construction. This endorsement excludes damages due to professional design services, so this proviso may be of little use to a design-builder attempting to cover errors and omissions from the design team.

The second endorsement offered by ISO Form CG 22 80 (Fig. 10.2) is called *Limited Exclusion-Contractor-Professional Liability Endorsement* and permits coverage for bodily injury or property damage from professional design services offered by design consultants working with a builder.

A third endorsement, CG 22 43 07 98 (Fig. 10.3), is called Exclusion for Engineers, Architects, or Surveyors Professional Liability.

COMMERCIAL GENERAL LIABILITY
CG 22 79 07 98

THIS ENDORSEMENT CHANGES THE POLICY. PLEASE READ IT CAREFULLY.

EXCLUSION – CONTRACTORS – PROFESSIONAL LIABILITY

This endorsement modifies insurance provided under the following:

COMMERCIAL GENERAL LIABILITY COVERAGE PART

The following exclusion is added to Paragraph **2.**, **Exclusions** of **Section I – Coverage A – Bodily Injury And Property Damage Liability** and Paragraph **2.**, **Exclusions** of **Section I – Coverage B – Personal And Advertising Injury Liability:**

1. This insurance does not apply to "bodily injury", "property damage" or "personal and advertising injury" arising out of the rendering of or failure to render any professional services by you or on your behalf, but only with respect to either or both of the following operations:

   a. Providing engineering, architectural or surveying services to others in your capacity as an engineer, architect or surveyor; and

   b. Providing, or hiring independent professionals to provide, engineering, architectural or surveying services in connection with construction work you perform.

2. Subject to Paragraph **3.** below, professional services include:

   a. Preparing, approving, or failing to prepare or approve, maps, shop drawings, opinions, reports, surveys, field orders, change orders, or drawings and specifications; and

   b. Supervisory or inspection activities performed as part of any related architectural or engineering activities.

3. Professional services do not include services within construction means, methods, techniques, sequences and procedures employed by you in connection with your operations in your capacity as a construction contractor.

CG 22 79 07 98

Figure 10.1 CGL endorsement 22-79–exclusion-contractors-professional liability. (*By permission via education license from Insurance Services Organization, Inc. (ISO), Jersey City, New Jersey.*)

COMMERCIAL GENERAL LIABILITY
CG 22 80 07 98

THIS ENDORSEMENT CHANGES THE POLICY. PLEASE READ IT CAREFULLY.

LIMITED EXCLUSION – CONTRACTORS – PROFESSIONAL LIABILITY

This endorsement modifies insurance provided under the following:

COMMERCIAL GENERAL LIABILITY COVERAGE PART

The following exclusion is added to Paragraph **2., Exclusions** of **Section I – Coverage A – Bodily Injury And Property Damage Liability** and Paragraph **2., Exclusions** of **Section I – Coverage B – Personal And Advertising Injury Liability:**

This insurance does not apply to "bodily injury", "property damage" or "personal and advertising injury" arising out of the rendering of or failure to render any professional services by you, but only with respect to your providing engineering, architectural or surveying services in your capacity as an engineer, architect or surveyor.

Professional services include:

1. Preparing, approving, or failing to prepare or approve, maps, shop drawings, opinions, reports, surveys, field orders, change orders, or drawings and specifications; and

2. Supervisory or inspection activities performed as part of any related architectural or engineering activities.

This exclusion does not apply to your operations in connection with construction work performed by you or on your behalf.

CG 22 80 07 98

Figure 10.2 CGL endorsement 22-80–limited exclusion-contractors-professional liability. (*By permission via educational license from Insurance Services Organization, Inc. (ISO), Jersey City, New Jersey.*)

COMMERCIAL GENERAL LIABILITY
CG 22 43 07 98

THIS ENDORSEMENT CHANGES THE POLICY. PLEASE READ IT CAREFULLY.

EXCLUSION – ENGINEERS, ARCHITECTS OR SURVEYORS PROFESSIONAL LIABILITY

This endorsement modifies insurance provided under the following:

COMMERCIAL GENERAL LIABILITY COVERAGE PART

The following exclusion is added to Paragraph **2., Exclusions** of **Section I – Coverage A – Bodily Injury And Property Damage Liability** and Paragraph **2., Exclusions** of **Section I – Coverage B – Personal And Advertising Injury Liability:**

This insurance does not apply to "bodily injury", "property damage" or "personal and advertising injury" arising out of the rendering of or failure to render any professional services by you or any engineer, architect or surveyor who is either employed by you or performing work on your behalf in such capacity.

Professional services include:

1. The preparing, approving, or failing to prepare or approve, maps, shop drawings, opinions, reports, surveys, field orders, change orders or drawings and specifications; and
2. Supervisory, inspection, architectural or engineering activities.

CG 22 43 07 98

Figure 10.3 CGL endorsement 22-43–exclusion-engineers, architects or surveyors professional liability. (*By permission via educational license from Insurance Services Organization, Inc. (ISO), Jersey City, New Jersey.*)

What Effect Does a Hold Harmless Clause Have?

The hold harmless clause has been inserted in many contracts by design-builders to limit the liability of the contractor or designer; however, these types of disclaimers may not produce their intended effect. Some states maintain strict provisions in their laws that limit the shifting of liability from one's negligence and failure to conform to these rules and regulations. These provisions may invalidate any hold harmless clauses.

A design-builder has liability exposure relating to their design and the subsequent construction; any deficiencies in design may continue through or become magnified during construction.

Risks associated with design

Ann Rudd Hickman of the International Risk Management Institute, Inc. stated that a contractor could not rely solely on a hold harmless clause or other contract provisions to protect them from design liability because:

1. A hold harmless provision can be unenforceable since many states have rather strict rules relating to the shift of liability for one's negligence to another party. Failure to conform to any such rule(s) may invalidate the hold harmless provision.
2. The design firm engaged at the time of design and/or construction may not be in business when the actual claim is filed by an owner. Professional liability insurance is written on a claims-made basis, therefore if the claim is made after the policy has expired, there is no coverage.
3. Architect/engineers' E&O insurance is limited to damages that result from professional negligence, which is somewhat different from the "standard of care" concept, but even if that provision applied, architects and engineers generally carry relatively low limits of insurance or high deductibles and may have limited tangible assets that could be seized to satisfy indemnity obligations.

A new focus on the increased or differing roles of insurance in the design-build process will concentrate on the following:

- Construction
- Design
- Environmental remediation

Insurance products in design-bid-build don't adequately address these new risks:

- Traditional professional liability insurance excludes design-build, most environmental issues, and equity interests.
- Traditional general liability insurance excludes professional errors and omissions coverage.

There is a need to coordinate insurance products to fit design-builder risks:

- Delete traditional exclusions.
- Purchase new insurance products to cover the design-builder's errors and omissions exposure.
- Wrap up insurance program.

The Standard of Care Standard

The courts do not expect a designer to prepare and be responsible for a perfect design, but they must prepare their design in a manner consistent with accepted "standards of care" and skills for their profession, and that is one reason why a standard of care provision is generally included in a design-build contract.

This adherence to an accepted standard of care will allow some leeway if there are slight shortcomings in the design. Although some owners will want to change this *generally accepted* standard of care provision to read *highest* standard of care, this should be avoided, for obvious reasons.

Risk is inherent in the design and construction process and risk must be managed *reasonably*—a word that is often used when rendering legal decisions.

As an example, let's look at the geotechnical consultant and the problem of differing site conditions or unknown subsurface conditions. If the owner hires the geotech, which is not unusual in design-build work, their exploration of the site should be representative of overall site conditions. Test borings are theoretically representative only of the area in which they were taken, i.e., the 6 or 8 inch boring or in close proximity thereto. However, if significant numbers of test borings reveal gravelly material down to a depth of 8 feet but one area of the site contains rock at a depth of 3 feet that was not evidenced by a test boring, who is responsible to remove the rock—the design-builder or the owner?

The design-builder could argue that their proposal, based on the owner's geotech, reasonably assumed all gravel materials and the presence of rock was an aberration and the cost to remove the rock must be paid by the owner.

Another twist to the liability issue

Insurance concerns for today's designers and constructors are no longer a simple matter of merely meeting personal and property liability requirements. Builder's risk insurance, carried either by owners or requested from contractors, is becoming more prevalent these days.

Contractors have historically depended on the E&O insurance of the designer maintained by the design firm to correct problems related to design shortcomings. But what about other problems that could multiply and take the form of consequential damages? An example of the design and construction of a heating, ventilating, airconditioning (HVAC) system that falls short of the owner's expectations illustrates the point.

The owner moves into a new factory building and finds that the HVAC system does not provide sufficient cooling (or heating) to allow their work force to operate productively. Although there may be additional design and construction costs to correct the situation, which will be borne by the design-build team, some of which may be reimbursable from CGL policies with endorsements, the owner has another claim; they had to shut down the factory, lay off 150 employees while the system was being modified and upgraded. The owner expected to be reimbursed for the lost wages and also for profit lost due to production downtime costs that probably greatly exceed the HVAC rework costs. The design-build team will suffer that loss, and if there is an indemnification clause in the contract with the design professionals, errors and omissions can't be applied.

The insurance industry has responded to problems like this and has developed errors and omissions coverage for the contingent liability when a contractor-led design-build team subcontracts the design. Owners will have an interest in how the design-builder handles matters of errors and omissions because these types of issues may somehow impact them.

Risk management through E&O insurance

E&O insurance requirements need to be clear and precise—what is the specific requirement for a specific project?

These types of policies should be retroactive to the commencement of prebid design activities providing coverage from the first dollar spent. There are no standard E&O policies, but there should be provisions that include:

- Cost overruns
- Delays associated with the design process
- Liquidated damages (LDs) in connection with delays, if LDs are in the owner contract
- Errors and omissions
- Coverage for defense costs
- Pollution liability

There may be several ways in which to mitigate errors and omissions insurance.

Option 1—Owner caps level of E&O insurance. Ultimately the cost of project insurance will be factored into the cost of the project and an owner willing to place a cap on E&O insurance coverage can, in effect, self-insure a portion of the project. This can be accomplished by inserting a provision in the owner-design-build contract limiting the design-builder's professional liability to the amount of insurance proceeds available from the design-builder's policy. Such a provision could read:

> To the fullest extent permitted by the governing law, the total liability in the aggregate of design-build and DB's officers, directors, employees, agents, and independent professional associates, to (Owner) for any and all injuries, claims, losses, expenses, or damages whatsoever arising out of or in any way connected to (the design-builder's) engineering, architectural or surveying services in the design-builder's capacity as an engineer, architect, or surveyor, shall be limited to the amount of insurance proceeds recoverable under the applicable errors and omissions policies for this project.

Option 2—The owner can prepare a release from design liability on completion of the project. The owner can insert a provision in the design-build contract releasing the design-builder from design liability upon some point in time after completion of the project. This would limit the design-builder's cost for extended liability and, in effect, make the owner self-insurable for that extended period.

Option 3—Use of a skip-over provision in the design-build contract. This provision would be affected at the front end of a project, by having the owner insert a provision that "skips over" the contractor and requires conventional E&O insurance to be provided by the design consultants. This, in effect, is the same type of insurance that is provided in the conventional design-bid-build process, which apparently has worked quite well in the past. This skip-over provision would read:

> In consideration of the Contractor entering into the Agreement, Owner hereby agrees that Contractor, or the Contractor's surety, shall not be liable or responsible in any manner whatsoever for any claims, damages, errors, or omissions arising out of the professional services to be performed by the Architect and/or Engineer as defined in the Agreement, or other design professionals under this Agreement, whether through indemnity or otherwise. Owner hereby agrees that Owner will not look to Contractor or contractor's surety for recourse as to any claims, errors, damages, or omission, and Owners' sole recourse shall be against the Architect and/or Engineer or other design professional performing such professional services. Contractor agrees to fully cooperate with Owner in pursuing its rights hereunder and under the Agreement including Without Limitation OR (i) assignment to the owner of any rights or remedies Contractor may have against the Architect and/or Engineer or other design professional relating to any such claims, damages, errors, or omissions.
>
> Other than recourse with respect to claims, damages, errors and omissions relating to the architect and/or engineer or other design professionals arising out of the professional services to be performed under this Agreement, this agreement shall not constitute a waiver of any other remedy which Owner may have against the Contractor for any other failure of contractor to perform in accordance with this Agreement or an other agreement or contract between owner and Contractor.

Option 4—Use separate contracts for design and construction. If separate prime contracts are used in establishing a design-build team, the design function with all its standard responsibilities, obligations, and liability exposure will fall to the design consultants in a more conventional manner. And if a contract for construction is prepared by the builder to provide all services generally accepted as those assigned to the contractor, the liability issues become more defined and localized. This arrangement can be established in the terms of the teaming agreement when the design-build entity is being formed.

The Necessity for Builder's Risk Insurance

Design-build teams will focus their insurance concerns on CGL insurance to cover property and bodily injury, but they also need to look at builder's risk insurance. This insurance, also known as *course of construction* insurance, provides coverage for loss or damage to the structure incurred during the construction process. Damage caused by a fire in one part of the building under construction will be covered by builder's risk. Although somewhat difficult to obtain in areas subjected to hurricanes and earthquakes, this type of policy, in general, is readily available to both project owners and builders.

Most builder's risk policies are written as "all risk," which means that the policy covers all risks except those expressly excluded. The other type of builder's risk policy is called *named peril*, a policy that covers only certain risks identified in the policy, such as wind, fire, or flood. These "named risk" policies can be supplemented by other insurance policies that include all risks other than those covered in the named risk policy. So if one party to the design-build team, say the owner, has a named risk policy, the design-builder, by opting for a policy with coverage other than the named risks will have, in effect, an "all risk" coverage.

Unless specified to the contrary, owners generally elect to furnish builder's risk insurance on a project; however, these policies may have large deductibles with the assumption that lesser amounts of damage, those not covered by the deductible, are expected to be covered by the design-builder. If the project is a turnkey, the design-builder, may, per contract, "own" the building until final payment and transfer of title is made to the owner, therefore the design-builder will be responsible for any losses due to damages such as fire, wind, water, and so forth.

It may be in the design-builder's interest to furnish builder's risk and include its costs in the proposal to the owner. Such a policy provided by the design-builder could have some advantages. These are listed as follows:

- The design-build team will be dealing with their own insurance company, possibly one with which they maintain other insurance coverage and receive more friendly attention in case of a claim.
- The proper coverage will be obtained and there will be no concern about less than adequate coverage provided by the owner.
- When a claim does occur, an owner's adjustor may not be as willing to provide a fair adjustment as the design-builder's insurance company, which will look forward to continuing business with the designer and contractor.

The Need for Waivers of Subrogation

Subrogation is the process by which a person or entity who pays a debt for another acquires the right of a creditor to that person for whom the debt had been paid. This means that after an insurance company pays a claim, they may pursue reimbursement from the party that was responsible for the loss. By including a waiver of subrogation in the design-builder's contract with the owner and obtaining a waiver from the builder's risk insurer, parties can insulate themselves from any subrogation claims by the builder's risk insurer.

Design errors can result in property damage claims, therefore including this type of waiver is especially important for design-builders.

Workers Compensation Insurance

Contractors are very familiar with the state's workers or workman compensation requirements and how a series of jobsite accidents can increase insurance costs dramatically. A poor accident record requires three years of good accident experience before rates will be reduced. Design consultants may not have as much experience with worker's compensation insurance, and the strong safety program that is required to keep jobsite accidents and injuries in check.

How premiums are established

Workers compensation premiums are determined by the following formula:

$$\text{WCIP} = \text{EMR} \times \text{manual rate} \times \text{payroll units}$$

Where WCIP = workers compensation insurance premiums and EMR = experience modification risk. This is the multiplier determined by previous insurance experience of the policyholder and is used to forecast future benefit payments to employees who have filed claims. Manual rate is a rate structure assigned to each type of work performed. Various trade crafts are classified into "families" based on their potential exposure to injury. Each family is assigned a four-digit number corresponding to their premium rate that takes into account worker's accident claims experience for that particular family of trades. Payroll units is a number determined by dividing the employer's annual direct labor costs by 100.

It is important for the design-build team to require certificates of worker's compensation insurance from each of their subcontractors. However, some states do allow workers, whose injuries were incurred as employees of subcontractors, the right to sue the design-build entity, and it is therefore important to be apprised of all subcontractor jobsite accidents by having these subcontractors furnish copies of all accident reports to the design-builder.

Lawsuits may have a long gestation period and without prompt notification of a subcontractor accident, the design-build team may be shocked when served with a legal notice three or four years down the road. So the collection of all jobsite injury and accident reports should become an integral part of the project's archival records.

Controlled Insurance Programs

A controlled insurance program, more commonly referred to as CIP or OCIP (owner controlled insurance program), has gained popularity in recent years as another method to control insurance costs while still maintaining the desired coverage.

A CIP is an insurance program that will cover the project's jobsite risk for the owner, the general contractor, the design-builder, design consultants, and subcontractors for such losses represented by worker's compensation, employer's liability, general liability, and builder's risk. The only parties to the construction process not covered are material and equipment suppliers. An OCIP program is essentially the same but one initiated and controlled by the owner.

The differences between the conventional insurance approach to insurance coverage and the CIP approach are rather straightforward (see Table 10.3).

A typical exhibit to a subcontractor agreement advising CIP and requiring the subcontractor to provide that insurance not included in the controlled insurance program is shown in Fig. 10.4.

OCIP and CIP programs are generally not cost-effective for small projects because the effect of combining the premiums of all contractors and subcontractors and other parties to the project may not result in significant savings to offset the added cost to administer the program. But on large construction projects, significant savings can accrue, and by combining all the premiums of the various parties on the site, the insured can present one large account for insurance companies to bid on, rather than a larger number of small accounts.

Using CIP on a project should be combined with a strong safety program backed up by an on-site safety supervisor in order to reap the benefits of reduced claims. The cost of the safety director, on these projects, will be more than offset by reduced claims and lost time due to accidents and injuries.

TABLE 10.3 Conventional and Controlled Insurance

Type of Coverage	Conventional Approach	CIP Approach
Workers compensation	Each contractor and subcontractor	Held for all parties*
CGL	Owner, each contractor, and subcontractor, design consultants	Held for all parties
Builders risk	Owner of contractor	Held for all parties
Auto liability	Each contractor, subcontractor, design consultants	Each contractor, subcontractor, and design consultant

*Some states offer workers compensation directly to contractors and subcontractors.

EXHIBIT B: Subcontractor, Sub-Subcontractor Provided Insurance

This project is an Owner Controlled Insurance Program (OCIP) Project. Therefore, much of the insurance necessary for the project is provided by the OCIP. There are, however, additional insurances requirements for "off site" liability exposures that are required in the OCIP manual. There are also additional insurance requirements under this Exhibit B.

Each Subcontractor of every tier shall purchase and maintain the following insurance (s) during the term of this Project:

Automobile Liability insurance for not less then:
$1,000,000. Bodily injury/Property Damage Combined Single Limit.

This Insurance must apply to all owned, leased, non-owned or hired vehicles to be used in the performance of the Work, and the policy shall include an Additional Insured Endorsement naming the Owner, its directors, officers, representatives, agents and employees, Construction Manager, and Developer as an Additional Insured with respect to their operations at the Project Site.

Coverage is primary and non-contributory with respects to Owner, its directors, officers, representatives, agents and employees, Construction Manager, and Developer.

NOTE: Automobiles are defined in accordance with the 1986 ISO insuring agreement. This definition includes, but is not limited to, a land motor vehicle, trailer or semi-trailer designed for travel on public roads, whether licensed or not (including any machinery or apparatus attached thereto).

Workers Compensation, including Employers' Liability Insurance with minimum limits of:
(a) Workers' Compensation- Statutory Limits with Other States Endorsements
(b) Employers Liability
$1,000,000. Bodily Injury with Accident – Each Accident
$1,000,000. Bodily Injury with Disease- Policy Limit
$1,000,000. Bodily Injury with Disease- Each Employee

To protect Subcontractor and Sub-subcontractor from and against all workers compensation claims arising from performance of work outside the project site under the contract.

Initials ______
Initials ______

Figure 10.4 Typical subcontract agreement exhibit when CIP Insurance is in effect.

General Liability Insurance for contract operations not physically occurring within the Project Site. Five (5) Year Completed Operations Coverage Extension with a limit of liability not less than:

$1,000,000 Per Occurrence
$1,000,000 Personal Injury and Advertising Injury
$2,000,000 General Aggregate (On a Per Project Basis)
$2,000,000 Products/Completed Operations Aggregate

The Coverage to be written on the ISO standardized CG 00 01 (10/01) or substitute form providing equivalent coverage. Such insurance shall cover liability arising from premises, operations, independent contractors, products-completed operations, personal and advertising injury, and liability assumed under an insured contract (including the tort liability of another assumed in a business contract). Such policy shall include an Additional Insured Endorsement naming the Owner, its directors, officers, agents, represents and employees, Construction Manager, and Developer as additional insured's.

IMPORTANT – The additional insured endorsement shall be maintained for a minimum period of at least (5) five years after end of the project completion. This endorsement wording should be equivalent either of the following ISO endorsements:

- CG 20 10 (11/85)
- Both CG 20 37 (10/01) and CG 20 10 (10/01)
- CG 20 26 (11/85)

Coverage is primary and non-contributory with respects to Owner, its directors, officers, representatives, agents and employees, Construction Manager, and Developer.

Wrap up exclusion to be removed five (5) years after completion of your work.

Excess Liability insurance written on an occurrence form for contract operations not physically occurring within the Project Site. Five (5) Year Completed Operations Coverage Extension with a limit of liability not less than:

Subcontractors of all tiers
$5,000,000 Per Occurrence
$5,000,000 Annual Aggregate

Initials ______
Initials ______

Figure 10.4 *(Continued)*

Excess Liability insurance will include coverage for Automobile Liability for the, Subcontractor, Sub-subcontractors while on-site and off-site.

Such policy shall include an Additional Insured Endorsement naming the Owner, its directors, officers, agents, represents and employees, Construction Manager, and Developer as additional insured's.

Contractors Equipment insurance, for all construction tools and equipment whether owned, leased, rented, borrowed or used on Work at the Project Site is the responsibility of the Subcontractors and Sub-subcontractors, the Owner or Construction Manager shall not be responsible for any loss or damage to tools and equipment. The Contractor's Equipment insurance shall include a waiver of subrogation against the Owner, its designee(s), Construction Manager, Developer, other contractor(s) and subcontractor(s) of all tiers to the extent of any loss or damage. If the Subcontractor or Sub-subcontractor does not purchase such insurance, he will hold harmless the Owner, its designee(s), broker(s), Construction Manager, Developer, other contractor(s) and subcontractor(s) of all tiers for damage to their tools and equipment.

Waiver of Subrogation

Subcontractors at every tier shall require all policies of insurance that are in any way related to the Work and that are secured and maintained by the Subcontractor and all tiers of Sub-subcontractors to include clauses providing that each underwriter shall waive all of its rights of recovery, under subrogation or otherwise against the Owner, their designee(s), broker(s), Construction Manager, Developer, contractor(s) and all tiers of subcontractors.

Construction Manager shall require Subcontractors and all tiers to waive the rights of recovery (as aforesaid waïver by Construction Manager) as stated above.

Notice of Cancellation/Termination

Each Certificate of Insurance shall provide that the insurer must give the Construction Manager at least 30 days prior written notice of cancellation and termination of the subcontractors and sub subcontractors coverage there under.

Insurance Carrier

Insurance Carrier must be licensed to do business in the State of Maryland (or whichever state in which work is being performed) and have a Best Rating of A VII or better.

Initials ______
Initials ______

Figure 10.4 *(Continued)*

Is CIP for DB?

Although the rationale for controlled insurance programs is ostensibly for cost savings, a design-build firm may find some other advantages when considering this method of obtaining insurance coverage. We have discussed the various types of policies for which contractors and design consultants are familiar with individually, but not collectively. One concern of a design-build firm, whether contractor led or architect/engineer led may be "Do we, as a new firm, have all our insurance requirements in hand? What about the civil or structural engineer, and what will our subcontractors bring to the table to augment our insurance needs?" These valid questions may be more easily handled if only one entity is responsible for gathering all the myriad insurance requirements and funneling them through one insurer to assure the design-build that they have adequately addressed their liability issues. The controlled insurance program may be the way to deal with this concern.

Steps to take when considering a CIP program are as follows:

1. *Conduct a feasibility study.* Determine the advantages and disadvantages of such a program. Are there any state or local statutes or regulations that govern or limit a program such as this? What added costs are involved—administrative, inspection, and supervisory—and what savings may accrue?
2. *Requests for proposals for insurance company bidders.* Solicit quotes from insurers via an RFP spelling out the types of coverage and the amounts of coverage required for the type and projected cost of the design-build project. In the RFP to the design consultant (if a contractor-led team) or to the builder (if an architect-led team) set forth the same coverage and limits, and request the proposed cost of insurance coverage to meet those needs. Request the same information from the subcontractors planned for the project, which may not be exactly accurate if these subcontractors have not been preselected by either an architect or a builder. But even if the subcontracted work is to be competitively bid, an order of magnitude of insurance costs can be gathered.
3. *Interview insurers.* Conduct interviews with a short list of insurers to include:
   - The background of the design-build firm
   - OCIP administrative services to be provided
   - Their approach to structuring the OCIP program
   - Availability of risk management assistance including loss control services
   - Availability of claim management services
   - Type of computerized risk management information system to be employed to determine compatibility
   - Fees for all services and premiums

The compelling reasons for considering CIP on a design-build project would include:

1. *Quality of coverage.* A CIP that ensures that the teams' coverage will be met since they are the ones that have specified exactly what their needs are.
2. The larger insurance package offered to the carrier ought to be rewarded with lower overall premiums.
3. *Increase insurance limits.* Many subcontractors can only provide small amounts of coverage with their general liability insurance, but by combining all policies, the higher limits so often required by owners today, can be offered.
4. *Stability of the insurer.* With some contractors having policies with offshore companies or smaller companies with less sophistication, one policy can be placed with an insurer that is not only financially sound, offering competitive rates, but also one that can provide risk management assessments, site inspections, and experience in the type of projects being undertaken.
5. Potential elimination of claims and disputes and subrogation between design-build team and insurers.
6. *Innovative program management.* Integrated risk management encompassing design, environmental remediation, force majeure, and builder's risk in a multidiscipline environment like design-build adds yet another dimension to the new way of doing business.

There are, as always, disadvantages to any system and CIP is not immune. These disadvantages are as follows:

1. Soliciting bids from subcontractors is somewhat more cumbersome since requesting, evaluating, and approving credits for their insurance must be done and, unless these subcontractors submit costs documented by their agents, these costs may not be accurate.
2. Documentation and reporting requirements will place a burden on an administrative staff that may be already overburdened.
3. A CIP usually concludes coverage for "completed operations" after a specified period of time, typically two to five years, while a contractor's liability exposure continues for a longer period of time.
4. Since auto insurance is excluded from CIPs, it may be difficult to separate a general liability claim from an auto claim.

The design-build process presents some new challenging liability issues for owners, contractors, subcontractors, architects, and engineers. In today's risk management field there are lots of options available to the design-build team to obtain proper coverage but the team individually and collectively needs to consult those experts in the field that can give them sufficient information to make an informed decision.

Note. With regard to Figs. 10.1, 10.2, and 10.3, ISO requires the following to be posted: Information which is copyrighted by and proprietary to Insurance

Chapter

11

The Legal Aspects of Design-Build

The legal aspect of design-build that sets it apart from other project delivery systems relates to both licensing and contractual issues. Many states require licenses for the practice of architecture and different types of licensing arrangements for contractors. Only recently, because of the increased interest in design-build in both the public and private sectors, state legislatures are implementing new laws to allow one firm to obtain a license to provide owners with both design and construction.

Legal issues for this increasingly popular project delivery system has also brought new contractual and liability issues to the forefront and because design-build is a relatively new approach, these issues are still a work-in-progress.

Contractors, designers, and owners, contemplating a new design-build venture need to be aware of the many legal concerns that will affect their decision to proceed and therefore, seeking legal advice must be an essential part of that decision-making process.

The roles and responsibilities of each party to the design-build process changes significantly from the design-bid-build approach. First and foremost is the determination of what kind of business entity must be formed in order to meld design with build. For this, several alternatives are available. These are listed as follows:

1. A joint venture (JV) agreement can be prepared, basically preserving the contractor's and the design firm's organizational structure while creating a new entity for the sole purpose of working on a specific project.
2. A new business or professional corporation combining personnel from previous design and contracting firms.
3. A limited liability corporation (LLC), a single entity, formed to contract with an owner.
4. A partnership.
5. A sole proprietorship.

Business Decisions

Sometimes, the choice of selecting a new entity will be limited or restricted by state laws, which prohibit a company providing architectural or engineering services from engaging in the contractor business. So, state laws need to be reviewed for that purpose.

A rather uncomplicated way of working together as a design-build team would be for either the designer or contractor, operating in their present organizational form, to merely issue a contract to the other firm for either design or construction work, in other words, subcontract either design or construction.

But this is just the start of a changed environment that must address further legal issues.

Once the choice of a new design-build business entity is selected, many other business responsibilities need to be addressed, questions asked, and decisions made, which are listed as follows:

- Who will be assigned responsibility for design and for errors and omissions?
- Who will determine the responsibility for personal and property insurance, and extended coverage?
- Which party will provide any bonds required?
- Who will be responsible for means and methods?
- Which party will provide a comprehensive safety plan and the appropriate supervisory staff?
- Who will be responsible for preparing the budget and ensure that design tracks the budget? Who pays for any redesign required to get back on budget?
- How will conflicts of interest between design and construction be resolved?
- How will the payments to the owner be prepared, monitored, collected, and disbursed?
- How will negotiations with vendors and subcontractors be conducted? How will final scope and price be established?
- Who will be responsible for delays during the design phase and construction phase? If any fines are levied by the owner, how will they be shared?
- How will capital contributions be apportioned at the commencement of the enterprise and distributed at the conclusion? What if additional capital is needed in interim? Who will provide it and in what proportion?
- How will the designer maintain some semblance of independence in their role toward satisfying the owner's demands as opposed to the demands of the design-build firm?

There are risk and reward issues to be addressed. These are as follows:

- Who will be responsible for cost overruns that are not reimbursed by the owner?
- Who will assume responsibility for dealing with differing and unknown site conditions?
- How will responsibility for delays be addressed if delays occur in the design or construction phase of the project, whether designer, contractor, or owner generated?
- How are claims and disputes to be resolved and related costs apportioned when not reimbursed by the owner?
- How will design and construction errors be handled after the warranty period but before the lapsing of applicable Statutes of Limitation?
- If there is a conflict of interest between the two parties how will it be equitably resolved?
- When owner-generated changes in either design or construction or both occur, how will costs be determined, apportioned, and reimbursed on receipt of funds?
- How will costs to manage both design and construction be apportioned within the design-build entity?

It is apparent from this short list that there are a multitude of concerns that must be presented, discussed, and resolved by the design team and the construction team in order to create a workable design-build team. Not only do these issues need to be raised and resolved, but also memorialized in the design-build agreement.

Liability Issues

New liability issues arise, simply due to the fact that the new entity will now assume liability for design and construction under a single point responsibility concept. One of the reasons owners look favorably upon the design-build concept is that, among other benefits, this arrangement fills the void that previously occurred between design responsibility and contractor responsibility.

Contractors working under a "build-only" contract with an owner were obliged to complete the work that was contained in the plans and specifications. Architects working under a similar arrangement where they provided an owner with those plans and specifications were really not expected to deliver a perfect set of documents, but the interpretation of what is considered perfect, reasonable, or within the doctrine of "standard of care" often led to lengthy lawsuits.

Liability under design-build

Liability under design-build is different in that the design-build firm is now responsible for design problems that affect the scope of the work, particularly when it comes to specialty contractor work. Under a conventional design-bid-build system, the specialty contractors, subcontractors, and experts in their chosen field, would frequently be the ones to bring coordination, errors, and omissions, and constructability issues to the attention of the general contractor, who, in turn, would pass them on to the owner, who would pass them on to the designers and everyone would say, "Who me?" That scenario would end up in finger pointing, sometimes contentious negotiations, and most likely increased costs to the owner. Design-build has changed much of this liability model.

Latent Defect Concerns

The combination of design and construction, however, raises several legal issues. One obvious issue is which party bears responsibility for serious design and/or construction defects occurring after the "contract warranty" period ends, when liability for such defects as structure failure may be governed by state or federal law? Isn't the design-builder responsible for systems that underperform the performance requirement set by the owner as included in the design-build contract?

Failure to meet these performance specifications rests with the design-build team and can't easily be dismissed under the standard of care provision, and may be the basis for an owner claiming breach of contract under the provisions of the Uniform Commercial Code (UCC). The UCC is basically a law that applies to product warranty, but as we shall see further in this chapter, it may also apply to design-build.

Other Liability Issues

When an architect-led design-build team is in place, the lead member, the architect may find themselves on more unfamiliar ground from a liability standpoint; responsible for overall jobsite safety including Occupational Safety and Health Administration (OSHA) citations and resultant penalties and payment obligations to subcontractors and vendors.

When an architect is a subcontractor to a contractor, in design-build, they may have risk of liability inherent in their contractual (subcontract) obligations, such as:

1. Performance in terms of preparing and submitting design documents per the schedule and also in a timely fashion.
2. When a subcontractor agreement is issued that includes the standard "pay-when-paid" clause, the architect, if payment is late from the owner, may be liable for payment.
3. If the architect's subcontract agreement with the design-builder includes a clause for payment of work-in-place that includes design documents, can the contractor use those design documents, when paid for, on another project, and if so, what kind of liability does this present for the architect?
4. When unanticipated costs occur due to inadequacies in the design documents, can the contractor look to the design consultant and their subconsultants for reimbursement of these costs?

5. If disputes and claims arise between the owner and design-builder and include design errors and omissions, will the design consultants have to wait until the design-builder settles with the owner in order to obtain any relief from monies withheld by the contractor pending resolution of the dispute or claim?

Licensing Issues

Many states require a license for the practice of architecture and engineering and a license to engage in business as a contractor. The design-build legal entity under consideration will need to take into account the requirements established in the state or states in which the design-build team is planning to operate. Some states currently have specific design-build licensing laws and each year more states are enacting these types of laws. Some of these laws have some unique disclosure requirements as witness those in Pennsylvania.

The Commonwealth of Pennsylvania recently enacted the Architect's Licensure Law, an amendment to Section 3 of the Act of 12/14/82. Until the passage of this amendment only architects licensed under the laws of the Commonwealth were allowed to practice architecture. Contractor-led design-build teams were therefore illegal before its passage and since designer-led teams were reluctant to engage in design-build projects for a number of reasons, design-build delivery in Pennsylvania was stifled. These new amendments defined the design-build entity as a single-source contract to provide a combination of design and construction services to a client and not untypical of the approach other states have taken.

The Act including a specific provision authorizing firms practicing architecture to provide design-build services as an architectural firm authorized under various subsections to offer design-build services consistent with Section 15(9) below:

> Section 15- Permitted Practices
>
> (9) Design-build services strictly in accordance with the following practices; a design-build entity not authorized to practice under section 13(a) through (I) may offer design-builds service, if the architectural services in the design-build process are provided in accordance with the following:
>
> (i) An architectural firm which has been authorized to practice in this Commonwealth under section 13 (a) through (i) shall independently contract with a design-build entity and is responsible for all material aspects of the practice of architectural as defined in section 3.
>
> (ii) At the time a design-build entity offers a written design-build proposal for a specific project the design-build entity shall give a written disclosure to the client stating an *architect will be engaged by and will be contractually responsible to the design-build entity and will not be responsible to the client.*
>
> (iii) The design-build entity shall agree that the architect will have direct supervision of the architectural work.
>
> (iv) The contract between the design-build entity and the client shall set forth the name of the architectural firm that will be contractually responsible to the design-build entity for providing architectural services.

An architect in Pennsylvania may become involved in a design-build venture in one of three ways in that state:

1. A designer-led design-build team
2. As a subcontractor to a contractor
3. Creation of a new design-build business entity

However, other states have slightly different requirements. In Arizona, a design-build firm does not have to be licensed to do construction work as long as the contractor is properly licensed. In Georgia, a general contractor may perform design-build work as long as the design is created by a properly licensed architectural firm.

Contractor's licensing laws

All states regulate the practice of architecture and engineering but have differing licensing requirements. With respect to contracting licenses, the requirements vary; Georgia has a requirement for the licensing of HVAC contractors, electrical contractors, utility contractors, and fire protection contractors, but no provision for general contractors; Alabama requires a general contractor license only for jobs exceeding $50,000; Virginia requires a written examination and a demonstration of sufficient knowledge, skills, and abilities to obtain a license depending on the specific class of contractor under consideration.

How the Law Looks at Design-Build

Designing to a budget and dealing with cost overruns are concerns that design consultants and contractors face many times.

Contracts between the architect and owner often do not properly address the potential of their competitively bid project exceeding the budget, even though such a clause does exist in a standard A.I.A. owner/architect contract. A standard clause in A.I.A. Document B-141—Standard Form of Architect's Services: Design and Contract Administration relating to situations where the budget has been exceeded after reviewing bids from the lowest qualified bidder, directs the owner to:

1. Give written approval of an increase in the budget.
2. Authorize rebidding or renegotiating of the project.
3. Terminate the process.
4. Cooperate in revising the project scope as required to reduce the cost of the work.

If the owner elects to proceed with item #4, Article 2.1.7.6 of this document requires:

> ... the architect, without additional compensation, shall modify the documents for which the Architect is responsible under this Agreement as necessary to comply with the budget for the Cost of the Work.

However, some B-141 contracts are short forms, containing only a table of articles that may exclude the redesign provisions of Article 2 above.

Many architects take exception to these types of contract clauses for good reason.

The design professional, in situations like this, often argues against accepting liability for cost overruns stating that too many world market forces are in play, affecting the budget requirements set by the owner.

Just look back to the late 1970s and early 1980s when the inflation spiral was increasing by about 1% a month. Also, look at the more recent upward movement in some basic construction materials; the doubling of steel prices between 2003 and 2005 and the increased costs of copper products, drywall, cement, and lumber.

But suppose that a design-build team has been contracted by an owner with a proviso that both design and budget requirements are to be met, the team members must determine how any cost overruns are to be distributed internally. Since the contractor is given the main responsibility for preparing estimates, it would be presumed that any overages would fall to them. However, a court in Georgia found an architect/engineer (A/E) partially liable to the contractor in a design-build entity for failure to provide accurate data which the JV used in its bid. The project was a power plant and the contractor sued the A/E team for breach of contractual and fiduciary duties within the framework of the JV stating that the contractor did not receive sufficient, accurate information upon which to base their bid and to make reasonable efforts to design the project within budgeted quantities. The contractor further stated that the A/E did not track quantities in its design and did not promptly notify the contractor that budgeted quantities would be exceeded.

The Georgia court found that the JV contract was ambiguous as it related to risk allocation and partially blamed the A/E for the damages incurred by the contractor.

Failure to Control the Design

An owner-architect agreement such as AIA's B-141 directs the architect to observe the work-in-progress, differentiating their supervisory role from one of inspection to one of observation and most standard contracts exempt the architect's authority over the contractor's "means and methods." But does this apply to the design-build process as well? This is yet another liability issue to be thrashed out between the design consultant and builder.

Other Legal Issues That Confront an Architect-Led Design-Build Team

The Architect's Guide to Design-Build Services published by the American Institute of Architects includes several legal cases involving design errors, breach of good faith and fair dealing, ownership of documents, and payment for design services.

Liability for design errors: statutes of limitations limits

In the case of Kishwaukee Community Health Services v. Hospital Building and Equipment, et al., 638 F.Supp.1492 (N.D.Ill 1986), the court ruled that the period from which liability for design errors commences in a design-build project starts at the completion of the project rather than at the completion of the design. The Court said that "components of a contract cannot be wrenched out of the contract for accrual purposes."

In some states a design consultant liability relating to design defects or errors ends after the design has been completed and the construction documents have been issued, but as we can see from the Kishwaukee ruling, the courts have taken a different approach when design and construction responsibility resides in one contract.

A court in New Jersey, a year earlier, appeared to confirm this interpretation. Welsh v. Engineering, Inc, 202 N.J. Super 387.495,A.2d 160 (1985), ruled that when design and construction are provided by one company, the completion of all construction signals the start of any action regarding either errors in design or defects in construction.

Ownership of documents

Who owns the construction documents while a design-build project is underway?

In Johnson v. Jones, 885 F.Supp.1008 (E.D.Mich 1995), the architect, Johnson was commissioned to design the renovation of Jones's residence and as the design progressed, Johnson proposed a design-build contract whereby he would also be responsible for construction. But three months into this arrangement, Jones fired Johnson and contacted another architect and builder to continue the work using Johnson's work produced to that point. Johnson sued Jones, claiming copyright infringement. On reviewing all the evidence, the court ruled that Johnson never intended to relinquish control of his design, neither when he agreed to prepare the drawings nor when he gave incomplete copies to Jones. The court did say that because Johnson repeatedly objected to contract language making Jones the owner of his work, Johnson intended to allow Jones to use his drawings for the renovation of her home only if he would complete those drawings, not if they were to be used by another architect to complete them.

A design-build firm, I.A.E. Inc., contracted with Shaver, a firm with experience in airport cargo and hanger design, and entered into an agreement to pay Shaver $10,000 for schematic drawings. This contract did not indicate any further business dealing between Shaver and I.A.E., and indeed, after completion and delivery of the schematic drawings, I.A.E. retained another architect to complete the design work. Shaver wrote to the airport authorities indicating that his drawings were copyrighted and shortly thereafter his lawyer sent a request for payment of $7000 for the assignment of copyright.

In I.A.E. Inc. v. Shaver, 74 F.3d 768 (7th Cir.1996), the U.S. Court of Appeals confirmed the district court's ruling that there was no copyright infringement and that Shaver had provided the design-builder with an implied license to use his schematic drawings. They based their ruling on the contents of the contract with the design-builder that stipulated that Shaver was to produce schematic drawings and that the contract did not infer or suggest that he would be retained as the architect of record for the completed project.

Other ownership and use considerations

Although an owner may pay for the printed or electronic version of the design, the stock-in-trade of the architect is the intellectual property they generate and have many reasons to protect. Several contract provisions can be used to address these issues so that owner and architect will be allowed to use the design documents for the purpose for which they were intended but restrict their use in future or other uses.

For example, an owner may not want their design-builder to use their design in promotional literature for a number of reasons: security, exclusivity, and the like, and this restriction can be easily covered in the contract. If there is a termination for cause, generally occurring if the design-builder has not lived up to their end of the bargain, plans developed to that date might rightfully belong to the owner. If there is termination for convenience, either because both parties were of a mind, then the ownership of design documents could result in either remaining with the design-builder or, on an agreed amount, be transferred to the owner. This again can be spelled out in the contract documents.

Design Error Liability

The design firm Skidmore, Owings and Merrill (SOM) was hired to design a high-rise complex known as the Newmark Building and they had represented to the owner that they could meet the owner's budget and exacting schedule. However, major defects in the design documents were found and required substantial changes resulting in increased costs. SOM sued the owner to recover their additional fees and the owner countersued stating that the added design fees were related to change orders to cover these defects. In Skidmore, Owings and Merrill v. Intrawest I Limited Partnership, the jury rejected SOM's argument that these additional costs would have been incorporated in the Guaranteed Maximum Price (GMP) if the design had been complete. The court said:

> Intrawest (the owner) bargained for complete designs that would allow it to establish the project's GMP....SOM knew that Intrawest had a tight budget so that significant design changes after the GMP was set would threaten the project's feasibility. The record shows that Intrawest would not have undertaken the project had it known the true extent of its cost.

Although design consultants must share some responsibility for design errors when they are part of the design-build team, the contractor side of that team must also share in some of that exposure if they failed to act in a professional manner. In C.I. Maddox, Inc. v. The Benham Group, Inc., 88 F.3d 592 (8th Cir.1996), the design-build team contracted to build a power plant in Joppa, Illinois. The design-builder was unable to complete the work and the owner "called" the bond. The bonding company paid $2.8 million toward completing the work. The job had gone badly and their design-builder sued their engineer and subcontractor for breach of contract and fraudulent and negligent misrepresentation. The court initially awarded the design-builder $5 million in damages but this was overturned on appeal, and reduced to $3.8 million, based on the appeals' court finding that there was insufficient contract language to shift construction liability and that the engineering firm had "no duty under the contract to act as insurance against Maddox's own carelessness."

The Implied Terms in a Design-Build Project

As the SOM court case revealed, the owner expected certain results from the architect they hired. They expected construction documents that were *reasonably complete*. Lawyers would probably call this an *implied* promise and although not specifically stated in a contract, there are certain things that are taken for granted—*implied*. These are:

- Good faith and fair dealing.
- An owner agrees to pass judgment on acceptable work within a reasonable period of time.
- An owner will review and either approve or reject a request for payment within a reasonable period of time.
- An owner will give sufficient information to a design-build team to allow it to design a reasonably complete program.
- A design-build team, once a design development scheme is approved by an owner, will not substantially deviate from that scheme in the preparation of final design documents.
- A design-build team has a certain fiduciary responsibility with respect to protecting the owner's interests.

The implied warranty issue has been dealt with in a number of court decisions.

Courts have upheld the theory that a design-build contractor can be held liable if their product fails to meet the specific purpose for which they were hired.

In Robertson Lumber Co. v. Stephens Farmer Cooperative Elevator, a contract was awarded to a design-build firm to construct a building that would safely store 100,000 bushels of wheat for one year; the building collapsed after being loaded with wheat. One of the principals upheld by the court was that there was an implied warranty to provide a building for a specific purpose and it failed the test.

The Kennedy v. Bowling case involved a design-builder who built a four-story warehouse on a fast track basis to store chemical. When the building needed extensive repairs, the court awarded the owners those repair costs under the theory of implied warranty for a specific purpose.

A Rhode Island design-builder was negotiating a contract with a local university for a fast-tracked sports facility and received authorization to proceed with this $7.15 million project although both parties could not agree on a final scope of work. The design-build firm actually completed the project without a signed contract in hand. During construction some change orders were generated for work not included in the original scope of work and after the university rejected the design-builder's claim, they sued them for $881,500. The university claimed that they owed the design-builder nothing for this change-order work because they had an implied contract for $7.15 million. An appeals court ruled that no implied contract had been created since both parties had been unable to arrive at an agreement upon the scope of work.

Then there is the responsibility of a contractor, or in this case a design-builder, to act in a professional manner—isn't that why the client hired them in the first place? In the case of Eichberger v. Follard, 169 Ill. App.3d 145,119 Ill. Dec 781,523 N.E. 2d 389 (1988), the builder blindly followed the plans and specifications and poured the foundation required by those plans and specs even though soil conditions dictated otherwise. The court ruled that the builder should have at least advised the owner that additional work was required, based on their past experience, and therefore the builder did not perform in a workmanlike manner—an *implied* responsibility.

Compliance with code responsibility

Is the design-builder responsible for adhering to all building codes? It would seem that an architect or a contractor cannot claim ignorance of building codes, and the case of Tips v. Hartland Developers, *Inc.*, 961 S.W.2d 618 (Tex.App.1998), upholds that premise—the theory of an implied covenant. The court said that in this particular design-build contract, contractors and not owners, were in the best position to know about building codes and that the implied covenant carries with it the supposition that the design-build team will comply with all building codes and provide a building that can be occupied for its intended purpose.

The Uniform Commercial Code

Known as the UCC, this law was generally thought to apply only to the sale of goods and products. For example, look on the reverse side of a bill of sale from any appliance manufacturer and you will see the provisions of UCC spelled out in great detail.

The courts have split on whether a design-builder is a "merchant of goods." One court decision ruled that the design-build subcontract for an airplane hanger was actually a sale of goods and therefore comes under the purview of the UCC. The court said in their ruling:

> Although Trident's alleged contract with Austin includes the erection of the hangar, the erection services are incidental to the sale of the steel and cladding for the hanger. Thus, the predominant thrust of this alleged agreement was for the sale of the steel and cladding for the hanger.

Design-builders need to follow these types of rulings closely because they may become part of a wider interpretation by the courts that design-build contracts may constitute the sale of a product and therefore fall under the provisions of the UCC.

Americans with Disabilities Act

Owners of new and remodeled projects have been sued for violations of the Americans with Disabilities Act (ADA) and design-builders may face a double whammy when it comes to responsibility for compliance with this Act.

There have been a few court cases challenging whether architects and contractors can be held liable for violation of what is basically a civil rights law.

Although design consultants have argued that they do not fall within the classes of potential parties referred to in Section 302 of the Act, and they provide only design but not construction, they claim that these types of violations do not apply to their profession. Further, Section 302 prohibits discrimination by parties who own, lease, or operate a public accommodation, which they say certainly doesn't apply to them.

Design-build companies may not stand on such firm ground since the courts have, in some instances, found architects and contractors liable under ADA, because of the design and construction language in Section 303.

Section 303 reads as follows:

> Section 3 – New Construction and Alterations of Public Accommodations and Commercial Facilities
>
> Application of term—except as provided in Subsection (b) as applied to public accommodations and commercial facilities, discrimination for purposes of section 302 (a) include:
>
> A failure to design and construct facilities for first occupancy later than 30 months after the date of enactment of this Act that are readily accessible to and usable by individuals with disabilities, except where an entity can demonstrate that is it structurally impartibly to meet the requirements of such subsection in accordance with the standards set forth or incorporated by reference in regulations issued under this title.

Writer's note: Subsection (b) deals with elevator installations.

In an Eighth Circuit Court of Appeals ruling, the court said that they applied the *design and construct* language in Section 303 "conjunctively" and found the defendants responsible for both design and construction of a hotel project.

The Fair Housing Act (FHA) also created a legal course of action against those who fail to design and construct "those dwellings" in a manner that allows them to be accessible by disabled persons. The FHA makes it unlawful to "discriminate in the sale or rent or to otherwise make unavailable or deny a dwelling to any buyer or renter because of a handicap of that buyer or renter." The same Eighth Circuit Court ruled that anyone possessing a significant degree of control over the final design and construction of a facility may be held liable for discrimination under Title III.

In United States v. Ellerbe Becket, Inc., 976 F.Supp.1262, 1267-68 (D.Minn 1997), the federal government extended liability to architects under Title III. The government, in their suit, claimed that Ellerbe Becket had shown a pattern of designing inaccessible sports arenas around the country and because the government alleged that the firm participated in not only design but also construction, the court did not have to decide whether participation in both functions was required for liability.

However, the Ninth Circuit Court in their rulings limited the potential class of defendants under Title III to those specifically cited in Section 302(a): owners, lessors, and operators of either pubic or commercial facilities.

Green Buildings—Avoiding Some Not-Too-Obvious Pitfalls

Green building design, green building components, and green building construction techniques are all relatively new. Just about seven years ago, in 1998, the U.S. Green Building Council issued their first version of the LEED rating system.

While conventional buildings must meet or exceed the local building code requirements, green building designers and contractors answer to a somewhat different code. These green building design-builders hold out the promise that they will employ, according to author Miriam Landman in her 1999 thesis at Tufts College "building design, construction methods, and materials that are resource efficient and that will not compromise the health of the environment or the health and well-being of the building's occupants."

Designers and contractors promising these goals, may open themselves to more legal liability.

Design firms may underestimate the amount of expertise required for some of these green and sustainable building projects. Relying on manufacturer's claims can be risky; some will overstate the case and others will not have had enough long-term testing to back up their claims. The limited warranties of new products and equipment are generally limited to one year and therefore limit the manufacturer's liability but maybe not the design-builder's. Nearly universal is a manufacturer's exclusion of consequential damages in their warranty program, and if a new heating or cooling system fails and a company must send many of their employees home and basically shut down their operation, the size of a claim for consequential damages may far exceed the costs to correct the mechanical problems.

Legal experts suggest having an owner sign-off on a new product, acknowledging that a manufacturer's performance claims may not be fully tested. They also suggest that the design-builder might explore a waiver of liability for the use of a new product. But in the real world, do you think an owner would buy into such a program?

Paybacks on Capital Costs That May or May Not Occur

Many "green" systems are more costly than conventional ones but hold out the promise of a payback within a certain period of time, after which the related expense such as water or electricity will be reduced, guaranteeing continued savings. With green building initial capital outlays between 7% and 10% higher than conventional design and construction, this promise of savings becomes one of the lures. But unless the design-build team advises the owner of the *possibility* that these cost savings may be less than originally stated or may kick in later than initially expected, they may have assumed more liability than they anticipated.

Building materials are subjected to provisions of the UCC, which sets a four-year limitation on product liability matters, but many limitations on a design professional's liability vary from six to ten years, leaving an exposure gap of two to six years if a product fails.

The Sick Building Concern

According to the National Institute for Occupational Safety and Health (NIOSH), 20% to 30% of all commercial buildings suffer from poor indoor air quality. Indoor air pollution sources have been identified as follows:

Inadequate ventilation—53%

Outdoor contaminants—19%

Indoor contaminants—15%

Unknown sources—13%

Although some green building HVAC systems are touted as reducing health-related issues, unless exact commissioning procedures are in place, these benefits may not accrue to the owner.

A design-builder would be wise to require the owner to engage a commissioning expert to oversee and vet that process.

Operating and maintenance procedures may be somewhat more complicated in order to retain the anticipated quality levels of these "green" systems, and owners need to be apprised of that fact.

The owner's commissioning agent should also be responsible for ensuring that the owner's maintenance crews are knowledgeable in the maintenance and operation of the building's equipment. Both these practices will greatly aid the owner and reduce the design-builder's potential for liability if any health related indoor air quality (issues) surface.

The Legal Implication of Electronic Records

As more and more designers and contractors turn to the electronic storage of everything from plans and specifications to daily logs and requests for information if, or when, a claim or dispute arises, these stored electrons may be called upon to diffuse or substantiate that dispute or claim. It is better that a design-build firm consult with their legal counsel to get the ground rules for electronic storage of construction documents just in case they are needed.

There are three basic sources of electronic documents that may be called upon. These are listed as follows:

1. Project documents generated during the normal course of construction, correspondence, schedules, and things such as Requests For Information (RFIs) and Requests For Proposal (RFPs), change-order requests, and the like.
2. Electronic mail—both outgoing and incoming.
3. Information obtained through the Internet.

These kinds of records, in legal parlance, are called out-of-court statements and are often interpreted by the courts as "hearsay evidence." An exception to this hearsay ruling is the business record, if it is being offered with a *sufficient foundation.* The American Bar Association in their Rule 803(6) spells out sufficient foundation.

> Records of Regularly Conducted Activity—A memorandum, report, record or data compilation, in any form, of acts, events, conditions, opinions or diagnoses, made at or near the time by, or from information transmitted by, a person with knowledge, if kept in the course of a regularly conducted business activity, and if it was the regular practice of that business activity to make the memorandum, report or record, or data compilation, all as shown by the testimony of the custodian or other qualified witness, or by certification that complies with Rule 902(11), Rule 902(12), or a statute permitting certification, unless the source of information or the method or circumstances of preparation indicate a lack of trustworthiness. The term "business" as used in this paragraph includes business, institution, association, profession, occupation and calling of every kind whether or not conducted for profit.

Discoverable Records

Discovery is the way in which litigants gather proof of their claim, a way for one party to examine documents in the other's possession in order to evaluate the other party's case or prepare for trial. Many of these documents are no longer paper but reside in an electronic format within a computer. As of 2001, 93% of all business documents in the United States were created electronically and approximately one-third of all electronically stored data are never printed out, as reported by KrollOnTrack, eDiscovery: Market Statistics (November 13, 2002).

Companies have an obligation to preserve documentation that can support or defuse a claim and most companies back up their electronic records with taped or CD versions. Companies also use archival record services and many employees within the company maintain their own back-up materials via a diskette or CD. One of the provisions of the Sarbanes-Oxley Act of July 2002 imposed criminal penalties of publicly traded companies that destroy or alter corporate documents with the intent to obstruct or influence a U.S. investigatory agency investigation or action or a bankruptcy under Chapter 11. This serves as a warning to those who may be tempted to alter their electronic records for self-serving purposes such as staying out of jail.

Obviously, electronic data are discoverable. As far back as 1985, a court noted that electronically stored information was recoverable under Rule 34 of the Federal Rules of Civil Procedure Act.

What is the proper foundation for a computer record? Courts generally accept computer printouts as electronic evidence. The California Evidence Code, effective January 1, 1999, recognized that computer printouts are an accurate representation of the computer information or computer program it purports to represent.

Electronic mail, or e-mail, has become an everyday method of communication in the design and construction business and, as has been evident in several Wall Street prosecutions in 2004–2005, this form of communication can contain a wealth of information to help or hinder a dispute or claim. Authentication of e-mail can be achieved by one of the several identifying markers —the sender or receiver's name and the company's name or e-mail address.

This entire subject of electronic storage, transmission, and retrieval of current and archival information will be an ever evolving topic for the courts to rule on for decades to come.

Chapter

12

Design-Build Contracts

A contract must be fair to all parties and when preparing a contract for the design and construction of a building project, the contract provisions should reflect the good faith and fairness of all those parties. Writing a contract that includes reasonable terms and conditions may be one of the most difficult parts of the design-build process because this is where the first elements of good faith and fairness will manifest itself. Provisions that tend to tilt heavily toward owners or design-builders to the exclusion of others as well as provisions where one party objects to something vehemently, but another party insists it cannot be changed is certainly not starting off on the right foot. A wise owner once told the writer, "You don't need contracts for the good guys, but since you sometimes don't know if the person you are dealing with is a good guy or a bad guy, I guess you need a contract." That still sounds like a good rationale for writing contracts.

Professional associations such as the American Institute of Architects, the Construction Management Association of America, the Associated General Contractors of America, and the Engineer's Joint Contract Documents Committee have all prepared standard contracts, all of which appear to include the same basic issues facing owners, design consultants, and contractors in much the same manner.

These standard documents serve a very valuable purpose besides their obvious purpose for existing; they provide a basic framework that the entire industry seems to accept and live by, even with the reams of modifications and exhibits that most of these "standard forms" finally evolve into.

Design-build contracts include provisions for design and construction but they neither have to be combined in one document nor be executed at the same time. Depending on the owner's building program, they may elect to use "bridging" as a first step in defining their project's scope and budget or opt for just some preliminary design work, and a "ball park" budget for a project with a long pipeline.

Exhibits—When More Clarity Is Required

A standard contract form can't respond to all the intricacies of a construction project and the use of exhibits is the method most frequently used to "customize" the standard form.

Although lump sum or stipulated sum contracts are often used in design-build projects, the cost plus a fee with a Guaranteed Maximum Price (GMP) format is equally popular. The GMP allows the design-builder to project a total cost for the job on less-than-100% complete documents and often affords an owner an "upset" price while holding out the potential for reduced costs due to buy-outs, by the refinement of the plans and specifications or, by *value engineering* proposals as subcontractors are brought on board.

The GMP contract format leaves a few questions open that can best be answered by an exhibit or two; such as what is a complete list of reimbursable costs within the GMP context; what costs will not be reimbursed; how certain costs are to be calculated; and how change orders are to be prepared, presented, evaluated and accepted, or modified or rejected.

Reimbursable Costs

Although some reimbursable costs appear to need no explanation, i.e., cost of labor, materials, and equipment incorporated in a project, on further investigation, a clearer explanation of what reimbursable equipment "costs" are may prevent some future owner questions. For example, contractor owned equipment—how should this be billed? A standard contract clarification frequently used is:

> The rental rates for any piece of equipment from the contractor's stock is not to exceed 75% of the published AED rate or the rental rate prevailing for similar equipment in the geographic area where the project is being built, whichever is lower.

The rates charged for equipment rental will conform to the following hourly/daily rate restrictions:

1. Rental for 4 hours or more—the daily rental rate may be charged instead of the hourly rate for less than a full day's rental.
2. If the equipment is used for three days or more during one week, the weekly rate shall be charged instead of the daily rate.
3. If the equipment is used for three weeks or more during a one-month period, the monthly rate shall be charged instead of the weekly rate.

Rental rates from third parties will be billed at the invoice rate plus the designated design-builder's overhead and profit rates as specified in Article (X) of the contract for construction.

Now all parties understand the reimbursement cost structure for equipment rentals.

Reimbursable design-build personnel costs

Costs for design-build personnel can range from hourly rates for tradesmen such as laborers, carpenters employed by the design-builder for on-site, self-performed work to hourly rates for managers and professional staff. It is advisable to include those hourly rates in a separate contract exhibit, say Exhibit A—Reimbursable Personnel Costs. Each type of tradesmen contemplated should be listed by title (laborer, laborer foreman, carpenter, apprentice, journeyman, foreman, and so on) and their corresponding rate, exclusive of overhead and profit, but including all labor burden costs. A similar exhibit can be prepared for construction, architectural and engineering managers, and professional staff, or these rates can be incorporated in the same exhibit for tradesmen.

The GMP contract will include in its general conditions division those costs for the project manager, project superintendent, and possibly project engineer actively engaged in the project, which are defined by the contract scope along with a section devoted to inspections and testing. In the case of a design-build project, jobsite visits may be required by various design consultants to respond to subcontractor queries, inspect for compliance with the design specifications, and address quality issues.

It may be a good idea to include hourly rates for manager and professional-type individuals such as estimator, purchasing agent, senior architect, junior architect, draftsman, senior and junior engineer, and so forth in this same exhibit.

If during the course of construction, the owner is considering adding work that may require the services of an estimator or purchasing person, or a sketch or two, and this work is proposed on the basis of time and materials, the rates for the various personnel involved in this work will have already been established.

Costs Not to Be Reimbursed

In some instances, it might be better to also list those costs that are not to be reimbursed, such as the following:

- Salaries and other compensation of design-builder's personnel stationed at the design-builder's principal office or at offices other than the jobsite except as specifically permitted by the definition of personnel costs in Exhibit X.
- Expenses of the contractor's principal offices.
- Overhead or general expenses other than the percentage specifically referred to in the appropriate section of the contract. (This is not the same as the contract allowable overhead and profit, but this refers to general office overhead required to run the entire design-build firm.)
- Costs incurred due to the fault or negligence of the design-builder or any subcontractor, or anyone directly or indirectly employed by any of them, or for whose acts any of them may be liable, including costs for correcting damaged work, disposal and replacement of materials and equipment incorrectly

ordered or supplied, and making good for damage to property not forming part of the work.

- Rental costs for equipment furnished by the design-builder that exceeds the costs expressed in the section of the contract that limits the design-builder's equipment costs.
- The cost of any items not specifically and expressly identified in this agreement as a permitted part of the cost of work.

Note. This is a catch-all phrase and all parties should carefully review the entire list of reimbursable costs prior to contract signing so that if any costs appear as work progresses, and the question of whether or not to reimburse emerges—"Just read the contract and resolve that question.

Dealing with Allowances and Alternates

It is rare in a project that some allowances and alternates are not included. An owner unsure of the quality level of certain components, or a design-builder unable to establish a firm price for a specialized component of work may elect to treat that item of work as an "allowance," but unless the terms and conditions of that allowance are detailed elsewhere in the contract, some explanation might be included in a separate exhibit for allowances. A section of the contract might include these requirements.

1. All amounts charged against an allowance item shall be documented by the design-builder. If it appears that the work will exceed the allowance, the owner is to be notified when it becomes apparent that those costs will be exceeded. The design-builder shall not proceed with the work until they receive written authorization from the owner to do so. The allowance items included in Exhibit Y include the cost to the design-builder for materials, labor, and equipment delivered to the jobsite, and all required taxes, less applicable trade discounts. Costs for unloading and handling at the jobsite are included as well as a portion of the design-builder's general conditions.

Note. Some confusion will arise if the design-builder's overhead and profit application pertaining to an allowance is not spelled out. Some contracts state that the design-builder's overhead and profit is *included* in the allowance item; other contracts stipulate that the design-builder's overhead and profit is *excluded.* It is best to clarify this item in the section on allowances, if not specified elsewhere in the contract.

Not all allowances, when reconciled, equal the amount set forth in the contract; some may result in higher or lesser costs, thus some method of price adjustment may be needed in the contract. Typical language is:

If actual costs are less than the allowance. If on reconciliation of the actual cost of the work to the allowance the costs are less than the amount of the allowance, all such savings will accrue to the owner and shall reduce the cost of work and the GMP via the issuance of a change order.

If actual costs exceed the allowance. If the design-builder anticipates that the cost of the allowance will exceed the amount of that allowance item, the design-builder is to notify the owner in not less than three days after becoming aware of these additional costs. The design-builder is to obtain the owner's written authorization to proceed with this additional work before doing so. If the design-builder exceeds the allowance item without first obtaining the owner's approval through a change order, then the design-builder shall be deemed to have waived any claim for additional expenditures associated with that work.

Note. Although the design-builder may not be able to increase the GMP because of failure to obtain written authorization to do so, is this not a cost of work as defined by other articles in the contract? If so, these costs can be applied as the cost of work as defined in the contract documents against the GMP, but these costs cannot increase the GMP.

The Contingency Account

Design-build contracts often include a contingency and this contingency should be structured such that it is suitable for the sole use of the design-builder. Most owners, prior to entering into a construction contract will have established their own contingency account, fully aware that budgets can be affected by the many unknown costs in their proposed capital facility expansion program.

Owners often place restrictions on the use of a contractor's contingency account stating that these funds must be used for a specific purpose such as:

1. Unforeseen subsurface conditions
2. Concealed conditions
3. Costs incurred due to default or bankruptcy of a subcontractor
4. Unanticipated spikes in cost of materials

Many contractors and design-builders will object to these restrictions opting for the following provision:

Included in the GMP is a line item entitled "Contingency," which represents the design-builder's best estimate of additional costs that may be incurred in order to perform the work that was not reasonably inferred from the plans and specifications when the GMP was developed, but which would be required to complete the work as set forth in the contract documents.

The design-builder will account for and document all costs charged against the contingency line item. If all or a part of the design-builder's contingency charges are less than the line item amount, then any savings will revert to the owner in the same manner as other savings are applied against the GMP.

Alternates

This is sometimes referred to as the owner's wish list and is comprised of added work that may or may not be elected, but for which costs had been previously established in the contract and listed in a separate exhibit.

There again, the contract should contain language that clarifies what is included in the alternate, i.e., all costs plus a portion of the design-builder's general conditions and applicable overhead and profit percentage (if that is the case). A typical contract article dealing with alternates will be similar to that specified below:

> The GMP includes several alternates which are listed in Exhibit X. Each alternate represents an item or items that the owner may add to, delete from, or modify or substitute for another item or items of work in completing some portion of the work. The owner shall be the sole determiner of whether or not to elect to incorporate any or all alternates. All alternates, when scope and cost are accepted by the owner will be affected by change order specifying how that change order will effect the GMP (either increase, decrease, or have no effect).

Note. One important statement needs to be added to any listing of alternates and that deals with the timing of their acceptance, rejection, or modification. Obviously an alternate relating to the substitution of carpet for resilient flooring, as an example, does not need to be dealt with until construction is well underway. But an alternate to add a vending area on the third floor of an office building must be elected during a certain phase of the mechanical, electrical, plumbing (MEP) rough-ins. A decision to accept this option at a later date will result in additional costs for both the design-builder and the owner. So it is relatively simple to add a statement alongside each alternate: "Decision to elect this alternate must be made by the owner not later than (date established by referencing the contract schedule). Any decision to authorize this alternate after that time may result in increased costs."

Everything Is Included in the Article

There are several ways in which to define the scope of work included in a design-build contract; an extensive inclusion list—the items of work specifically included in the project and an extensive exclusion and qualification list. The purpose of both, as supplements to design development drawings, is simply to spell out, sometimes in extreme detail (App. 12.1), all items of work as recognized by both owner and design-builder. But some owners may wish to tighten the rope a little more by adding an article similar to this:

> The intent of the contract documents is to include all of the work required to complete the project except as specifically excluded. It is acknowledged that as of the date of the contract the plans and specifications are not complete, but define the scope and nature of the work and are sufficient to establish the contract sum. No adjustment shall be made in the contract sum if, as a prudent design-builder, design-builder should have been aware of or anticipated such additional work as may be required to produce a first-class (office building or whatever the nature of the project is).

The writer, at one point, was presented with an owner's contract draft that stated that no requests for contract scope and contract sum increases would be considered unless the owner's request resulted in enlarging the cubic footage of the building! Obviously this was a clause that was unreasonable and was therefore rejected.

Change Orders

It is rare that a project doesn't generate change orders. Although one advantage to owners considering design-build is its history of decreasing the potential for change orders, the opportunity still exists for owner requested improvements and enhancements. There is a widespread notion among project owners that contractors make bundles of money on change orders; but most contractors would be very happy never to see a change order.

Clarification of the change order process will lessen any owner concerns about what constitutes reasonable costs.

Inserting a specific procedure for the development and submission of a change order will establish the rules right upfront and avoid prolonged discussion if, and when, there is a need to prepare and issue a change order.

The design-builder must include sufficient information to allow an owner to fully comprehend the nature and related costs that constitute a change in scope. They should be able to review, evaluate, and comment promptly without the need to get additional backup information from a subcontractor or supplier; this is a prudent thing to do. This will greatly speed up the review process.

Subcontracted costs

First of all, the contract ought to include language that limits the subcontractors' overhead and profit as well as that of their second and third tier subcontractors and must include the same restrictions in the subcontractor agreements:

> For all subcontracted work, the subcontractor will be limited to no more than 15% representing their overhead and profit. Lower-tier subcontractors are allowed 10%; however, the overall allowable limits for overhead and profit submitted by a subcontractor cannot exceed 25% (or some other agreed upon upper limit on total overhead and profit).

The writer has seen change orders where a series of prime, second, and third tier subcontractors add as much as 60% to 75% to the cost of the work *before* the design-builder adds their fee. Owners reviewing these change order requests can get very upset and point fingers at the design-builder for being lax in their handling and preparation and subcontractor costs.

Those subcontractor proposals responding to a request for additional work should include enough detail to allow an owner to intelligently understand the nature and costs of this work. When a design-build submits insufficient information for an owner to review, they should rightfully be accused of not doing their job.

A subcontractor-submitted proposal ought to include the following:

1. Costs delineated by breakdowns for labor, materials, and equipment.
2. Labor costs should be further defined by listing the hourly rate and number of hours required for the work for each division of tradesmen.
3. Materials should be accompanied by a quote from a vendor or breakdown from the subcontractor's estimate.
4. Equipment should be listed at the appropriate rental rate.
5. A quantity definition, if applicable should be included, i.e., lineal or square footage of work, cubic yards of concrete or excavation, and number of pieces.
6. Overhead and profit should be listed separately, as should applicable taxes, permits, and insurance.

The design-builder, in preparing their Proposal Change Order (PCO) or Change Order Request (COR) to an owner needs to furnish the following additional information:

1. A brief statement explaining the nature of the change (Fig. 12.1 is a good example).
2. Attach all supporting documents, person initiating the change request (owner, design consultant, subcontractor, and building official); include a portion of any revised drawings reflecting the change or a narrative requesting change, including the scope of work.
3. If the scope of work has increased or decreased, state the condition prior to the requested change and the effect of the proposed change, i.e., if railings are added, state "Railings between Column 9 and 10 per Dwg A-5.6 measure 32 feet. Owner requests railing between Column 9 and 9.5 only, thereby reducing length approximately 16 feet."
4. Include subcontract proposal as defined above for content.
5. Equipment—indicate if design-builder owned or rented from third party, approximate number of hours/days required and applicable rates, idle and active rates, move-in and move-out costs, if applicable.
6. Include whether this work will increase, decrease, or have no effect on contract time.

If work is to be performed on the basis of time and materials (T&M) add the following:

1. The design-build supervisor is to obtain daily tickets for all T&M work self-performed to include worker's trade category, number of hours worked, and task performed. Ticket to be signed by the design-builder supervisor.
2. For all subcontracted work, daily tickets from the subcontractor listing tradesmen, number of hours of work for each, and task performed and signed by

CHANGE ORDER REQUEST
No. 00028

TITLE: Reroute Chilled Water Risers **DATE:** 4/7/2005

PROJECT: **JOB:**

TO: **CONTRACT No:** 00001-01

RE: **To:** **From:** **Number:**

DESCRIPTION OF PROPOSAL

FOR FINAL APPROVAL 6/18/05

Inc is pleased to submit for approval under REVISED WBI JM# 98.1 to reroute the CWS&R lines from the riser near the elevator as shown on drawing H-7 to near the 85 line as shown on the sketch SKH-7 received dated 2/11/05 and per the e-mail dated 2/10/05. No Filed Instruction was issued for this work. The total REVISED cost of this Change Order Request is for the amount of Nineteen Thousand Three Hundred and Fifty Six Dollars ($19,356.00). Attached please find Inc. Cost Summary Sheet dated 6/18/05 and the necessary attachments from each subcontractor involved to complete this scope of work.

Inc is pleased to submit for approval the Clubhouse Project Change Order Request #28/WBI JM# 98 to reroute the CWS&R lines from the riser near the elevator as shown on drawing H-7 to near the 85 line as shown on the sketch SKH-7 received dated 2/11/05 and per the e-mail dated 2/10/05. No Filed Instruction was issued for this work.

The additional pipe run was required because the location of the riser as shown on the contract drawings was determined unacceptable by Engineering and then explored other routing possibilities. This change request is for the additional 50' of pipe run that was required from where the riser was shown to where the riser was actually installed as directed. Has proceeded with this work in good faith to complete this work with the parameters of the project schedule.

The total cost of this Change Order Request is for Twenty One Thousand Four Hundred and Seventy Eight Dollars ($21,478.00). Attached please find Cost Summary Sheet dated 4/7/05 and the Engineering proposal. Also included is the WBI independent review of the Cox request.

Please review and sign this change order request so it can be included in the next WBI/Red Sox Change Order.

Unit Cost:	**$0.00**
Unit Tax:	**$0.00**
Lump Sum:	**$19,356.00**
Lump Tax:	**$0.00**
Total:	**#19,356.00**

APPROVAL:

By: ____________________ **By:** ____________________

Date: ____________________ **Date:** ____________________

Figure 12.1 Change Order Request (COR) containing brief explanation of change.

the subcontractor, foreman, or supervisor. This ticket is to be signed by the design-builder supervisor each day T&M work is performed.

3. Receiving receipts for all materials used for the extra work are to be signed by the design-build supervisory and must accompany the COR.
4. Receiving tickets for equipment and tickets when equipment is turned in or leaves site will be required, signed by the design-build supervisor acknowledging both active and idle equipment time and work tasks performed.

Winter Conditions

Another exhibit might be devoted to winter conditions for those design-builders working in a geographic area where freezing temperatures during winter months are fairly common. Because of its very nature, design-builders often exclude costs for winter conditions (temporary heat and temporary protection of surfaces) altogether or include an amount as an allowance item.

In either case to insure an owner where winter conditions apply, all such costs will need to be verified and a statement in the contract and a procedure for documenting these types of costs may forestall any disagreements later on when reimbursement for such costs are requested by the design-builder.

Documentation to support winter conditions costs is listed as follows:

1. Prepare daily tickets, supported by entries in the superintendent's daily log.
2. Indicate the operation taking place that requires winter conditions.
3. Provide a log with temperature readings taken at 7:00 a.m., noon, and at 2:00 p.m. and a brief description of weather conditions, such as, snow flurries, over cast, and the like.
4. Provide daily tickets for all labor employed in providing for winter conditions work and indicate actual task(s) being performed.
5. Provide a list of materials required and what they are being used for.
6. All such tickets should be signed by the site superintendent.
7. Submit weekly cost estimates to the owner to keep them apprised of the current costs and the projected costs, if they can be reasonably predicted.

An Owner Viewpoint about Design-Build Contract Provisions

Viewing a situation from the other party's perspective can often shed new light on the other's concerns, and design-builders doing so may better understand several owner concerns. The design-build process represents a departure from other forms of project delivery systems in both institutional and contractual variations and owners embarking on this type of project need to consult with their legal counsel and discuss how these changes will affect the contract they are going to enter into.

The institutional or cultural changes have mainly to do with the one-point responsibility and the shifting of the architect's role from one of an owner's ombudsman to one of a design-build team member. Although the roles of the architect and engineer in this form of project formulation and consummation are different from their roles in a design-bid-build project, their professionalism and integrity are very much on the line, and possibly more so. Architects or engineers will abrogate their responsibilities in dealing fairly and honestly with an owner while acting as a partner with the builder. No contractor looking to remain in business for any extended period of time will place their reputation on the line by doing anything unethical in this partnership with design consultants to bring a capital facility project on line.

We assume that owners, having selected design-build, either from a not-too-satisfactory experience with design-bid-build or having learned about the team approach are more receptive to testing this collaborative effort.

But that said, contracts must still be prepared, which hopefully will reduce the ambiguities that frequently lead to misunderstandings and not contain provisions that raise the hackles of the design-build team.

One of the concerns frequently voiced by an owner considering design-build has to do with the supposed lack of checks and balances that an independent architect provides in the more conventional design-bid-build approach. These concerns can be partially allayed by having the owner engage a consultant versed in both design and construction processes. The chapter on construction managers discusses the virtues of taking that owner's representative route. There are provisions or modifications that can be included in the owner/design builder contract which can reduce some of these concerns.

Who Owns the Drawings?

When a two-part contract such as the AIA A191—Standard Form of Agreement between Owner and Design-Builder is used, Part 1 deals primarily with design and Part 2 relates to the construction portion. But what if an owner is not comfortable with the design-build team during this phase and wishes to terminate the agreement, and let's say the termination is mutual; who owns the design development drawings produced to the point of termination?

Article 3 of AIA A191 entitled *Ownership and Use of Documents and Electronic Data* states that if Part 2 is not executed, the owner does not have the right to use the drawings or the specifications furnished by the design-builder without their written permission, even if they have paid the design-builder for those documents. This may seem unfair, but looking at it from the designer's standpoint, they don't want their partially completed drawings out there for someone else to complete or scavenge. Contract language may be able to assuage those owner concerns in several ways. These are listed as follows:

1. Article 3 can just be stricken from the agreement.
2. Change the language to read that if the owner pays the design-build team for the documents produced to that point of payment, the documents become the property of the owner.

3. Language can be added so that if Part 2 is not elected and the design-build team consists of a builder with a subcontract agreement with design consultants, the contract with the designers can be assigned to the owner.

Understanding Dispute and Claims Resolution Options

Article 6 in the A191 contract contains provisions to resolve disputes as most other professional organization forms do, and this article calls for mediation as a first step and binding arbitration as the next and final step in settling a dispute. Some contracts require mediation as a first step, followed by nonbinding arbitration as a second step, and litigation as the final step. An owner's lawyer may point out problems sometimes associated with binding arbitration, that arbitrators who are beholden to neither party tend to see faults in each party's arguments and split the claim, sometimes unevenly, not making an award solely to one party.

Mediation only works if both parties to the dispute are truly open to understanding each other's claim and are primarily interested in settling their differences and moving on. Anyone can leave a mediation proceeding at any time, so there is little to be lost by trying this first step in dispute resolution. The costs are not great and the time spent with the mediator may be no more than 4 to 5 hours.

Arbitration has its proponents and opponents. Initially conceived as a method of resolving disputes quickly by having a rather informal meeting, controlled by a panel versed in the topic to be discussed, lacking the formality and several restrictions of a legal hearing, such as strict rules of evidence, allowing hearsay evidence, these procedures have evolved into quasi-judicial hearings where each side brings their own team of lawyers to the fray and where proceedings can last for years.

An owner may decide to elect litigation if the dispute or claim exceeds a certain value. But with the cost of legal representation, including the discovery process and trial preparation expenses, the cost to litigate is often much more than the value of the claim.

The Schedule of Values Line Item Issue

There are many questions about the management of a GMP contract that some owners don't fully understand and a design-builder working with a first time GMP project owner may need to explain some of its intricacies. The matter of a reimbursable cost when a change order is rejected, as previously discussed, is just one. Another GMP principal that often leads to misunderstandings is the fallibility of line-item costs as displayed by the schedule of values, which is usually the basis for future requisitions. For example, if costs for masonry as the project proceeds exceeds the line-item value assigned in the schedule of values and duplicated in the requisition, can those excess costs be recouped, if by doing so they exceed 100% of their stated value?

Rarely will the final cost of any item of work match the value assigned to it at the beginning of the project and therefore an adjustment of actual versus proposed costs must be anticipated.

A typical article in the contract ought to read to the effect that overages in one line item can be applied against underages in another line item as long as the GMP contract sum is not affected. When the validity of this concept is fully understood by all, there should be no problem in using it.

The Importance of Substantial Completion

A milestone schedule will have been submitted prior to contract signing and when reviewed and approved by the owner, it becomes a contract requirement. Construction schedules by their very nature are fluid; some activities will take less time and some will require more time to complete; some late deliveries will not have the impact originally anticipated, but will have no net impact on timely completion. Increases in scope of work may impact the initial completion date and should be documented in the change order pertaining to that scope change. But completion is slightly different from substantial completion.

Article 4 of AIA A191 states that "substantial completion shall be achieved on or before the date established in Article 14" (other provisions).

The term substantial completion means that the building is ready for the use for which it was intended but this does not mean that the building will be 100% complete. Oftentimes, owners wish to begin some phased occupancy of their building to set up equipment, bring in furniture that has been stored elsewhere, and begin stocking supply cabinets. Substantial completion requires that the design-builder, besides completing the work to the point where the building can function as planned, must obtain a certificate of occupancy before an owner can actually occupy the building. Transient personnel bringing in equipment or stocking paper in the duplicating machine does not really constitute "occupancy" in the eyes of most building officials, but a design-builder has other concerns, liabilities, and property damage.

Both these issues can be dealt with by modifying the substantial occupancy portion of the contract.

A statement can be inserted, allowing the owner phased access to the building prior to the point where it reaches substantial completion. The design-builder is concerned about damage caused by the owner or their vendors. A moving company can badly scar an expensive stone floor in the vestibule, damage drywall partitions and painted surfaces, and so forth. A common approach to this kind of problem is for an owner to request access to a certain portion of the building, and if the design-builder agrees, representatives from both parties will walk that area, note any damaged surfaces or areas, note any incomplete work, and in general, agree on the condition of the space before turning it over to the owner, not for occupancy but to move in furniture, fixtures, or equipment. Any damages discovered after this move that were not noted in the turnover walk through will then become the responsibility of the owner to repair. As far as personal liability issues are concerned, both parties can develop and insert some form of "hold harmless" clause to deal with that issue. An owner should also realize what their responsibilities are upon signing a certificate of substantial completion. Owners need to consider that substantial completion means the building is ready for the use for which it was intended and that someone can function fairly well in an office with sufficient lighting and conditioned air but no carpet on the floor.

Although the design-builder may still be working in the building installing flooring or finishing the painting, all utility costs will be transferred from the design-builder to the owner at that point. Unless stated otherwise, the design-builder's insurance coverage will cease as will their responsibility for any equipment maintenance (warranty work is still in effect), and payment of retainage less any hold-backs for incomplete or nonconforming work will be requested from the owner.

The Standard Form Contracts

Several professional trade associations and organizations have created standard contract forms, some specifically for design-build work and others that are generic and can be used in that fashion. These "standard" forms provide the framework for modifying individual project contracts and incorporate the specific concerns of the individuals who created them.

The American Institute of Architects contracts

The American Institute of Architects (AIA) publishes some of the most widely used construction contract forms in the country and they have prepared contract forms specifically for design-build work. In July 2004, a new family of design-build documents was introduced by the AIA. Critics of existing documents voiced concerns about previous AIA design-build documents and wanted the AIA to respond to these concerns.

- Owners and design-builders wanted only one contract and not two like the previous A191 document.
- Owners did not want to pay for a preliminary design and then have to pay an unacceptably high price to be able to use that design.
- Some owners were able to develop their own project criteria agenda or hire their own consultants for that purpose and therefore a separate preliminary phase type document was no longer a necessity.
- Some owners expressed a desire to bid these design-build projects competitively and needed a document that fit that need.

The following new documents were created to meet these owner demands:

AIA B142-2004—Standard Form of Agreement Between Owner and Consultant Where the Owner is Contemplating the Design-Build Method of Project Delivery. Using this document an owner can retain a consultant to establish their design-build program and the owner, via this document, has the irrevocable license to use the intellectual property created by the consultant.

AIA A141-2004—Standard Form of Agreement Between Owner and Design-Builder. A two-part agreement is no longer necessary when using this form and allows for either a fixed-price contract or a GMP-type contract. This contract format places stricter responsibility on the part of the design-builder in that the design-builder can't hide behind the "professional standard of care" provision as to the adequacy of the design. This strict liability approach distinguishes this new document from the other organization's contracts. This document, however, places more responsibility on the owner in that it requires them to review and approve the criteria package prepared by their consultants and then approve conformance of that criteria package to the final design created by the design-builder.

AIA A142-2004—Standard Form of Agreement Between Design-Builder and Contractor. This contract allows the design-builder to choose to work with only one prime contractor or retain subcontractors as separate primes.

AIA B143-2004—Standard Form of Agreement Between Design-Builder and Architect. This contract form contains an exhibit, Exhibit B that is a checklist of services from which the parties can select all or only some of the normal architectural services. This document also allows the design-builder to select those construction services that are to be provided by the architect.

One of the more significant changes in these 2004 edition documents deals with dispute resolution. In the owner/design-builder agreement there are provisions to employ the use of a "neutral" to be the initial arbiter of disputes on a project. This neutral will be jointly selected and that individual will serve in a capacity similar to that of the architect in A201—General Conditions—Article 4—Administration of the Contract, which can be extensive if fully applied. These services will also add costs to the project.

The older editions of AIA design-build projects that may still be in effect are as follows:

AIA A191—Contract Between Owner and Design-Builder. The two-part contract that was a standard in the industry.

AIA A491—Contract Between Design-Build and Contractor. A form that will have application by an architect planning to contract with a contractor to form an architect-led design–build team.

AIA B901—Contract Between Design-Builder and Architect. This can be used as the basis of an agreement between a builder and an architect desirous of creating a contractor-led design-build team.

AIA C801—Joint Venture Agreement. To be used when a joint venture (JV) between a contractor and architect/engineer is being considered for purposes of creating design-build work.

Construction Management contracts

The Construction Management Association of America (CMAA) has developed several construction management type contracts that are used for design-build work. Some of these forms are listed in Chap. 7, all 2005 editions, and they include the following:

CM A-1—Standard Form of Agreement Between Owner and Construction Manager (Construction Manager as Owner's Agent)

CM A-2—Standard Form of Contract Between Owner and Contractor

CM A-3—General Conditions of the Construction Contract

CM A-4—Standard Form of Agreement Between Owner and Designer

The Associated General Contractors of America contracts

The Associated General Contractors (AGC) of America, basically a builder's organization, developed a number of well-thought-out contracts specifically for the design-build process.

AGC Standard Form of Teaming Agreement for Design-Build Project (App. 12.1) is a model document for establishing the relationship between contractor and architect/engineer in that it provides the matrix for defining the relationship and responsibilities of all parties by filling in the blanks and adding complete documents as part of its Article 6.

AGC Document 410—Standard Form of Design-Build Agreement Between Owner and Design-Builder (Where the Basis of Payment is the Cost of the Work Plus a Fee with a GMP) – App. 12.2. This contract format spells out, in simple language, the obligations and responsibilities of the design-builder and the owner.

Figure 12.2 is a checklist of owner and design-builder responsibilities and identifies the specific article in the AGC 410 document where each of these topics is listed. This is a convenient form to use when finalizing contract terms and conditions with the owner and can act as a handy checklist when any of those events subsequently require notifying the owner or reminding the owner of some items that need to be attended to.

Figure 12.3, developed as an exhibit to the contract, is another well-thought-out document that sets forth a dispute resolution menu on which the parties to the contract can elect any of the five following methods to resolve disputes:

1. Create a Dispute Resolution Board (DRB).
2. Submit to advisory arbitration.
3. Conduct a mini-trial.
4. Agree to binding arbitration.
5. Consider litigation as a course of action.

POST AGREEMENT SUBMITTALS AND ADMINISTRATIVE OBLIGATIONS AGC 410

Paragraph	Responsibility		Task (asterisk indicates task is optional)	Completed Task
	Owner	Design-Builder		
3.1.1		X	Preliminary Evaluation of Project Feasibility	
3.1.2		X	Preliminary Schedule of Work	
	X		Approve milestone dates in schedule	
		X	Recommend schedule adjustments, if needed	
3.1.3		X	Preliminary Estimate; adjustments, if needed	
3.1.4		X	Schematic Design Documents	
	X		Approve Schematic Design Documents	
		X	Identify material deviations in Sch. Design Docs, if any	
3.1.5		X	Obtain Planning Permits, if needed	
3.1.6		X	Design Development Documents	
	X		Approve Design Development Documents	
		X	Identify material deviations from earlier documents, if any	
		X	Update schedule and estimate	
3.1.7		X	Construction Documents	
	X		Approve Construction Documents	
		X	Identify material deviations from Design Dev. Docs., if any	
3.1.8.5		X	Obtain from A/E & Subs: Property rights and rights of use	
3.2.1		X	GMP Proposal	
3.2.3	X		Written comments, if any	
3.2.4	X		Accept GMP Proposal, if appropriate	
3.2.6	X		Authorize Pre-GMP work*	
3.3.1	X		Notice to Proceed	
		X	Pre-GMP list of documents applicable to authorized work, if any	
	X		Approve pre-GMP list & incorporate in Notice to Proceed	
3.3.3		X	Give all legally required notices	
3.3.4		X	Obtain building permits	
3.3.5		X	Keep detailed accounts & preserve for 3 years	
3.3.6		X	Provide periodic written reports	
3.3.7		X	Monitor actual costs; report at agreeable intervals	
3.3.9		X	Final marked up as-built drawings	
3.4		X	Prepare Schedule of Work	
	X		Approve Schedule of Work	
3.5.3		X	Identify safety representative	
		X	Immediately provide written accident reports, if any	
3.5.4		X	Provide copies of legally-required notices	
3.6.2		X	Report Hazardous Materials, if any, to owner/government	
3.6.4	X		Retain independent testing lab re: hazardous materials	
3.8.3		X	Obtain Certificates of Inspection, testing or approval	
3.8.4		X	Deliver warranties and manuals to Owner	
3.10	X	X	Define Additional Services before performed	
4.1.2.1	X		Info. re: physical characteristics of site	
4.1.2.2	X		Inspection & testing services during construction	

AGC DOCUMENT NO. 410 • STANDARD FORM OF DESIGN-BUILD AGREEMENT AND GENERAL CONDITIONS BETWEEN OWNER AND DESIGN-BUILDER (Where the Basis of Payment is the Cost of the Work Plus a Fee with a Guaranteed Maximum Price)

Figure 12.2 AGC checklist of owner, design-builder responsibilities.

Paragraph	Responsibility		Task (asterisk indicates task is optional)	Completed Task
	Owner	Design-Builder		
4.1.3	X		Evidence of sufficient funds committed to Project	
		X	Notice/stop work without evidence of funds*	
4.2.1	X		Provide Owner's Program	
4.3.2	X		Prompt written notice of defects/errors in requirements or Work	
4.4.3	X		Notify of Changes in Owner's Representative	
5.1	X		Propose subcontractors*	
5.3		X	In subcontracts, provide assignability	
5.4		X	Bind all subs and suppliers to this Agreement	
6.2.2	X		Direct commencement of work before insurance is effective*	
7.1.5		X	Application for Payment (Design Phase)	
	X		Within 15 days, accept or reject Application for Payment	
	X		Pay accepted amounts within 15 days of acceptance of Application	
7.1.6		X	Notice to Stop Work if Owner doesn't pay*	
8.2.10	X		Approve insurance and surety bonds	
9.1.1	X	X	Execute written Change Orders	
9.2.1	X		Issue Work Change Directives*	
9.2.2		X	Submit costs for work per Work Change Directive	
9.3.2		X	Promptly give written notice of minor changes made	
9.4		X	Within 21 days of occurrence, give notice of unknown conditions	
9.5.1.4		X	Maintain itemized account of expenses & savings	
9.6		X	Within 21 days of occurrence, make claim for add'l cost or time	
		X	Within 21 days, make claim for design costs of work not pursued	
10.1.1		X	Application for Progress Payments	
		X	Submit bills of sale and applicable insurance	
		X	Statement of funds disbursed from last payment	
10.1.2	X		Within 10 days, accept or reject App. for Payment (w/reasons)	
	X		Pay accepted amount within 15 days of acceptance of App. for Pymt.	
10.1.3		X	Notice to Stop Work if Owner doesn't pay when due*	
10.3	X		Written reasons for disapproving Application for Payment	
10.4.1	X	X	Certificate of Substantial Completion	
10.5.1	X		Before final payment, request proof bills paid*	
11.2.1		X	Obtain and maintain required insurance	
11.2.4		X	File Certificates of Insurance with Owner before Commencement	
11.4	X		Obtain Liability insurance and provide Certificate of Insurance	
11.5.1	X		Obtain and maintain Property Insurance	
11.5.5		X	Require Owner to provide copy of property insurance policies*	
	X		Give written notice before work if Owner will not have prop. insurance	
11.7.1		X	In subcontracts, require subrogation waivers	
12.1.1	X		Order suspension for convenience*	
12.2.1	X		Undertake work upon 5 days notice if non-performance*	
12.2.2	X		Terminate contract on 5 additional days notice*	
12.4.1		X	Terminate for cause on 5 days notice*	
12.4.2		X	Give 5 day notice and terminate for nonpayment*	
13.2	either	either	File mediation request*	
13.4		X	In subcontracts, require consolidation of cases	
Exhibit 1	either	either	File arbitration demand within reasonable time, if an option*	

AGC DOCUMENT NO. 410 • STANDARD FORM OF DESIGN-BUILD AGREEMENT AND GENERAL CONDITIONS BETWEEN OWNER AND DESIGN-BUILDER (Where the Basis of Payment is the Cost of the Work Plus a Fee with a Guaranteed Maximum Price)

Figure 12.2 *(Continued)*

dated ______________________________ ♦

AGC DOCUMENT NO. 410
STANDARD FORM OF DESIGN-BUILD AGREEMENT AND GENERAL CONDITIONS BETWEEN OWNER AND DESIGN-BUILDER
(Where the Basis of Payment is the Cost of the Work Plus a Fee with a Guaranteed Maximum Price)

DISPUTE RESOLUTION MENU

Pursuant to Paragraph 13.3, if neither direct discussions nor mediation successfully resolve the dispute, the parties agree that the following shall be used to resolve the dispute.

(Check the appropriate selection(s). These procedures can be used singularly, or progressively, as agreed to by the parties.)

___ **Dispute Review Board** The Dispute Review Board is composed of one member selected by the Owner, one selected by the Design-Builder, and a third member selected by the Owner and Design-Builder selected members. This Board shall be selected by the time construction commences, shall meet periodically, and shall make advisory decisions which may be introduced into evidence at any subsequent dispute resolution process. If a Dispute Review Board is selected, it is understood its review will precede mediation. ♦

___ **Advisory Arbitration** Advisory Arbitration shall be pursuant to the Construction Industry Rules of the American Arbitration Association. ♦

___ **Mini Trial** Each party, in the presence of senior management, shall submit its position to a mutually selected individual who shall make a non-binding recommendation to the parties. Such advisory decision may be introduced into evidence at any subsequent dispute resolution process.

___ **Binding Arbitration** Binding Arbitration shall be pursuant to the Construction Industry Rules of the American Arbitration Association unless the parties mutually agree otherwise. A written demand for arbitration shall be filed with the American Arbitration Association and the other party to the Agreement within a reasonable time after the dispute or claim has arisen, but in no event after the applicable statute of limitations for a legal or equitable proceeding would have run. The location of the arbitration proceedings shall be at the office of the American Arbitration Association nearest the Project, unless the parties agree otherwise. The arbitration award shall be final. Notwithstanding Paragraph 14.2, this agreement to arbitrate shall be governed by the Federal Arbitration Act and judgment upon the award may be confirmed in any court having jurisdiction. ♦

___ **Litigation** Action may be filed in the appropriate state or federal court located in the jurisdiction in which the Project is located. ♦

AGC DOCUMENT NO. 410 • STANDARD FORM OF DESIGN-BUILD AGREEMENT AND GENERAL CONDITIONS BETWEEN OWNER AND DESIGN-BUILDER (Where the Basis of Payment is the Cost of the Work Plus a Fee with a Guaranteed Maximum Price)

Figure 12.3 Contract exhibit listing various dispute resolution options.

The Engineers Joint Contract Documents Committee

Engineering-driven design-build projects need their own form of contract, and the Engineers Joint Contract Documents Committee (EJCDC) has a full range listed as follows:

1. Owner-consultant agreement for owners that require an independent planning service or administrative assistance.
2. Agreement for preliminary services where it may be premature to commit to a full-blown design-build program.
3. Documents that provide guidance in preparing Requests For Proposals (RFPs).
4. Standard general conditions for design-build services.
5. Standard general conditions for construction services if provided by a subcontractor to a design-builder.
6. Performance and payment bonds specifically for use in design-build projects:

   D-500—Standard Form of Agreement Between Owner and Owner's Consultant for Professional Services on Design-Build, 2002 Edition

   D-505—Standard Form of Subagreement Between Design-Builder and Engineer for Professional Services, 2002 Edition

   D-510—Standard Form of Agreement Between Owner and Design-Builder for Preliminary Services, 2002 Edition

   D-520—Suggest Form of Agreement Between Owner and Design-Builder on the Basis of a Stipulated Price, 2002 Edition

   D-525—Suggest Form of Agreement Between Owner and Design-Build on the Basis of Cost Plus, 2002 Edition

   D-700—Standard General Conditions of the Contract Between Owner and Design-Builder, 2002 Edition

Typical steps to follow when considering EJCDC design-build documents are as follows:

- Evaluate the design-build concept and compare with other project delivery options.
- Evaluate the capabilities, professional and administrative, of in-house personnel to determine whether outside consultants may be required.
- Confirm that these EJCDC prepared documents are the appropriate ones for the project being considered.
- Define, evaluate, and review the level of detail that will be required in the design development or conceptual design documents.
- Develop the performance parameters such as time and cost.
- Consider whether a limited preliminary agreement with the design-builder is the appropriate step or another form of agreement is more appropriate.

- After award, develop the necessary conceptual documents that will be a precursor to an RFP.
- Develop and issue the RFP.
- Prequalify proposers.
- Commence the review, evaluation, and award processes.

The posting in bold letters on the face of AIA contract forms should be heeded by all who enter into a contract: "this document has important legal consequences, consulting with an attorney is encouraged with respect to its use, completion, or modification."

Appendix 12.1: Teaming Agreement prepared by the Associated General Contractors of America (AGC)

THE ASSOCIATED GENERAL CONTRACTORS OF AMERICA

AGC DOCUMENT NO. 499
STANDARD FORM OF TEAMING AGREEMENT FOR DESIGN-BUILD PROJECT

This Agreement is made this ________________ day of ______________________ in the year __________, ♦

by and between

TEAM LEADER ____ ♦
(Name and Address)

and **TEAM MEMBER** ____ ♦
(Name and Address)

and **TEAM MEMBER** (if applicable) ____ ♦
(Name and Address)

and **TEAM MEMBER** (if applicable) ____ ♦
(Name and Address)

the parties collectively referred to as the **TEAM** for services in connection with the following **PROJECT**

____ ♦
(Name, Location and Brief Description)

for **OWNER** ____ ♦
(Name and Address)

ARTICLE 1

TEAM RELATIONSHIP AND RESPONSIBILITIES

1.1 This Agreement shall define the respective responsibilities of the Team Members for the preparation of responses to the Owner's request for qualifications and request for proposals for the Project. Each Team Member agrees to proceed with this Agreement on the basis of mutual trust, good faith and fair dealing and to use its best efforts in the preparation of the statement of qualifications and proposal for the Project, as required by the Owner, and any contract arising from the proposal.

1.2 The Team Leader, ____________________ ◆
____________________,
shall provide overall direction and leadership for the Team and be the conduit for all communication with the Owner. In addition the Team Leader shall provide expertise in the areas of (a) construction management and construction; (b) the procurement of equipment, materials and supplies; (c) the coordination and tracking of equipment and materials shipping and receiving; (d) construction scheduling, budgeting and materials tracking; and (e) administrative support. The Team Leader's representative shall be: __________ ◆
____________________.

1.3 The principal design professional is Team Member,
____________________ ◆
who shall perform the following design and engineering services required for the Project: __________ ◆
____
____
____
____
____
____________________.

In addition this Team Member shall coordinate the design activities of the remaining design professionals, if any. This Team Member's representative shall be: __________ ◆
____
____
____________________.

1.4 Team Member, ____________________ ◆
____________________,
shall provide expertise in the following areas:
____________________ ◆
____
____
____
____
____
____
____________________.

This Team Member's representative shall be: __________ ◆
____
____________________.

1.5 Team Member, ____________________ ◆
____________________,
shall provide expertise in the following areas:
____________________ ◆
____
____
____
____
____
____
____________________.

This Team Member's representative shall be: __________ ◆
____
____________________.

1.6 Each Team Member shall be responsible for its own costs and expenses incurred in the preparation of materials for the statement of qualifications and the proposal and in the negotiation of any contracts arising from the proposal, except as specifically described herein:
____________________ ◆
____
____
____
____
____
____
____
____
____
____
____________________.

Any stipends provided by the Owner to the Team shall be shared on the following basis:
____________________ ◆
____
____
____
____
____
____
____
____
____________________.

1.7 **EXCLUSIVITY** No Team Member shall participate in Owner's selection process except as a member of the Team, or participate in the submission of a competing statement of qualifications or proposal, except as otherwise mutually agreed by all Team Members.

ARTICLE 2

STATEMENT OF QUALIFICATIONS AND PROPOSAL

2.1 The Team Members shall use their best efforts to prepare a statement of qualifications in response to the request of the Owner. Each Team Member shall submit to the Team Leader appropriate data and information concerning its area or areas of professional expertise. Each Team Member shall make available appropriate and qualified personnel to work on its portion of the statement of qualifications in the time frame proscribed, and shall provide reasonable assistance to the Team Leader in preparation of the statement of qualifications.

2.2 The Team Leader shall integrate the information provided by the Team Members, prepare the statement of qualifications and submit it to the Owner. The Team Leader has responsibility for the form and content of the statement of qualifications and agrees to consult with each Team Member, before submission to the Owner, on all matters concerning such Team Member's area of professional expertise. The Team Leader shall represent accurately the qualifications and professional expertise of each Team Member as stated in the submitted materials.

2.3 If requested by the Owner, the Team Members shall prepare and submit a proposal for the Project to the Owner. Each Team Member shall support the Team Leader with a level of effort and personnel, licensed as required by law, sufficient to complete and submit the proposal in the time frame allowed by the Owner. A clear and concise statement of the division of responsibilities between the Team Members will be prepared by the Team Leader. The Team Leader shall make all final determinations as to the form and content of the proposal. The Team Leader shall use its best efforts, after the Team has qualified for the Project, to obtain the contract award, and each Team Member shall assist in such efforts as the Team Leader may reasonably request.

ARTICLE 3

CONFIDENTIAL INFORMATION

3.1 The Team Members may receive from one another Confidential Information, including proprietary information, as is necessary to prepare the statement of qualifications and the proposal. Confidential Information shall be designated as such in writing by the Team Member supplying such information. If required by the Team Member supplying the Confidential Information, a Team Member receiving such information shall execute an appropriate confidentiality agreement. A Team Member receiving Confidential Information shall not use such information or disclose it to third parties except as is consistent with the terms of any executed confidentiality agreement and for the purposes of preparing the statement of qualifications, the proposal and in performing any contract awarded to the Team as a result of the proposal, or as required by law. Unless otherwise provided by the terms of an executed confidentiality agreement, if a contract is not awarded to the Team or upon the termination or completion of an contract awarded to the Team, each Team Member will return any Confidential Information supplied to it.

ARTICLE 4

OWNERSHIP OF DOCUMENTS

4.1 Each Team Member shall retain ownership of property rights, including copyrights, to all documents, drawings, specifications, electronic data and information prepared, provided or procured by it in furtherance of this Agreement or any contract awarded as a result of a successful proposal. In the event the Owner chooses to award a contract to the Team Leader on the condition that a Team Member not be involved in the Project, that Team Member shall transfer in writing to the Team Leader, upon the payment of an amount to be negotiated by the parties in good faith, ownership of the property rights, except copyright, of all documents, drawings, specifications, electronic data and information prepared, provided or procured by the Team Member pursuant to this Agreement and shall grant to the Team Leader a license for this Project alone, in accordance with Paragraph 4.2.

4.2 The Team Leader may use, reproduce and make derivative works from such documents in the performance of any contract. The Team Leader's use of such documents shall be at the Team Leader's sole risk, except that the Team Member shall be obligated to indemnify the Team Leader for any claims of royalty, patent or copyright infringement arising out of the selection of any patented or copyrighted materials, methods or systems by the Team Member.

ARTICLE 5

POST AWARD CONSIDERATIONS

5.1 Following notice from the Owner that the Team has been awarded a contract, the Team Leader shall prepare and submit to the Team Members a proposal for a Project-specific agreement of association among them. (Such agreement may take the form of a design-builder/subcontractor agreement, a joint venture agreement, a limited partnership agreement or an operating agreement for a limited liability company.) The Team Members shall negotiate in good faith such Project-specific agreement of association so that a written agreement may be executed by the Team Members on a schedule as determined by the Team Leader or by the Owner, if required by the request for proposal. The Team Leader shall use its best efforts, with the cooperation of all Team Members, to negotiate and achieve a written contract with the Owner for the Project.

ARTICLE 6

OTHER PROVISIONS ♦

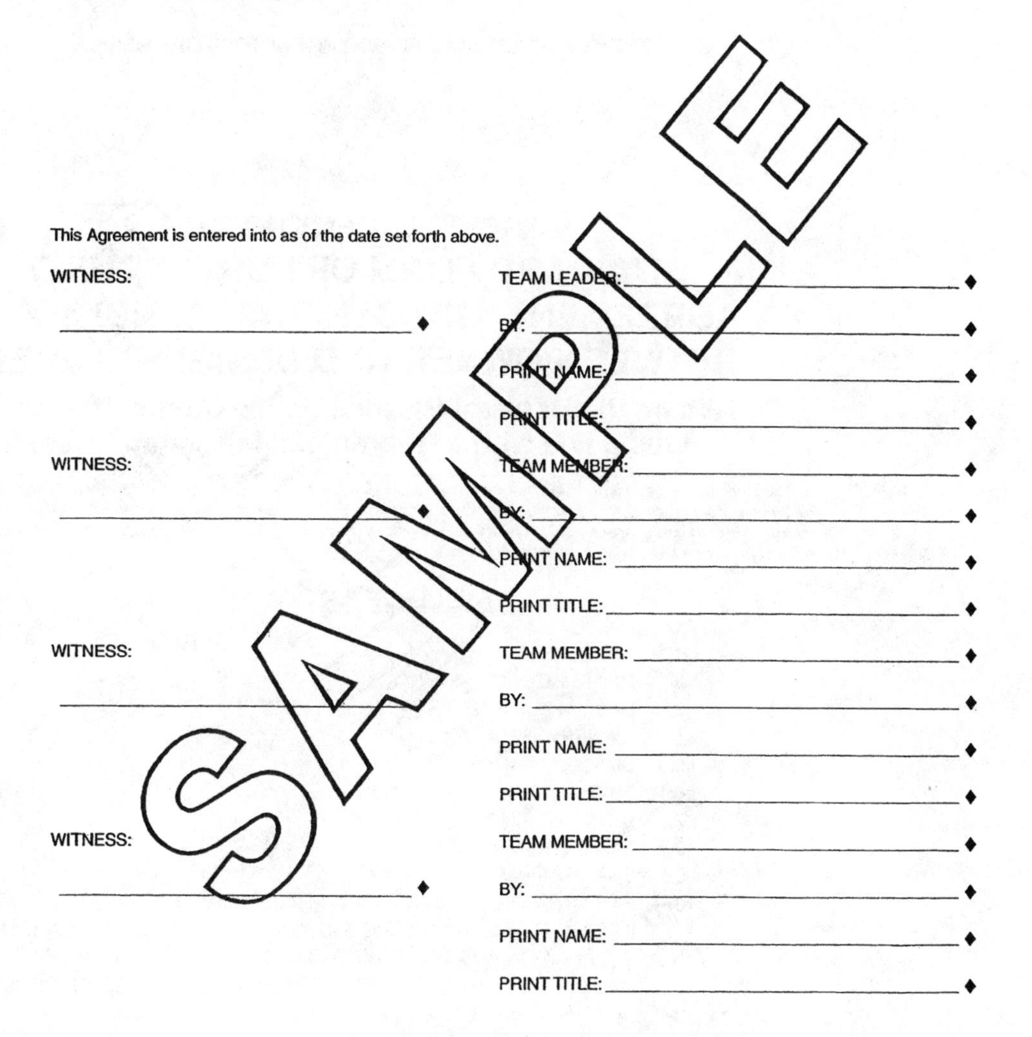

This Agreement is entered into as of the date set forth above.

WITNESS:

________________________ ♦

TEAM LEADER: ________________________ ♦

BY: ________________________ ♦

PRINT NAME: ________________________ ♦

PRINT TITLE: ________________________ ♦

WITNESS:

________________________ ♦

TEAM MEMBER: ________________________ ♦

BY: ________________________ ♦

PRINT NAME: ________________________ ♦

PRINT TITLE: ________________________ ♦

WITNESS:

____

TEAM MEMBER: ________________________ ♦

BY: ________________________ ♦

PRINT NAME: ________________________ ♦

PRINT TITLE: ________________________ ♦

WITNESS:

________________________ ♦

TEAM MEMBER: ________________________ ♦

BY: ________________________ ♦

PRINT NAME: ________________________ ♦

PRINT TITLE: ________________________ ♦

Appendix 12.2: AGC Document 410—Standard Form of Design-Build Agreement between Owner and Design-Builder

THE ASSOCIATED GENERAL CONTRACTORS OF AMERICA

AGC DOCUMENT NO. 410
STANDARD FORM OF DESIGN-BUILD AGREEMENT AND GENERAL CONDITIONS BETWEEN OWNER AND DESIGN-BUILDER
(Where the Basis of Payment is the Cost of the Work Plus a Fee with a Guaranteed Maximum Price)

This standard form agreement was developed with the advice and cooperation of the AGC Private Industry Advisory Council, a number of Fortune 500 owners' design and construction managers who have been meeting with AGC contractors to discuss issues of mutual concern. AGC gratefully acknowledges the contributions of these owners' staff who participated in this effort to produce a basic agreement for construction.

TABLE OF ARTICLES

This Agreement has important legal and insurance consequences. Consultation with an attorney and insurance consultant is encouraged with respect to its completion or modification.

AMENDMENT NO. 1
TO

AGC DOCUMENT NO. 410
STANDARD FORM OF DESIGN-BUILD AGREEMENT AND GENERAL CONDITIONS BETWEEN OWNER AND DESIGN-BUILDER
(Where the Basis of Payment is the Cost of the Work Plus a Fee with a Guaranteed Maximum Price)

Pursuant to Paragraph 3.2 of the Agreement dated ________________________________ ♦
between the Owner, ________________________________ ♦
and the Design-Builder, ________________________________ ♦
for ________________________________ (the Project), ♦
the Owner and the Design-Builder desire to establish a Guaranteed Maximum Price ("GMP") for the Work. Therefore, the Owner and the Design-Builder agree as follows:

ARTICLE 1

GUARANTEED MAXIMUM PRICE

The Design-Builder's GMP for the Work, including the Cost of the Work as defined in Article 8 and the Design-Builder's Fee as set forth in Paragraph 7.3, is ________________________________ ♦
Dollars ($ ________________________________) ♦

The GMP is for the performance of the Work in accordance with the documents listed below, which are part of the Agreement.

APPENDIX ___ Drawings and Specifications, including Addenda, if any, dated ________________, ________________ pages. ♦

APPENDIX ___ Allowance Items, dated ________________, ________________ pages. ♦

APPENDIX ___ Assumptions and Clarifications, dated ________________, ________________ pages. ♦

APPENDIX ___ Schedule of Work, dated ________________, ________________ pages. ♦

APPENDIX ___ Alternate Prices, dated ________________, ________________ pages. ♦

APPENDIX ___ Unit Prices, dated ________________, ________________ pages. ♦

APPENDIX ___ Additional Services included, dated ________________, ________________ pages. ♦

AGC DOCUMENT NO. 410
STANDARD FORM OF DESIGN-BUILD
AGREEMENT AND GENERAL CONDITIONS
BETWEEN OWNER AND DESIGN-BUILDER
(Where the Basis of Payment is the Cost of the Work
Plus a Fee with a Guaranteed Maximum Price)

ARTICLE 1

AGREEMENT

This Agreement is made this ______________ day of ______________ in the year __________, ◆

by and between the

OWNER
(Name and Address) ◆

and the
DESIGN-BUILDER
(Name and Address) ◆

for services in connection with the following
PROJECT
(Name, location and brief description) ◆

Notice to the parties shall be given at the above addresses.

ARTICLE 2

DATE OF SUBSTANTIAL COMPLETION

The Date of Substantial Completion of the Work is ____. ◆

ARTICLE 3

DATE OF FINAL COMPLETION

The Date of Final Completion of the Work is: ____ ◆
or within ________________________ (_______) days after the Date of Substantial Completion, ◆
subject to adjustments as provided for in the Contract Documents.

This Amendment is entered into as of ____. ◆

OWNER: ______________________________ ◆

ATTEST: ______________________________ ◆ BY: ______________________________ ◆

PRINT NAME: ______________________________ ◆

PRINT TITLE: ______________________________ ◆

DESIGN-BUILDER: ______________________________ ◆

ATTEST: ______________________________ ◆ BY: ______________________________ ◆

PRINT NAME: ______________________________ ◆

PRINT TITLE: ______________________________ ◆

ARTICLE 2

GENERAL PROVISIONS

2.1 TEAM RELATIONSHIP The Owner and the Design-Builder agree to proceed with the Project on the basis of trust, good faith and fair dealing and shall take all actions reasonably necessary to perform this Agreement in an economical and timely manner, including consideration of design modifications and alternative materials or equipment that will permit the Work to be constructed within the Guaranteed Maximum Price (GMP) and by the Dates of Substantial Completion and Final Completion if they are established by Amendment No. 1. The Design-Builder agrees to procure or furnish, as permitted by the law of the state where the project is located, the design phase services and construction phase services as set forth below.

2.1.1 The Design-Builder represents that it is an independent contractor and that it is familiar with the type of work it is undertaking.

2.1.2 Neither the Design-Builder nor any of its agents or employees shall act on behalf of or in the name of the Owner unless authorized in writing by the Owner's Representative.

2.2 ARCHITECT/ENGINEER Architectural and engineering services shall be procured from licensed independent design professionals retained by the Design-Builder or furnished by licensed employees of the Design-Builder, or as permitted by the law of the state where the Project is located. The standard of care for architectural and engineering services performed under this Agreement shall be the care and skill ordinarily used by members of the architectural and engineering professions practicing under similar conditions at the same time and locality. The person or entity providing architectural and engineering services shall be referred to as the Architect/Engineer. If the Architect/Engineer is an independent design professional, the architectural and engineering services shall be procured pursuant to a separate agreement between the Design-Builder and the Architect/Engineer. The Architect/Engineer for the Project is

____ .

2.3 EXTENT OF AGREEMENT This Agreement is solely for the benefit of the parties, represents the entire and integrated agreement between the parties, and supersedes all prior negotiations, representations or agreements, either written or oral. The Owner and the Design-Builder agree to look solely to each other with respect to the performance of the Agreement. The Agreement and each and every provision is for the exclusive benefit of the Owner and the Design-Builder and not for the benefit of any third party nor any third party beneficiary, except to the extent expressly provided in the Agreement.

2.4 DEFINITIONS

.1 The *Contract Documents* consist of:

a. Change Orders and written amendments to this Agreement including exhibits and appendices, signed by both the Owner and the Design-Builder, including Amendment No. 1 if executed;

b. this Agreement except for the existing Contract Documents set forth in item e. below;

c. the most current documents approved by the Owner pursuant to Subparagraph 3.1.4, 3.1.6 or 3.1.7 ;

d. the information provided by the Owner pursuant to Clause 4.1.2.1;

e. the Contract Documents in existence at the time of execution of this Agreement which are set forth in Article 15;

f. the Owner's Program provided pursuant to Subparagraph 4.1.1;

In case of any inconsistency, conflict or ambiguity among the Contract Documents, the documents shall govern in the order in which they are listed above.

.2 The term *day* shall mean calendar day, unless otherwise specifically defined.

.3 *Design-Builder's Fee* means the compensation paid to the Design-Builder for salaries and other mandatory or customary compensation of the Design-Builder's employees at its principal and branch offices except employees listed in Subparagraph 8.2.2, general and administrative expenses of the Design-Builder's principal and branch offices other than the field office, and the Design-Builder's capital expenses, including interest on the Design-Builder's capital employed for the Work, and profit.

.4 *Defective Work* is any portion of the Work not in conformance with the Contract Documents as more fully described in Paragraph 3.8.

.5 The term *Fast-track* means accelerated scheduling which involves commencing construction prior to the completion of drawings and specifications and then using means such as bid packages and efficient coordination to compress the overall schedule.

with regard to access, traffic, drainage, parking, building placement and other considerations affecting the building, the environment and energy use, as well as information regarding applicable governmental laws, regulations and requirements. The Design-Builder shall also propose alternative architectural, civil, structural, mechanical, electrical and other systems for review by the Owner, to determine the most desirable approach on the basis of cost, technology, quality and speed of delivery. The Design-Builder will also review existing test reports but will not undertake any independent testing nor be required to furnish types of information derived from such testing in its Preliminary Evaluation. Based upon its review and verification of the Owner's Program and other relevant information the Design-Builder shall provide a Preliminary Evaluation of the Project's feasibility for the Owner's acceptance. The Design-Builder's Preliminary Evaluation shall specifically identify any deviations from the Owner's Program.

3.1.2 PRELIMINARY SCHEDULE The Design-Builder shall prepare a preliminary schedule of the Work. The Owner shall provide written approval of milestone dates established in the preliminary schedule of the Work. The schedule shall show the activities of the Owner, the Architect/Engineer and the Design-Builder necessary to meet the Owner's completion requirements. The schedule shall be updated periodically with the level of detail for each schedule update reflecting the information then available. If an update indicates that a previously approved schedule will not be met, the Design-Builder shall recommend corrective action to the Owner in writing.

3.1.3 PRELIMINARY ESTIMATE When sufficient Project information has been identified, the Design-Builder shall prepare for the Owner's acceptance a preliminary estimate utilizing area, volume or similar conceptual estimating techniques. The estimate shall be updated periodically with the level of detail for each estimate update reflecting the information then available. If the preliminary estimate or any update exceeds the Owner's budget, the Design-Builder shall make recommendations to the Owner.

3.1.4 SCHEMATIC DESIGN DOCUMENTS The Design-Builder shall submit for the Owner's written approval Schematic Design Documents, based on the agreed upon Preliminary Evaluation. Schematic Design Documents shall include drawings, outline specifications and other conceptual documents illustrating the Project's basic elements, scale, and their relationship to the Worksite. One set of these documents shall be furnished to the Owner. When the Design-Builder submits the Schematic Design Documents the Design-Builder shall identify in writing all material changes and deviations that have taken place from the Design-Builder's Preliminary Evaluation, schedule and estimate. The Design-Builder shall update the preliminary schedule and estimate based on the Schematic Design Documents.

3.1.5 PLANNING PERMITS The Design-Builder shall obtain and the Owner shall pay for all planning permits necessary for the construction of the Project.

3.1.6 DESIGN DEVELOPMENT DOCUMENTS The Design-Builder shall submit for the Owner's written approval Design Development Documents based on the approved Schematic Design Documents. The Design Development Documents shall further define the Project including drawings and outline specifications fixing and describing the Project size and character as to site utilization, and other appropriate elements incorporating the structural, architectural, mechanical and electrical systems. One set of these documents shall be furnished to the Owner. When the Design-Builder submits the Design Development Documents, the Design-Builder shall identify in writing all material changes and deviations that have taken place from the Schematic Design Documents. The Design-Builder shall update the schedule and estimate based on the Design Development Documents.

3.1.7 CONSTRUCTION DOCUMENTS The Design-Builder shall submit for the Owner's written approval Construction Documents based on the approved Design Development Documents. The Construction Documents shall set forth in detail the requirements for construction of the Work, and shall consist of drawings and specifications based upon codes, laws and regulations enacted at the time of their preparation. When the Design-Builder submits the Construction Documents, the Design-Builder shall identify in writing all material changes and deviations that have taken place from the Design Development Documents. Construction shall be in accordance with these approved Construction Documents. One set of these documents shall be furnished to the Owner prior to commencement of construction. If a GMP has not been established, the Design-Builder shall prepare a further update of the schedule and estimate based on the Construction Documents.

3.1.8 OWNERSHIP OF DOCUMENTS Upon the making of payment pursuant to Paragraph 10.5, the Owner shall receive ownership of the property rights, except for copyrights, of all documents, drawings, specifications, electronic data and information prepared, provided or procured by the Design-Builder, its Architect/Engineer, Subcontractors and consultants and distributed to the Owner for this Project. ("Design-Build Documents")

.1 If this Agreement is terminated pursuant to Paragraph 12.2, the Owner shall receive ownership of the property rights, except for copyrights, of the Design-Build Documents upon payment for all Work performed in accordance with this Agreement, at which time the Owner shall have the right to use, reproduce and make derivative works from the Design-Build Documents to complete the Work.

.2 If this Agreement is terminated pursuant to Paragraph 12.3, the Owner shall receive ownership of the property rights, except for copyrights, of the Design-Build Documents upon payment of all sums provided in Paragraph 12.3, at which time the Owner shall have the right to use, reproduce and make derivative works from the Design-Build Documents to complete the Work.

.3 The Owner may use, reproduce and make derivative works from the Design-Build Documents for subsequent renovation and remodeling of the Work, but shall not use, reproduce or make derivative works from the Design-Build Documents for other projects without the written authorization of the Design-Builder, who shall not unreasonably withhold consent.

.4 The Owner's use of the Design-Build Documents without the Design-Builder's involvement or on other projects is at the Owner's sole risk, except for the Design-Builder's indemnification obligation pursuant to Paragraph 3.7, and the Owner shall defend, indemnify and hold harmless the Design-Builder, its Architect/Engineer, Subcontractors and consultants, and the agents, officers, directors and employees of each of them from and against any and all claims, damages, losses, costs and expenses, including but not limited to attorney's fees, costs and expenses incurred in connection with any dispute resolution process, arising out of or resulting from the Owner's use of the Design-Build Documents.

.5 The Design-Builder shall obtain from its Architect/Engineer, Subcontractors and consultants property rights and rights of use that correspond to the rights given by the Design-Builder to the Owner in this Agreement.

3.2 GUARANTEED MAXIMUM PRICE (GMP)

3.2.1 GMP PROPOSAL At such time as the Owner and the Design-Builder jointly agree, the Design-Builder shall submit a GMP Proposal in a format acceptable to the Owner. Unless the parties mutually agree otherwise, the GMP shall be the sum of the estimated Cost of the Work as defined in Article 8 and the Design-Builder's Fee as defined in Article 7. The GMP is subject to modification as provided in Article 9.

3.2.1.1 If the Design-Build Documents are not complete at the time the GMP Proposal is submitted to the Owner, the Design-Builder shall provide in the GMP for further development of the Design-Build Documents consistent with the Owner's Program. Such further development does not include changes in scope, systems, kinds and quality of materials, finishes or equipment, all of which if required, shall be incorporated by Change Order.

3.2.2 BASIS OF GUARANTEED MAXIMUM PRICE The Design-Builder shall include with the GMP Proposal a written statement of its basis, which shall include:

.1 a list of the drawings and specifications, including all addenda, which were used in preparation of the GMP Proposal;

.2 a list of allowances and a statement of their basis;

.3 a list of the assumptions and clarifications made by the Design-Builder in the preparation of the GMP Proposal to supplement the information contained in the drawings and specifications;

.4 the Date of Substantial Completion and/or the Date of Final Completion upon which the proposed GMP is based, and the Schedule of Work upon which the Date of Substantial Completion and/or the Date of Final Completion is based;

.5 a schedule of applicable alternate prices;

.6 a schedule of applicable unit prices;

.7 a statement of Additional Services included, if any;

.8 the time limit for acceptance of the GMP proposal;

.9 the Design-Builder's Contingency as provided in Subparagraph 3.2.7;

.10 a statement of any work to be self-performed by the Design-Builder; and

.11 a statement identifying all patented or copyrighted materials, methods or systems selected by the Design-Builder and incorporated in the Work that are likely to require the payment of royalties or license fees.

3.2.3 REVIEW AND ADJUSTMENT TO GMP PROPOSAL The Design-Builder shall meet with the Owner to review the GMP Proposal. In the event that the Owner has any comments relative to the GMP Proposal, or finds any inconsistencies or inaccuracies in the information presented, it shall give prompt written notice of such comments or findings to the Design-Builder, who shall make appropriate adjustments to the GMP, its basis or both.

3.2.4 ACCEPTANCE OF GMP PROPOSAL Upon acceptance by the Owner of the GMP Proposal, as may be amended by the Design-Builder in accordance with Subparagraph 3.2.3, the GMP and its basis shall be set forth in Amendment No. 1. The GMP and the Date of Substantial Completion and/or the Date of Final Completion shall be subject to modification in Article 9.

3.2.5 FAILURE TO ACCEPT THE GMP PROPOSAL Unless the Owner accepts the GMP Proposal in writing on or before the date specified in the GMP Proposal for such acceptance and so notifies the Design-Builder, the GMP Proposal shall not be effective. If the Owner fails to accept the GMP Proposal, or rejects the GMP Proposal, the Owner shall have the right to:

.1 Suggest modifications to the GMP Proposal. If such modifications are accepted in writing by Design-Builder, the GMP Proposal shall be deemed accepted in accordance with Subparagraph 3.2.4;

.2 Direct the Design-Builder to proceed on the basis of reimbursement as provided in Articles 7 and 8 without a GMP, in which case all references in this Agreement to the GMP shall not be applicable; or

.3 Terminate the Agreement for convenience in accordance with Paragraph 12.3.

In the absence of a GMP the parties may establish a Date of Substantial Completion and/or a Date of Final Completion.

3.2.6 PRE-GMP WORK Prior to the Owner's acceptance of the GMP Proposal, the Design-Builder shall not incur any cost to be reimbursed as part of the Cost of the Work, except as provided in this Agreement or as the Owner may specifically authorize in writing.

3.2.7 DESIGN-BUILDER'S CONTINGENCY The GMP Proposal will contain, as part of the estimated Cost of the Work, the Design-Builder's Contingency, a sum mutually agreed upon and monitored by the Design-Builder and the Owner for use at the Design-Builder's discretion to cover costs which are properly reimbursable as a Cost of the Work but are not the basis for a Change Order.

3.3 CONSTRUCTION PHASE SERVICES

3.3.1 The Construction Phase will commence upon the issuance by the Owner of a written notice to proceed with construction. If construction commences prior to execution of Amendment No. 1, the Design-Builder shall prepare for the Owner's written approval a list of the documents that are applicable to the part of the Work which the Owner has authorized, which list shall be included in the Owner's written notice to proceed.

3.3.2 In order to complete the Work, the Design-Builder shall provide all necessary construction supervision, inspection, construction equipment, labor, materials, tools, and subcontracted items.

3.3.3 The Design-Builder shall give all notices and comply with all laws and ordinances legally enacted at the date of execution of the Agreement which govern the proper performance of the Work.

3.3.4 The Design-Builder shall obtain and the Owner shall pay for the building permits necessary for the construction of the Project.

3.3.5 The Design-Builder shall keep such full and detailed accounts as are necessary for proper financial management under this Agreement. The Owner shall be afforded access to all the Design-Builder's records, books, correspondence, instructions, drawings, receipts, vouchers, memoranda and similar data relating to this Agreement. The Design-Builder shall preserve all such records for a period of three years after the final payment or longer where required by law.

3.3.6 The Design-Builder shall provide periodic written reports to the Owner on the progress of the Work in such detail as is required by the Owner and as agreed to by the Owner and the Design-Builder.

3.3.7 The Design-Builder shall develop a system of cost reporting for the Work, including regular monitoring of actual costs for activities in progress and estimates for uncompleted tasks and proposed changes in the Work. The reports shall be presented to the Owner at mutually agreeable intervals.

3.3.8 The Design-Builder shall regularly remove debris and waste materials at the Worksite resulting from the Work. Prior to discontinuing Work in an area, the Design-Builder shall clean the area and remove all rubbish and its construction equipment, tools, machinery, waste and surplus materials. The Design-Builder shall minimize and confine dust and debris resulting from construction activities. At the completion of the Work, the Design-Builder shall remove from the Worksite all construction equipment, tools, surplus materials, waste materials and debris.

3.3.9 The Design-Builder shall prepare and submit to the Owner:

final marked up as-built drawings

updated electronic data
(Cross-out one of the above) ♦

in general documenting how the various elements of the Work including changes were actually constructed or installed, or as defined by the parties by attachment to this Agreement.

3.4 SCHEDULE OF THE WORK The Design-Builder shall prepare and submit a Schedule of Work for the Owner's acceptance and written approval as to milestone dates. This schedule shall indicate the dates for the start and completion of the various stages of the Work, including the dates when information and approvals are required from the Owner. The Schedule shall be revised as required by the conditions of the Work.

3.5 SAFETY OF PERSONS AND PROPERTY

3.5.1 SAFETY PRECAUTIONS AND PROGRAMS The Design-Builder shall have overall responsibility for safety precautions and programs in the performance of the Work. While the provisions of this Paragraph establish the responsibility for safety between the Owner and the Design-Builder, they do not relieve Subcontractors of their responsibility for the safety of persons or property in the performance of their work, nor for compliance with the provisions of applicable laws and regulations.

3.5.2 The Design-Builder shall seek to avoid injury, loss or damage to persons or property by taking reasonable steps to protect:

.1 its employees and other persons at the Worksite;

.2 materials, supplies and equipment stored at the Worksite for use in performance of the Work; and

.3 the Project and all property located at the Worksite and adjacent to work areas, whether or not said property or structures are part of the Project or involved in the Work.

3.5.3 DESIGN-BUILDER'S SAFETY REPRESENTATIVE The Design-Builder shall designate an individual at the Worksite in the employ of the Design-Builder who shall act as the Design-Builder's designated safety representative with a duty to prevent accidents. Unless otherwise identified by the Design-Builder in writing to the Owner, the designated safety representative shall be the Design-Builder's project superintendent. The Design-Builder will report immediately in writing all accidents and injuries occurring at the Worksite to the Owner. When the Design-Builder is required to file an accident report with a public authority, the Design-Builder shall furnish a copy of the report to the Owner.

3.5.4 The Design-Builder shall provide the Owner with copies of all notices required of the Design-Builder by law or regulation. The Design-Builder's safety program shall comply with the requirements of governmental and quasi-governmental authorities having jurisdiction over the Work.

3.5.5 Damage or loss not insured under property insurance which may arise from the performance of the Work, to the extent of the negligence attributed to such acts or omissions of the Design-Builder, or anyone for whose acts the Design-Builder may be liable, shall be promptly remedied by the Design-Builder. Damage or loss attributable to the acts or omissions of the Owner or Others and not to the Design-Builder shall be promptly remedied by the Owner.

3.5.6 If the Owner deems any part of the Work or Worksite unsafe, the Owner, without assuming responsibility for the Design-Builder's safety program, may require the Design-Builder to stop performance of the Work or take corrective measures satisfactory to the Owner, or both. If the Design-Builder does not adopt corrective measures, the Owner may perform them and reduce by the costs of the corrective measures the amount of the GMP, or in the absence of a GMP, the Cost of the Work as provided in Article 8. The Design-Builder agrees to make no claim for damages, for an increase in the GMP, compensation for Design Phase Services, the Design-Builder's Fee and/or the Date of Substantial Completion and/or the Date of Final Completion based on the Design-Builder's compliance with the Owner's reasonable request.

3.6 HAZARDOUS MATERIALS

3.6.1 A Hazardous Material is any substance or material identified now or in the future as hazardous under any federal, state or local law or regulation, or any other substance or material which may be considered hazardous or otherwise subject to statutory or regulatory requirements governing handling, disposal and/or clean-up. The Design-Builder shall not be obligated to commence or continue work until all Hazardous Material discovered at the Worksite has been removed, rendered or determined to be harmless by the Owner as certified by an independent testing laboratory approved by the appropriate government agency.

3.6.2 If after the commencement of the Work, Hazardous Material is discovered at the Project, the Design-Builder shall be entitled to immediately stop Work in the affected area. The Design-Builder shall report the condition to the Owner and, if required, the government agency with jurisdiction.

3.6.3 The Design-Builder shall not be required to perform any Work relating to or in the area of Hazardous Material without written mutual agreement.

3.6.4 The Owner shall be responsible for retaining an independent testing laboratory to determine the nature of the material encountered and whether it is a Hazardous Material requiring corrective measures and/or remedial action. Such measures shall be the sole responsibility of the Owner, and shall be performed in a manner minimizing any adverse effects upon the Work of the Design-Builder. The Design-

Builder shall resume Work in the area affected by any Hazardous Material only upon written agreement between the parties after the Hazardous Material has been removed or rendered harmless and only after approval, if necessary, of the governmental agency or agencies with jurisdiction.

3.6.5 If the Design-Builder incurs additional costs and/or is delayed due to the presence or remediation of Hazardous Material, the Design-Builder shall be entitled to an equitable adjustment in the GMP, compensation for Design Phase Services, the Design-Builder's Fee and/or the Date of Substantial Completion and/or the Date of Final Completion.

3.6.6 Provided the Design-Builder, its Subcontractors, Material Suppliers and Subsubcontractors, and the agents, officers, directors and employees of each of them, have not, acting under their own authority, knowingly entered upon any portion of the Work containing Hazardous Materials, and to the extent not caused by the negligent acts or omissions of the Design-Builder, its Subcontractors, Material Suppliers and Subsubcontractors, and the agents, officers, directors and employees of each of them, the Owner shall defend, indemnify and hold harmless the Design-Builder, its Subcontractors and Subsubcontractors, and the agents, officers, directors and employees of each of them, from and against any and all direct claims, damages, losses, costs and expenses, including but not limited to attorney's fees, costs and expenses incurred in connection with any dispute resolution process, arising out of or relating to the performance of the Work in any area affected by Hazardous Material. To the fullest extent permitted by law, such indemnification shall apply regardless of the fault, negligence, breach of warranty or contract, or strict liability of the Owner.

3.6.7 Material Safety Data (MSD) sheets as required by law and pertaining to materials or substances used or consumed in the performance of the Work, whether obtained by the Design-Builder, Subcontractors, the Owner or Others, shall be maintained at the Project by the Design-Builder and made available to the Owner and Subcontractors.

3.6.8 During the Design-Builder's performance of the Work, the Design-Builder shall be responsible for the proper handling of all materials brought to the Worksite by the Design-Builder. Upon the issuance of the Certificate of Substantial Completion, the Owner shall be responsible under this Paragraph for materials and substances brought to the site by the Design-Builder if such materials or substances are required by the Contract Documents.

3.6.9 The terms of this Paragraph 3.6 shall survive the completion of the Work under this Agreement and/or any termination of this Agreement.

3.7 ROYALTIES, PATENTS AND COPYRIGHTS The Design-Builder shall pay all royalties and license fees which may be due on the inclusion of any patented or copyrighted materials, methods or systems selected by the Design-Builder and incorporated in the Work. The Design-Builder shall defend, indemnify and hold the Owner harmless from all suits or claims for infringement of any patent rights or copyrights arising out of such selection. The Owner agrees to defend, indemnify and hold the Design-Builder harmless from all suits or claims of infringement of any patent rights or copyrights arising out of any patented or copyrighted materials, methods or systems specified by the Owner.

3.8 WARRANTIES AND COMPLETION

3.8.1 The Design-Builder warrants that all materials and equipment furnished under the Construction Phase of this Agreement will be new unless otherwise specified, of good quality, in conformance with the Contract Documents, and free from defective workmanship and materials. Warranties shall commence on the Date of Substantial Completion of the Work or of a designated portion. The Design-Builder agrees to correct all construction performed under this Agreement which is defective in workmanship or materials within a period of one year from the Date of Substantial Completion or for such longer periods of time as may be set forth with respect to specific warranties required by the Contract Documents.

3.8.2 To the extent products, equipment, systems or materials incorporated in the Work are specified and purchased by the Owner, they shall be covered exclusively by the warranty of the manufacturer. There are no warranties which extend beyond the description on the face of any such warranty. To the extent products, equipment, systems or materials incorporated in the Work are specified by the Owner but purchased by the Design-Builder and are inconsistent with selection criteria that otherwise would have been followed by the Design-Builder, the Design-Builder shall assist the Owner in pursuing warranty claims. **ALL OTHER WARRANTIES EXPRESSED OR IMPLIED INCLUDING THE WARRANTY OF MERCHANTABILITY AND THE WARRANTY OF FITNESS FOR A PARTICULAR PURPOSE ARE EXPRESSLY DISCLAIMED.**

3.8.3 The Design-Builder shall secure required certificates of inspection, testing or approval and deliver them to the Owner.

3.8.4 The Design-Builder shall collect all written warranties and equipment manuals and deliver them to the Owner in a format directed by the Owner.

3.8.5 With the assistance of the Owner's maintenance personnel, the Design-Builder shall direct the checkout of utilities and start up operations, and adjusting and balancing of systems and equipment for readiness.

3.9 CONFIDENTIALITY The Design-Builder shall treat as confidential and not disclose to third persons, except Subcontractors, Subsubcontractors and the Architect/Engineer as is necessary for the performance of the Work, or use for its own benefit any of the Owner's developments, confidential information, know-how, discoveries, production methods and the like that may be disclosed to the Design-Builder or which the Design-Builder may acquire in connection with the Work. The Owner shall treat as confidential information all of the Design-Builder's estimating systems and historical and parameter cost data that may be disclosed to the Owner in connection with the performance of this Agreement.

3.10 ADDITIONAL SERVICES The Design-Builder shall provide or procure the following Additional Services upon the request of the Owner. A written agreement between the Owner and the Design-Builder shall define the extent of such Additional Services before they are performed by the Design-Builder. If a GMP has been established for the Work or any portion of the Work, such Additional Services shall be considered a Change in the Work, unless they are specifically included in the statement of the basis of the GMP as set forth in Amendment No. 1.

.1 Development of the Owner's Program, establishing the Project budget, investigating sources of financing, general business planning and other information and documentation as may be required to establish the feasibility of the Project.

.2 Consultations, negotiations, and documentation supporting the procurement of Project financing.

.3 Surveys, site evaluations, legal descriptions and aerial photographs.

.4 Appraisals of existing equipment, existing properties, new equipment and developed properties.

.5 Soils, subsurface and environmental studies, reports and investigations required for submission to governmental authorities or others having jurisdiction over the Project.

.6 Consultations and representations before governmental authorities or others having jurisdiction over the Project other than normal assistance in securing building permits.

.7 Investigation or making measured drawings of existing conditions or the reasonably required verification of Owner-provided drawings and information.

.8 Artistic renderings, models and mockups of the Project or any part of the Project or the Work.

.9 Inventories of existing furniture, fixtures, furnishings and equipment which might be under consideration for incorporation into the Work.

.10 Interior design and related services including procurement and placement of furniture, furnishings, artwork and decorations.

.11 Making revisions to the Schematic Design, Design Development, Construction Documents or documents forming the basis of the GMP after they have been approved by the Owner, and which are due to causes beyond the control of the Design-Builder. Causes beyond the control of the Design-Builder do not include acts or omissions on the part of Subcontractors, Material Suppliers, Subsubcontractors or the Architect/Engineer.

.12 Design, coordination, management, expediting and other services supporting the procurement of materials to be obtained, or work to be performed, by the Owner, including but not limited to telephone systems, computer wiring networks, sound systems, alarms, security systems and other specialty systems which are not a part of the Work.

.13 Estimates, proposals, appraisals, consultations, negotiations and services in connection with the repair or replacement of an insured loss, provided such repair or replacement did not result from the negligence of the Design-Builder.

.14 The premium portion of overtime work ordered by the Owner, including productivity impact costs, other than that required by the Design-Builder to maintain the Schedule of Work.

.15 Out-of-town travel by the Architect/Engineer in connection with the Work, except between the Architect/Engineer's office, the Design-Builder's office, the Owner's office and the Worksite.

.16 Obtaining service contractors and training maintenance personnel, assisting and consulting in the use of systems and equipment after the initial start up.

.17 Services for tenant or rental spaces not a part of this Agreement.

.18 Services requested by the Owner or required by the Work which are not specified in the Contract Documents and which are not normally part of generally accepted design and construction practice.

.19 Serving or preparing to serve as an expert witness in connection with any proceeding, legal or otherwise, regarding the Project.

.20 Document reproduction exceeding the limits provided for in this Agreement.

3.11 DESIGN-BUILDER'S REPRESENTATIVE The Design-Builder shall designate a person who shall be the Design-Builder's authorized representative. The Design-Builder's Representative is ______________________

______________________________________. ◆

ARTICLE 4

OWNER'S RESPONSIBILITIES

4.1 INFORMATION AND SERVICES PROVIDED BY OWNER

4.1.1 The Owner shall provide full information in a timely manner regarding requirements for the Project, including the Owner's Program and other relevant information.

4.1.2 The Owner shall provide:

.1 all available information describing the physical characteristics of the site, including surveys, site evaluations, legal descriptions, existing conditions, subsurface and environmental studies, reports and investigations;

.2 inspection and testing services during construction as required by law or as mutually agreed; and

.3 unless otherwise provided in the Contract Documents, necessary approvals, site plan review, rezoning, easements and assessments, fees and charges required for the construction, use, occupancy or renovation of permanent structures, including legal and other required services.

4.1.3 The Owner shall provide reasonable evidence satisfactory to the Design-Builder, prior to commencing the Work and during the progress of the Work, that sufficient funds are available and committed for the entire cost of the Project, including a reasonable allowance for changes in the Work as may be approved in the course of the Work. Unless such reasonable evidence is provided, the Design-Builder shall not be required to commence or continue the Work. The Design-Builder may stop Work after seven (7) days written notice to the Owner if such evidence is not presented within a reasonable time. The failure of the Design-Builder to insist upon the providing of this evidence at any one time shall not be a waiver of the Owner's obligation to make payments pursuant to this Agreement, nor shall it be a waiver of the Design-Builder's right to require that such evidence be provided at a later date.

4.1.4 The Design-Builder shall be entitled to rely on the completeness and accuracy of the information and services required by this Paragraph 4.1.

4.2 RESPONSIBILITIES DURING DESIGN PHASE

4.2.1 The Owner shall provide the Owner's Program at the inception of the Design Phase and shall review and timely approve in writing schedules, estimates, Preliminary Estimate, Schematic Design Documents, Design Development Documents and Construction Documents furnished during the Design Phase as set forth in Paragraph 3.1, and the GMP Proposal as set forth in Paragraph 3.2.

4.3 RESPONSIBILITIES DURING CONSTRUCTION PHASE

4.3.1 The Owner shall review the Schedule of the Work as set forth in Paragraph 3.4 and timely approve the milestone dates set forth.

4.3.2 If the Owner becomes aware of any error, omission or failure to meet the requirements of the Contract Documents or any fault or defect in the Work, the Owner shall give prompt written notice to the Design-Builder.

4.3.3 The Owner shall communicate with the Design-Builder's Subcontractors, Material Suppliers and the Architect/Engineer only through or in the presence of the Design-Builder. The Owner shall have no contractual obligations to Subcontractors, suppliers, or the Architect/Engineer.

4.3.4 The Owner shall provide insurance for the Project as provided in Article 11.

4.4 OWNER'S REPRESENTATIVE The Owner's Representative is ______________________

______________________________________. ◆

The Representative:

.1 shall be fully acquainted with the Project;

.2 agrees to furnish the information and services required of the Owner pursuant to Paragraph 4.1 so as not to delay the Design-Builder's Work; and

.3 shall have authority to bind the Owner in all matters requiring the Owner's approval, authorization or written notice. If the Owner changes its representative or the representative's authority as listed above, the Owner shall notify the Design-Builder in writing in advance.

4.5 TAX EXEMPTION If in accordance with the Owner's direction the Design-Builder claims an exemption for taxes, the Owner shall defend, indemnify and hold the Design-Builder harmless for all liability, penalty, interest, fine, tax assessment, attorney's fees or other expense or cost incurred by the Design-Builder as a result of any action taken by the Design-Builder in accordance with the Owner's direction.

ARTICLE 5

SUBCONTRACTS

Work not performed by the Design-Builder with its own forces shall be performed by Subcontractors or the Architect/Engineer.

5.1 RETAINING SUBCONTRACTORS The Design-Builder shall not retain any subcontractor to whom the Owner has a reasonable and timely objection, provided that the Owner agrees to compensate the Design-Builder for any additional costs incurred by the Design-Builder as a result of such objection. The Owner may propose subcontractors to be considered by the Design-Builder. The Design-Builder shall not be required to retain any subcontractor to whom the Design-Builder has a reasonable objection.

5.2 MANAGEMENT OF SUBCONTRACTORS The Design-Builder shall be responsible for the management of the Subcontractors in the performance of their work.

5.3 ASSIGNMENT OF SUBCONTRACT AGREEMENTS The Design-Builder shall provide for assignment of subcontract agreements in the event that the Owner terminates this Agreement for cause as provided in Paragraph 12.2. Following such termination, the Owner shall notify in writing those Subcontractors whose assignments will be accepted, subject to the rights of sureties.

5.4 BINDING OF SUBCONTRACTORS AND MATERIAL SUPPLIERS The Design-Builder agrees to bind every Subcontractor and Material Supplier (and require every Subcontractor to so bind its Subsubcontractors and Material Suppliers) to all the provisions of this Agreement and the Contract Documents as they apply to the Subcontractor's and Material Supplier's portions of the Work.

5.5 LABOR RELATIONS (Insert here or attach as exhibit as necessary any conditions, obligations or requirements relative to labor relations and their effect on the Project. Legal counsel is recommended.) ◆

ARTICLE 6

TIME

6.1 DATE OF COMMENCEMENT The Date of Commencement is the effective date of this Agreement as first written in Article 1 unless otherwise set forth below: (Insert here any special provisions concerning Notices to Proceed and the Date of Commencement.) ◆

The Work shall proceed in general accordance with the Schedule of Work as such schedule may be amended from time to time, subject, however, to other provisions of this Agreement.

6.2 SUBSTANTIAL/FINAL COMPLETION Unless the parties agree otherwise, the Date of Substantial Completion and/or the Date of Final Completion shall be established in Amendment No. 1 to this Agreement subject to adjustments as provided for in the Contract Documents. The Owner and the Design-Builder may agree not to establish such dates, or in the alternative, to establish one but not the other of the two dates. If such dates are not established upon the execution of this Agreement, at such time as a GMP is accepted a Date of Substantial Completion and/or Date of Final Completion of the Work shall be established in Amendment No. 1. If a GMP is not established and the parties desire to establish a Date of Substantial Completion and/or Date of Final Completion, it shall be set forth in Amendment No. 1.

6.2.1 Time limits stated in the Contract Documents are of the essence.

6.2.2 Unless instructed by the Owner in writing, the Design-Builder shall not knowingly commence the Work before the effective date of insurance that is required to be provided by the Design-Builder or the Owner.

6.3 DELAYS IN THE WORK

6.3.1 If causes beyond the Design-Builder's control delay the progress of the Work, then the GMP, compensation for Design Phase Services, the Design-Builder's Fee and/or the Date of Substantial Completion and/or the Date of Final Completion shall be modified by Change Order as appropriate. Such causes shall include but not be limited to: changes ordered in the Work, acts or omissions of the Owner or Others, the Owner preventing the Design-Builder from performing the Work pending dispute resolution, Hazardous Materials or differing site conditions. Causes beyond the control of the Design-Builder do not include acts or omissions on the part of the Design-Builder, Subcontractors, Subsubcontractors, Material Suppliers or the Architect/Engineer.

6.3.2 To the extent a delay in the progress of the Work is caused by adverse weather conditions not reasonably anticipated, fire, unusual transportation delays, general labor disputes impacting the Project but not specifically related to the Worksite, governmental agencies, or unavoidable accidents or circumstances, the Design-Builder shall only be entitled to its actual costs without fee and an extension of the Date of Substantial Completion and/or the Date of Final Completion.

6.3.3 In the event delays to the Project are encountered for any reason, the parties agree to undertake reasonable steps to mitigate the effect of such delays.

ARTICLE 7

COMPENSATION

7.1 DESIGN PHASE COMPENSATION

7.1.1 To the extent required by applicable law, the cost of services performed directly by the Architect/Engineer is computed separately and is independent from the Design-Builder's compensation for work or services performed directly by the Design-Builder; these costs shall be shown as separate items on applications for payment. If an Architect/Engineer is retained by the Design-Builder, the payments to the Architect/Engineer shall be as detailed in a separate agreement between the Design-Builder and the Architect/Engineer.

7.1.2 The Owner shall compensate the Design-Builder for services performed during the Design Phase as described in Paragraph 3.1, including preparation of a GMP Proposal, if applicable, as described in Paragraph 3.2, as follows:
(State whether a stipulated sum, actual cost, or other basis. If a stipulated sum, state what portion of the sum shall be payable each month.) ♦

7.1.3 Compensation for Design Phase Services, as part of the Work, shall include the Design-Builder's Fee as established in Paragraph 7.3, paid in proportion to the services performed, subject to adjustment as provided in Paragraph 7.4.

7.1.4 Compensation for Design Phase Services shall be equitably adjusted if such services extend beyond ______ ______________________ from the date of this ♦ Agreement for reasons beyond the reasonable control of the Design-Builder or as provided in Paragraph 9.1. For changes in Design Phase services, compensation shall be adjusted as follows: ♦

7.1.5 Within fifteen (15) days after receipt of each monthly application for payment, the Owner shall give written notice to the Design-Builder of the Owner's acceptance or rejection, in whole or in part, of such application for payment. Within fifteen (15) days after accepting such application, the Owner shall pay directly to the Design-Builder the appropriate amount for which application for payment is made, less amounts previously paid by the Owner. If such application is rejected in whole or in part, the Owner shall indicate the reasons for its rejection. If the Owner and the Design-Builder cannot agree on a revised amount then, within fifteen (15) days after its initial rejection in part of such application, the Owner shall pay directly to the Design-Builder the appropriate amount for those items not rejected by the Owner for which application for payment is made, less amounts previously paid by the Owner. Those items rejected by the Owner shall be due and payable when the reasons for the rejection have been removed.

7.1.6 If the Owner fails to pay the Design-Builder at the time payment of any amount becomes due, then the Design-Builder may, at any time thereafter, upon serving written notice that the Work will be stopped within seven (7) days after receipt of the notice by the Owner, and after such seven (7) day period, stop the Work until payment of the amount owing has been received.

7.1.7 Payments due pursuant to Subparagraph 7.1.5, may bear interest from the date payment is due at the prime rate prevailing at the location of Project.

7.2 CONSTRUCTION PHASE COMPENSATION

7.2.1 The Owner shall compensate the Design-Builder for Work performed following the commencement of the Construction Phase on the following basis:

.1 the Cost of the Work as allowed in Article 8; and

.2 the Design-Builder's Fee paid in proportion to the services performed subject to adjustment as provided in Paragraph 7.4.

7.2.2 The compensation to be paid under this Paragraph 7.2 shall be limited to the GMP established in Amendment No. 1, as the GMP may be adjusted under Article 9.

7.2.3 Payment for Construction Phase Services shall be as set forth in Article 10. If Design Phase Services continue to be provided after construction has commenced, the Design-Builder shall continue to be compensated as provided in Paragraph 7.1, or as mutually agreed.

7.3 DESIGN-BUILDER'S FEE The Design-Builder's Fee shall be as follows, subject to adjustment as provided in Paragraph 7.4:
(State whether a stipulated sum or other basis. If a stipulated sum, state what portion of the sum shall be payable each month.)

◆

7.4 ADJUSTMENT IN THE DESIGN-BUILDER'S FEE Adjustment in the Design-Builder's Fee shall be made as follows:

.1 for changes in the Work as provided in Article 9, the Design-Builder's Fee shall be adjusted as follows:

◆

.2 for delays in the Work not caused by the Design-Builder, except as provided in Subparagraph 6.3.2, there will be an equitable adjustment in the Design-Builder's Fee to compensate the Design-Builder for increased expenses; and

.3 if the Design-Builder is placed in charge of managing the replacement of an insured or uninsured loss, the Design-Builder shall be paid an additional fee in the same proportion that the Design-Builder's Fee bears to the estimated Cost of the Work for the replacement.

ARTICLE 8

COST OF THE WORK

The Owner agrees to pay the Design-Builder for the Cost of the Work as defined in this Article. This payment shall be in addition to the Design-Builder's Fee stipulated in Paragraph 7.3.

8.1 COST ITEMS FOR DESIGN PHASE SERVICES

8.1.1 Compensation for Design Phase Services as provided in Paragraph 7.1.

8.2 COST ITEMS FOR CONSTRUCTION PHASE SERVICES

8.2.1 Wages paid for labor in the direct employ of the Design-Builder in the performance of the Work.

8.2.2 Salaries of the Design-Builder's employees when stationed at the field office, in whatever capacity employed, employees engaged on the road expediting the production or transportation of material and equipment, and employees from the principal or branch office performing the functions listed below:

◆

8.2.3 Cost of all employee benefits and taxes including but not limited to workers' compensation, unemployment compensation, Social Security, health, welfare, retirement and other fringe benefits as required by law, labor agreements, or paid under the Design-Builder's standard personnel policy, insofar as such costs are paid to employees of the Design-Builder who are included in the Cost of the Work under Subparagraphs 8.2.1 and 8.2.2.

8.2.4 Reasonable transportation, travel, hotel and moving expenses of the Design-Builder's personnel incurred in connection with the Work.

8.2.5 Cost of all materials, supplies and equipment incorporated in the Work, including costs of inspection and testing if not provided by the Owner, transportation, storage and handling.

8.2.6 Payments made by the Design-Builder to Subcontractors for work performed under this Agreement.

8.2.7 Fees and expenses for design services procured or furnished by the Design-Builder except as provided by the Architect/Engineer and compensated in Paragraph 7.1.

8.2.8 Cost, including transportation and maintenance of all materials, supplies, equipment, temporary facilities and hand tools not owned by the workers that are used or consumed in the performance of the Work, less salvage value and/or residual value; and cost less salvage value on such items used, but not consumed that remain the property of the Design-Builder.

8.2.9 Rental charges of all necessary machinery and equipment, exclusive of hand tools owned by workers, used at the Worksite, whether rented from the Design-Builder or Others, including installation, repair and replacement, dismantling, removal, maintenance, transportation and delivery costs. Rental from unrelated third parties shall be reimbursed at actual cost. Rentals from the Design-Builder or its affiliates, subsidiaries or related parties shall be reimbursed at the prevailing rates in the locality of the Worksite up to eighty-five percent (85%) of the value of the piece of equipment.

8.2.10 Cost of the premiums for all insurance and surety bonds which the Design-Builder is required to procure or deems necessary, and approved by the Owner.

8.2.11 Sales, use, gross receipts or other taxes, tariffs or duties related to the Work for which the Design-Builder is liable.

8.2.12 Permits, fees, licenses, tests, royalties, damages for infringement of patents and/or copyrights, including costs of defending related suits for which the Design-Builder is not responsible as set forth in Paragraph 3.7, and deposits lost for causes other than the Design-Builder's negligence.

8.2.13 Losses, expenses or damages to the extent not compensated by insurance or otherwise, and the cost of corrective work and/or redesign during the Construction Phase and for a period of one year following the Date of Substantial Completion, provided that such corrective work and/or redesign did not arise from the negligence of the Design-Builder.

8.2.14 All costs associated with establishing, equipping, operating, maintaining and demobilizing the field office.

8.2.15 Reproduction costs, photographs, cost of telegrams, facsimile transmissions, long distance telephone calls, data processing services, postage, express delivery charges, telephone service at the Worksite and reasonable petty cash expenses at the field office.

8.2.16 All water, power and fuel costs necessary for the Work.

8.2.17 Cost of removal of all non-hazardous substances, debris and waste materials.

8.2.18 Costs incurred due to an emergency affecting the safety of persons and/or property.

8.2.19 Legal, mediation and arbitration fees and costs, other than those arising from disputes between the Owner and the Design-Builder, reasonably and properly resulting from the Design-Builder's performance of the Work.

8.2.20 All costs directly incurred in the performance of the Work or in connection with the Project, and not included in the Design-Builder's Fee as set forth in Article 7, which are reasonably inferable from the Contract Documents as necessary to produce the intended results.

8.3 DISCOUNTS All discounts for prompt payment shall accrue to the Owner to the extent such payments are made directly by the Owner. To the extent payments are made with funds of the Design-Builder, all cash discounts shall accrue to the Design-Builder. All trade discounts, rebates and refunds, and all returns from sale of surplus materials and equipment, shall be credited to the Cost of the Work.

ARTICLE 9

CHANGES IN THE WORK

Changes in the Work which are within the general scope of this Agreement may be accomplished, without invalidating this Agreement, by Change Order, Work Change Directive, or a minor change in the work, subject to the limitations stated in the Contract Documents.

9.1 CHANGE ORDER

9.1.1 The Design-Builder may request and/or the Owner, without invalidating this Agreement, may order changes in the Work within the general scope of the Contract Documents consisting of additions, deletions or other revisions to the GMP or the estimated cost of the work, compensation for Design Phase Services, the Design-Builder's Fee and/or the Date of Substantial Completion and/or the Date of Final Completion being adjusted accordingly. All such changes in the Work shall be authorized by applicable Change Order, and shall be performed under the applicable conditions of the Contract Documents.

9.1.2 Each adjustment in the GMP and/or estimated Cost of the Work resulting from a Change Order shall clearly separate the amount attributable to compensation for Design Phase Services, other Cost of the Work and the Design-Builder's Fee, with the Design-Builder's Fee not to exceed ______________ percent (________%).

9.1.3 The Owner and the Design-Builder shall negotiate in good faith an appropriate adjustment to the GMP or the estimated Cost of the Work, compensation for Design Phase Services, the Design-Builder's Fee and/or the Date of Substantial Completion and/or the Date of Final Completion and shall conclude these negotiations as expeditiously as possible. Acceptance of the Change Order and any adjustment in the GMP, the estimated Cost of the Work, compensation for Design Phase Services, the Design-Builder's Fee and/or the Date of Substantial Completion and/or the Date of Final Completion shall not be unreasonably withheld.

9.2 WORK CHANGE DIRECTIVES

9.2.1 The Owner may issue a written Work Change Directive directing a change in the Work prior to reaching agreement with the Design-Builder on the adjustment, if any, in the GMP, estimated Cost of the Work, the Design-Builder's Fee, the Date of Substantial Completion and/or the Date of Final Completion, and if appropriate, the compensation for Design Phase Services.

9.2.2 The Owner and the Design-Builder shall negotiate expeditiously and in good faith for appropriate adjustments, as applicable, to the GMP, estimated Cost of the Work, the Design-Builder's Fee, the Date of Substantial Completion and/or the Date of Final Completion, and if appropriate the compensation for Design Phase Services, arising out of Work Change Directives. As the changed work is completed, the Design Builder shall submit its costs for such work with its application for payment beginning with the next application for payment within thirty (30) days of the issuance of the Work Change Directive. Pending final determination of cost to the Owner, amounts not in dispute may be included in applications for payment and shall be paid by Owner.

9.2.3 If the Owner and the Design-Builder agree upon the adjustments in the GMP, estimated Cost of the Work, the Design-Builder's Fee, the Date of Substantial Completion and/or the Date of Final Completion, and if appropriate the compensation for Design Phase Services, for a change in the Work directed by a Work Change Directive, such agreement shall be the subject of an appropriate Change Order. The Change Order shall include all outstanding Change Directives issued since the last Change Order.

9.3 MINOR CHANGES IN THE WORK

9.3.1 The Design-Builder may make minor changes in the design and construction of the Project consistent with the intent of the Contract Documents which do not involve an adjustment in the GMP, estimated Cost of the Work, the Design-Builder's Fee, the Date of Substantial Completion and/or the Date of Final Completion, and do not materially and adversely affect the design of the Project, the quality of any of the materials or equipment specified in the Contract Documents, the performance of any materials, equipment or systems specified in the Contract Documents, or the quality of workmanship required by the Contract Documents.

9.3.2 The Design-Builder shall promptly inform the Owner in writing of any such changes and shall record such changes on the Design-Build Documents maintained by the Design-Builder.

9.4 UNKNOWN CONDITIONS If in the performance of the Work the Design-Builder finds latent, concealed or subsurface physical conditions which materially differ from the conditions the Design-Builder reasonably anticipated, or if physical conditions are materially different from those normally encountered and generally recognized as inherent in the kind of work provided for in this Agreement, then the GMP, estimated Cost of the Work, the Design-Builder's Fee, the Date of Substantial Completion and/or the Date of Final Completion, and if appropriate the compensation for Design Phase Services, shall be equitably adjusted by Change Order within a reasonable time after the conditions are first observed. The Design-Builder shall provide the Owner with written notice within the time period set forth in Paragraph 9.6.

9.5 DETERMINATION OF COST

9.5.1 An increase or decrease in the GMP and/or estimated Cost of the Work resulting from a change in the Work shall be determined by one or more of the following methods:

.1 unit prices set forth in this Agreement or as subsequently agreed;

.2 a mutually accepted, itemized lump sum;

.3 costs determined as defined in Paragraph 7.2 and Article 8 and a mutually acceptable Design-Builder's Fee as determined in Subparagraph 7.4.1; or

.4 if an increase or decrease cannot be agreed to as set forth in Clauses 9.5.1.1 through 9.5.1.3 above, and the Owner issues a Work Change Directive, the cost of the change in the Work shall be determined by the reasonable actual expense and savings of the performance of the Work resulting from the change. If there is a net increase in the GMP, the Design-Builder's Fee shall be adjusted as set forth in Subparagraph 7.4.1. In case of a net decrease in the GMP, the Design-Builder's Fee shall not be adjusted unless ten percent (10%) or more of the Project is deleted. The Design-Builder shall maintain a documented, itemized accounting evidencing the expenses and savings.

9.5.2 If unit prices are indicated in the Contract Documents or are subsequently agreed to by the parties, but the character or quantity of such unit items as originally contemplated is so different in a proposed Change Order that the original unit prices will cause substantial inequity to the Owner or the Design-Builder, such unit prices shall be equitably adjusted.

9.5.3 If the Owner and the Design-Builder disagree as to whether work required by the Owner is within the scope of the Work, the Design-Builder shall furnish the Owner with an estimate of the costs to perform the disputed work in accordance with the Owner's interpretations. If the Owner issues a written order for the Design-Builder to proceed, the Design-Builder shall perform the disputed work and the Owner shall pay the Design-Builder fifty percent (50%) of its estimated cost to perform the work. In such event, both parties reserve their rights as to whether the work was within the scope of the Work. The Owner's payment does not prejudice its right to be reimbursed should it be determined that the disputed work was within the scope of Work. The Design-Builder's receipt of payment for the disputed work does not prejudice its right to receive full payment for the disputed work should it be determined that the disputed work is not within the scope of the Work.

9.6 CLAIMS FOR ADDITIONAL COST OR TIME For any claim for an increase in the GMP, estimated Cost of the Work, the Design-Builder's Fee and the Date of Substantial Completion and/or the Date of Final Completion, and if appropriate the compensation for Design Phase Services, the Design-Builder shall give the Owner written notice of the claim within twenty-one (21) days after the occurrence giving rise to the claim or within twenty-one (21) days after the Design-Builder first recognizes the condition giving rise to the claim, whichever is later. Except in an emergency, notice shall be given before proceeding with the Work. Claims for design and estimating costs incurred in connection with possible changes requested by the Owner, but which do not proceed, shall be made within twenty-one (21) days after the decision is made not to proceed. Any change in the GMP, estimated Cost of the Work, the Design-Builder's Fee, the Date of Substantial Completion and/or the Date of Final Completion, and if appropriate the compensation for Design Phase Services, resulting from such claim shall be authorized by Change Order.

9.7 EMERGENCIES In any emergency affecting the safety of persons and/or property, the Design-Builder shall act, at its discretion, to prevent threatened damage, injury or loss. Any change in the GMP, estimated Cost of the Work, the Design-Builder's Fee, the Date of Substantial Completion and/or the Date of Final Completion, and if appropriate the compensation for Design Phase Services, on account of emergency work shall be determined as provided in this Article.

9.8 CHANGES IN LAW In the event any changes in laws or regulations affecting the performance of the Work are enacted after either the date of this Agreement or the date a GMP Proposal is accepted by the Owner and set forth in Amendment No. 1 to this Agreement, whichever occurs later, the GMP, estimated Cost of the Work, the Design-Builder's Fee, the Date of Substantial Completion and/or the Date of Final Completion, and if appropriate the compensation for Design Phase Services, shall be equitably adjusted by Change Order.

ARTICLE 10

PAYMENT FOR CONSTRUCTION PHASE SERVICES

10.1 PROGRESS PAYMENTS

10.1.1 On the ________________ day of each month ♦ after the Construction Phase has commenced, the Design-Builder shall submit to the Owner an application for payment consisting of the Cost of the Work performed up to the ____ ________________ day of the month, including the cost ♦ of material suitably stored on the Worksite or at other locations approved by the Owner, along with a proportionate share of the Design-Builder's Fee. Approval of payment applications for such stored materials shall be conditioned upon submission by the Design-Builder of bills of sale and applicable insurance or such other procedures satisfactory to the Owner to establish the Owner's title to such materials, or otherwise to protect the Owner's interest, including transportation to the site. Prior to submission of the next application for payment, the Design-Builder shall furnish to the Owner a statement accounting for the disbursement of funds received under the previous application. The extent of such statement shall be as agreed upon between the Owner and the Design-Builder.

10.1.2 Within ten (10) days after receipt of each monthly application for payment, the Owner shall give written notice to the Design-Builder of the Owner's acceptance or rejection, in whole or in part, of such application for payment. Within fifteen (15) days after accepting such application, the Owner shall pay directly to the Design-Builder the appropriate amount for which application for payment is made, less amounts previously paid by the Owner. If such application is rejected in whole or in part, the Owner shall indicate the reasons for its rejection. If the Owner and the Design-Builder cannot agree on a revised amount then, within fifteen (15) days after its initial rejection in part of such application, the Owner shall pay directly to the Design-Builder the appropriate amount for those items not rejected by the Owner for which application for payment is made, less amounts previously paid by the Owner. Those items rejected by the Owner shall be due and payable when the reasons for the rejection have been removed.

10.1.3 If the Owner fails to pay the Design-Builder at the time payment of any amount becomes due, then the Design-Builder may, at any time thereafter, upon serving written notice that the Work will be stopped within seven (7) days after receipt of the notice by the Owner, and after such seven day period, stop the Work until payment of the amount owing has been received.

10.1.4 Payments due but unpaid pursuant to Subparagraph 10.1.2, less any amount retained pursuant to Paragraphs 10.2 and 10.3 may bear interest from the date payment is due at the prime rate prevailing at the place of the Project.

10.1.5 The Design-Builder warrants and guarantees that title to all Work, materials and equipment covered by an application for payment, whether incorporated in the Project or not, will pass to the Owner upon receipt of such payment by the Design-Builder free and clear of all liens, claims, security interests or encumbrances, hereinafter referred to as "liens."

10.1.6 The Owner's progress payment, occupancy or use of the Project, whether in whole or in part, shall not be deemed an acceptance of any Work not conforming to the requirements of the Contract Documents.

10.1.7 Upon Substantial Completion of the Work, the Owner shall pay the Design-Builder the unpaid balance of the Cost of the Work, compensation for Design Phase Services and the Design-Builder's Fee, less one-hundred-fifty percent (150%) of the cost of completing any unfinished items as agreed to between the Owner and the Design-Builder as to extent and time for completion. The Owner thereafter shall pay the Design-Builder monthly the amount retained for unfinished items as each item is completed.

10.2 RETAINAGE From each progress payment made prior to the time Substantial Completion of the Work has been reached, the Owner shall retain ________________ ♦ percent (____________ %), if required, of the amount otherwise due after deduction of any amounts as provided in Paragraph 10.3 of this Agreement. If the Owner chooses to use this retainage provision:

.1 at the time the Work is fifty percent (50%) complete and thereafter, the Owner may choose to withhold no more retainage and pay the Design-Builder the full amount of what is due on account of subsequent progress payments;

.2 once each early finishing trade Subcontractor has completed its work and that work has been accepted by the Owner, the Owner may release final retention on such work;

.3 in lieu of retainage, the Design-Builder may furnish securities, acceptable to the Owner, to be held by the Owner. The interest on such securities shall accrue to the Design-Builder;

.4 the Owner may, in its sole discretion, reduce the amount to be retained at any time.

10.3 ADJUSTMENT OF DESIGN-BUILDER'S APPLICATION FOR PAYMENT The Owner may adjust or reject an application for payment or nullify a previously approved Design-Builder application for payment, in whole or in part, as may reasonably be necessary to protect the Owner from loss or damage based upon the following, to the extent that the Design-Builder is responsible under this Agreement:

.1 the Design-Builder's repeated failure to perform the Work as required by the Contract Documents;

.2 loss or damage arising out of or relating to this Agreement and caused by the Design-Builder to the Owner or Others to whom the Owner may be liable;

.3 the Design-Builder's failure to properly pay the Architect/Engineer, Subcontractors or Material Suppliers for labor, materials, equipment or supplies furnished in connection with the Work, provided that the Owner is making payments to the Design-Builder in accordance with the terms of this Agreement;

.4 Defective Work not corrected in a timely fashion;

.5 reasonable evidence of delay in performance of the Work such that the Work will not be completed by the Date of Substantial Completion and/or the Date of Final Completion, and that the unpaid balance of the GMP is not sufficient to offset any direct damages that may be sustained by the Owner as a result of the anticipated delay caused by the Design-Builder; and

.6 reasonable evidence demonstrating that the unpaid balance of the GMP is insufficient to fund the cost to complete the Work.

The Owner shall give written notice to the Design-Builder at the time of disapproving or nullifying all or part of an application for payment of the specific reasons. When the above reasons for disapproving or nullifying an application for payment are removed, payment will be made for the amount previously withheld.

10.4 OWNER OCCUPANCY OR USE OF COMPLETED OR PARTIALLY COMPLETED WORK

10.4.1 Portions of the Work that are completed or partially completed may be used or occupied by the Owner when (a) the portion of the Work is designated in a Certificate of Substantial Completion, (b) appropriate insurer(s) and/or sureties consent to the occupancy or use, and (c) appropriate public authorities authorize the occupancy or use. Such partial occupancy or use shall constitute Substantial Completion of that portion of the Work. The Design-Builder shall not unreasonably withhold consent to partial occupancy or use. The Owner shall not unreasonably refuse to accept partial occupancy or use, provided such partial occupancy or use is of value to the Owner.

10.5 FINAL PAYMENT

10.5.1 Final Payment, consisting of the unpaid balance of the Cost of the Work, compensation for Design Phase Services and the Design-Builder's Fee, shall be due and payable when the work is fully completed. Before issuance of final payment, the Owner may request satisfactory evidence that all payrolls, material bills and other indebtedness connected with the Work have been paid or otherwise satisfied.

10.5.2 In making final payment the Owner waives all claims except for:

.1 outstanding liens;

.2 improper workmanship or defective materials appearing within one year after the Date of Substantial Completion;

.3 work not in conformance with the Contract Documents; and

.4 terms of any special warranties required by the Contract Documents.

10.5.3 In accepting final payment, the Design-Builder waives all claims except those previously made in writing and which remain unsettled.

ARTICLE 11

INDEMNITY, INSURANCE, BONDS, AND WAIVER OF SUBROGATION

11.1 INDEMNITY

11.1.1 To the fullest extent permitted by law, the Design-Builder shall defend, indemnify and hold harmless the Owner, Owner's officers, directors, members, consultants, agents and employees from all claims for bodily injury and property damage (other than to the Work itself and other property required to be insured under Paragraph 11.5 owned by or in the custody of the owner), that may arise from the performance of the Work, to the extent of the negligence attributed to such acts or omissions by the Design-Builder, Subcontractors or anyone employed directly or indirectly by any of them or by anyone for whose acts any of them may be liable. The Design-Builder shall not be required to defend, indemnify or hold harmless the Owner, Owner's officers, directors, members, consultants, agents and employees for any acts, omissions or negligence of the Owner, the Owner's officers, directors, members, consultants, employees, agents or separate contractors.

11.1.2 To the fullest extent permitted by law, the Owner shall defend, indemnify and hold harmless the Design-Builder, its officers, directors or members, Subcontractors or anyone employed directly or indirectly by any of them or anyone for whose acts any of them may be liable from all claims for bodily injury and property damage, other than property insured under Paragraph 11.5, that may arise from the performance of work by Others, to the extent of the negligence attributed to such acts or omissions by Others.

11.2 DESIGN-BUILDER'S LIABILITY INSURANCE

11.2.1 The Design-Builder shall obtain and maintain insurance coverage for the following claims which may arise out of the performance of this Agreement, whether resulting from the Design-Builder's operations or from the operations of any Subcontractor, anyone in the employ of any of them, or by an individual or entity for whose acts they may be liable:

.1 workers' compensation, disability and other employee benefit claims under acts applicable to the Work;

.2 under applicable employer's liability law, bodily injury, occupational sickness, disease or death claims of the Design-Builder's employees;

.3 bodily injury, sickness, disease or death claims for damages to persons not employed by the Design-Builder;

.4 personal injury liability claims for damages directly or indirectly related to the person's employment by the Design-Builder or for damages to any other person;

.5 claims for physical injury to tangible property, including all resulting loss of use of that property, to property other than the Work itself and property insured under Paragraph 11.5;

.6 bodily injury, death or property damage claims resulting from motor vehicle liability in the use, maintenance or ownership of any motor vehicle; and

.7 contractual liability claims involving the Design-Builder's obligations under Subparagraph 11.1.1.

11.2.2 The Design-Builder's Commercial General and Automobile Liability Insurance as required by Subparagraph 11.2.1 shall be written for not less than the following limits of liability:

.1 Commercial General Liability Insurance

a. Each Occurrence Limit
$____________ ♦

b. General Aggregate
$____________ ♦

c. Products/Completed Operations Aggregate
$____________ ♦

d. Personal and Advertising Injury Limit
$____________ ♦

.2 Comprehensive Automobile Liability Insurance

a. Combined Single Limit Bodily Injury and Property Damage
$____________ ♦
Each Occurrence

b. Bodily Injury
$____________ ♦
Each Person

$____________ ♦
Each Occurrence

c. Property Damage
$____________ ♦
Each Occurrence

11.2.3 Commercial General Liability Insurance may be arranged under a single policy for the full limits required or by a combination of underlying policies and an Excess or Umbrella Liability policy.

11.2.4 The policies shall contain a provision that coverage will not be canceled or not renewed until at least thirty (30) days' prior written notice has been given to the Owner. Certificates of insurance showing required coverage to be in force shall be filed with the Owner prior to commencement of the Work.

11.2.5 Products and Completed Operations insurance shall be maintained for a minimum period of __________ ♦ __________ year(s) after either ninety (90) days following the Date of Substantial Completion or final payment, whichever is earlier.

11.3 PROFESSIONAL LIABILITY INSURANCE The Design Builder shall obtain, either itself or through the Architect/Engineer, professional liability insurance for claims arising from the negligent performance of professional services under this Agreement, which shall be:

General Office Coverage

Project Specific Professional Liability Insurance (Cross-out one of the above), ♦

written for not less than $___________________ ♦ per claim and in the aggregate with a deductible not to exceed $___________________. The Professional Liability Insurance shall include prior acts coverage sufficient to cover all services rendered by the Architect/Engineer. This coverage shall be continued in effect for ____________ year(s) after the Date of Substantial ♦ Completion.

11.4 OWNER'S LIABILITY INSURANCE The Owner shall be responsible for obtaining and maintaining its own liability insurance. Insurance for claims arising out of the performance of this Agreement may be purchased and maintained at the Owner's discretion. The Owner shall provide the Design-Builder with a certificate of insurance at the request of the Design-Builder.

11.5 INSURANCE TO PROTECT PROJECT

11.5.1 The Owner shall obtain and maintain "All Risk" Builder's Risk insurance in a form acceptable to the Design-Builder upon the entire Project for the full cost of replacement at the time of any loss. This insurance shall include as named insureds the Owner, the Design-Builder, the Architect/Engineer, Subcontractors, Material Suppliers and Subsubcontractors. This insurance shall include "all risk" insurance for physical loss or damage including without duplication of coverage at least theft, vandalism, malicious mischief, transit, materials stored off site, collapse, falsework, temporary buildings, debris removal, flood, earthquake, testing, and damage resulting from defective design, workmanship or material. The Owner shall increase limits of coverage, if necessary, to reflect estimated replacement cost. The insurance policy shall be written without a co-insurance clause. The Owner shall be solely responsible for any deductible amounts.

11.5.2 If the Owner occupies or uses a portion of the Project prior to its Substantial Completion, such occupancy or use shall not commence prior to a time mutually agreed to by the Owner and the Design-Builder. Permission for partial occupancy from the insurance company shall be included as standard in the property insurance policy, to ensure that this insurance shall not be canceled or lapsed on account of partial occupancy. Consent of the Design-Builder to such early occupancy or use shall not be unreasonably withheld.

11.5.3 The Owner shall obtain and maintain boiler and machinery insurance as necessary. The interests of the Owner, the Design-Builder, the Architect/Engineer, Subcontractors, Material Suppliers and Subsubcontractors shall be protected under this coverage.

11.5.4 The Owner shall purchase and maintain insurance to protect the Owner, the Design-Builder, the Architect/Engineer, Subcontractors, Material Suppliers and Subsubcontractors against loss of use of the Owner's property due to those perils insured pursuant to Paragraph 11.5. Such policy will provide coverage for expediting expenses of materials, continuing overhead of the Owner and the Design-Builder, the Architect/Engineer, Subcontractors, Material Suppliers and Subsubcontractors, necessary labor expense including overtime, loss of income by the Owner and other determined exposures. Exposures of the Owner, the Design-Builder, the Architect/Engineer, Subcontractors and Subsubcontractors, shall be determined by mutual agreement with separate limits of coverage fixed for each item.

11.5.5 The Owner shall provide the Design-Builder with a copy of all property insurance policies before an exposure to loss may occur. Copies of any subsequent endorsements shall be furnished to the Design-Builder. The Design-Builder shall be given thirty (30) days notice of cancellation, non-renewal, or any endorsements restricting or reducing coverage. The Owner shall give written notice to the Design-Builder before commencement of the Work if the Owner will not be obtaining property insurance. In that case, the Design-Builder may obtain insurance in order to protect its interest in the Work as well as the interest of the Architect/Engineer, Subcontractors, Material Suppliers and Subsubcontractors in the Work. The cost of this insurance shall be a Cost of the Work pursuant to Article 8, and the GMP shall be increased by Change Order. If the Design-Builder is damaged by the failure of the Owner to purchase or maintain property insurance or to so notify the Design-Builder, the Owner shall bear all reasonable costs incurred by the Design-Builder arising from the damage.

11.5.6 The Owner shall have the right to self-insure against the risks covered in Subparagraphs 11.5.1 and 11.5.4 upon providing evidence satisfactory to the Design-Builder of the ability to so self-insure.

11.6 PROPERTY INSURANCE LOSS ADJUSTMENT

11.6.1 Any insured loss shall be adjusted with the Owner and the Design-Builder and made payable to the Owner and Design-Builder as trustees for the insureds, as their interests may appear, subject to any applicable mortgagee clause.

11.6.2 Upon the occurrence of an insured loss, monies received will be deposited in a separate account and the trustees shall make distribution in accordance with the agreement of the parties in interest, or in the absence of such agreement, in accordance with a dispute resolution award pursuant to Article 13. If the trustees are unable to agree between themselves on the settlement of the loss, such dispute shall also be submitted for resolution pursuant to Article 13.

11.7 WAIVER OF SUBROGATION

11.7.1 The Owner and the Design-Builder waive all rights against each other, the Architect/Engineer, and any of their respective employees, agents, consultants, Subcontractors, Material Suppliers and Subsubcontractors for damages covered by the insurance provided pursuant to Paragraph 11.5 to the extent they are covered by that insurance, except such rights as they may have to the proceeds of such insurance held by the Owner and the Design-Builder as trustees. The Design-Builder shall require similar waivers from the Architect/Engineer and all Subcontractors, and shall require each of them to include similar waivers in their subsubcontracts and consulting agreements.

11.7.2 The Owner waives subrogation against the Design-Builder, the Architect/Engineer, Subcontractors, Material Suppliers and Subsubcontractors on all property and consequential loss policies carried by the Owner on adjacent properties and under property and consequential loss policies purchased for the Project after its completion.

11.7.3 The policies shall also be endorsed to state that the carrier waives any right of subrogation against the Design-Builder, the Architect/Engineer, Subcontractors, Material Suppliers, or Subsubcontractors.

11.8 MUTUAL WAIVER OF CONSEQUENTIAL DAMAGES The Owner and the Design-Builder agree to waive all claims against the other for all consequential damages that may arise out of or relate to this Agreement. The Owner agrees to waive damages including but not limited to the Owner's loss of use of the Property, all rental expenses incurred, loss of services of employees, or loss of reputation. The Design-Builder agrees to waive damages including but not limited to the loss of business, loss of financing, principal office overhead and profits, loss of profits not related to this Project, or loss of reputation. This paragraph shall not be construed to preclude contractual provisions for liquidated damages when such provisions relate to direct damages only. The provisions of this paragraph shall govern the termination of this Agreement and shall survive such termination.

11.9 BONDING

11.9.1 Performance and Payment Bonds

are

are not

(Cross-out one of the above) ♦

required of the Design-Builder. Such bonds shall be issued by a surety licensed in the state of the location of the Project and must be acceptable to the Owner.

11.9.2 Such Performance Bond shall be issued in the penal sum equal to one-hundred percent (100%) of the

GMP (if there is no GMP, then the agreed estimated cost of the Project, including design and construction).

agreed estimated construction cost of the Project.

(Cross-out one of the above) ♦

Such Performance Bond shall cover the cost to complete the Work, but shall not cover any damages of the type specified to be covered by the insurance pursuant to Paragraph 11.2 and Paragraph 11.3, whether or not such insurance is provided or is in an amount sufficient to cover such damages.

11.9.3 The penal sum of the Payment Bond shall equal the penal sum of the Performance Bond.

ARTICLE 12

SUSPENSION AND TERMINATION OF THE AGREEMENT AND OWNER'S RIGHT TO PERFORM DESIGN-BUILDER'S RESPONSIBILITIES

12.1 SUSPENSION BY THE OWNER FOR CONVENIENCE

12.1.1 The Owner may order the Design-Builder in writing to suspend, delay or interrupt all or any part of the Work without cause for such period of time as the Owner may determine to be appropriate for its convenience.

12.1.2 Adjustments caused by suspension, delay or interruption shall be made for increases in the GMP, compensation for Design Phase Services, the Design-Builder's Fee and/or the Date of Substantial Completion and/or the Date of Final Completion. No adjustment shall be made if the Design-Builder is or otherwise would have been responsible for the suspension, delay or interruption of the Work, or if another provision of this Agreement is applied to render an equitable adjustment.

12.2 OWNER'S RIGHT TO PERFORM DESIGN-BUILDER'S OBLIGATIONS AND TERMINATION BY THE OWNER FOR CAUSE

12.2.1 If the Design-Builder persistently fails to perform any of its obligations under this Agreement, the Owner may, after five (5) days' written notice, during which period the Design-Builder fails to perform such obligation, undertake to perform such obligations. The GMP shall be reduced by the cost to the Owner of performing such obligations.

12.2.2 Upon an additional five (5) days' written notice to the Design-Builder and the Design-Builder's surety, if any, the Owner may terminate this Agreement for any of the following reasons:

.1 if the Design-Builder persistently utilizes improper materials and/or inadequately skilled workers;

.2 if the Design-Builder does not make proper payment to laborers, Material Suppliers or Subcontractors, provided that the Owner is making payments to the Design-Builder in accordance with the terms of this Agreement;

.3 if the Design-Builder persistently fails to abide by the orders, regulations, rules, ordinances or laws of governmental authorities having jurisdiction; or

.4 if the Design-Builder otherwise materially breaches this Agreement.

If the Design-Builder fails to cure or commence and continue to cure within the five (5) days, the Owner, without prejudice to any other right or remedy, may take possession of the Worksite and complete the Work utilizing any reasonable means. In this event, the Design-Builder shall not have a right to further payment until the Work is completed.

12.2.3 If the Design-Builder files a petition under the Bankruptcy Code, this Agreement shall terminate if the Design-Builder or the Design-Builder's trustee rejects the Agreement or, if there has been a default, the Design-Builder is unable to give adequate assurance that the Design-Builder will perform as required by this Agreement or otherwise is unable to comply with the requirements for assuming this Agreement under the applicable provisions of the Bankruptcy Code.

12.2.4 In the event the Owner exercises its rights under Subparagraph 12.2.1 or 12.2.2, upon the request of the Design-Builder the Owner shall provide a detailed accounting of the cost incurred by the Owner.

12.3 TERMINATION BY OWNER WITHOUT CAUSE

If the Owner terminates this Agreement other than as set forth in Paragraph 12.2, the Owner shall pay the Design-Builder for all Work executed and for all proven loss, cost or expense in connection with the Work, plus all demobilization costs. In addition, the Design-Builder shall be paid an amount calculated as set forth below:

.1 If the Owner terminates this Agreement prior to commencement of the Construction Phase, the Design-Builder shall be paid for the Design-Builder's Design Phase services provided to date as set forth in Subparagraph 7.1.2 and 7.1.3, and a premium as set forth below:

(Insert here the amount agreed to by the parties) ♦

.2 If the Owner terminates this Agreement after commencement of the Construction Phase, the Design-Builder shall be paid for the Construction Phase Services provided to date pursuant to Subparagraph 7.2.1 and a premium as set forth below:

(Insert here the amount agreed to by the parties) ♦

.3 The Owner shall also pay to the Design-Builder fair compensation, either by purchase or rental at the election of the Owner, for all equipment retained. The Owner shall assume and become liable for obligations, commitments and unsettled claims that the Design-Builder has previously undertaken or incurred in good faith in connection with the Work or as a result of the termination of this Agreement. As a condition of receiving the payments provided under this Article 12, the Design-Builder shall cooperate with the Owner by taking all steps necessary to accomplish the legal assignment of the Design-Builder's rights and benefits to the Owner, including the execution and delivery of required papers.

12.4 TERMINATION BY THE DESIGN-BUILDER

12.4.1 Upon five (5) days' written notice to the Owner, the Design-Builder may terminate this Agreement for any of the following reasons:

.1 if the Work has been stopped for a sixty (60) day period

a. under court order or order of other governmental authorities having jurisdiction; or

b. as a result of the declaration of a national emergency or other governmental act during which, through no act or fault of the Design-Builder, materials are not available;

.2 if the Work is suspended by the Owner for sixty (60) consecutive days;

.3 if the Owner fails to furnish reasonable evidence that sufficient funds are available and committed for the entire cost of the Project in accordance with Subparagraph 4.1.3 of this Agreement.

12.4.2 If the Owner has for thirty (30) days failed to pay the Design-Builder pursuant to Subparagraph 10.1.2, the Design-Builder may give written notice of its intent to terminate this Agreement. If the Design-Builder does not receive payment within five (5) days of giving written notice to the Owner, then upon five (5) days' additional written notice to the Owner, the Design-Builder may terminate this Agreement.

12.4.3 Upon termination by the Design-Builder in accordance with this Subparagraph, the Design-Builder shall be entitled to recover from the Owner payment for all Work executed and for all proven loss, cost or expense in connection with the Work, plus all demobilization costs and reasonable damages. In addition, the Design-Builder shall be paid an amount calculated as set forth either in Subparagraph 12.3.1 or 12.3.2, depending on when the termination occurs, and Subparagraph 12.3.3.

ARTICLE 13

DISPUTE RESOLUTION

13.1 WORK CONTINUANCE AND PAYMENT Unless otherwise agreed in writing, the Design-Builder shall continue the Work and maintain the approved schedules during all dispute resolution proceedings. If the Design-Builder continues to perform, the Owner shall continue to make payments in accordance with the Agreement.

13.2 INITIAL DISPUTE RESOLUTION If a dispute arises out of or relates to this Agreement or its breach, the parties shall endeavor to settle the dispute first through direct discussions. If the dispute cannot be settled through direct discussions, the parties shall endeavor to settle the dispute by mediation under the Construction Industry Mediation Rules of the American Arbitration Association before recourse to the dispute resolution procedures contained in this Agreement. The location of the mediation shall be the location of the Project. Once one party files a request for mediation with the other contracting party and with the American Arbitration Association, the parties agree to conclude such mediation within sixty (60) days of filing of the request. Either party may terminate the mediation at any time after the first session, but the decision to terminate must be delivered in person by the party's representative to the other party's representative and the mediator.

13.3 EXHIBIT NO. 1 If the dispute cannot be settled by mediation within sixty (60) days, the parties shall submit the dispute to any dispute resolution process set forth in Exhibit No. 1 to this Agreement.

13.4 MULTIPARTY PROCEEDING The parties agree that all parties necessary to resolve a claim shall be parties to the same dispute resolution proceeding. Appropriate provisions shall be included in all other contracts relating to the Work to provide for the consolidation of such dispute resolution proceedings.

13.5 COST OF DISPUTE RESOLUTION The prevailing party in any dispute arising out of or relating to this Agreement or its breach that is resolved by the dispute resolution process set forth in Exhibit No. 1 to this Agreement shall be entitled to recover from the other party those reasonable attorneys fees, costs and expenses incurred by the prevailing party in connection with such dispute resolution process after direct discussions and mediation.

13.6 LIEN RIGHTS Nothing in this Article shall limit any rights or remedies not expressly waived by the Design-Builder which the Design-Builder may have under lien laws.

ARTICLE 14

MISCELLANEOUS PROVISIONS

14.1 ASSIGNMENT Neither the Owner nor the Design-Builder shall assign its interest in this Agreement without the written consent of the other except as to the assignment of proceeds. The terms and conditions of this Agreement shall be binding upon both parties, their partners, successors, assigns and legal representatives. Neither party to this Agreement shall assign the Agreement as a whole without written consent of the other except that the Owner may assign the Agreement to a wholly-owned subsidiary of the Owner when the Owner has fully indemnified the Design-Builder or to an institutional lender providing construction financing for the Project as long as the assignment is no less favorable to the Design-Builder than this Agreement. In the event of such assignment, the Design-Builder shall execute all consents reasonably required. In such event, the wholly-owned subsidiary or lender shall assume the Owner's rights and obligations under the Contract Documents. If either party attempts to make such an assignment, that party shall nevertheless remain legally responsible for all obligations under the Agreement, unless otherwise agreed by the other party.

14.2 GOVERNING LAW This Agreement shall be governed by the law in effect at the location of the Project.

14.3 SEVERABILITY The partial or complete invalidity of any one or more provisions of this Agreement shall not affect the validity or continuing force and effect of any other provision.

14.4 NO WAIVER OF PERFORMANCE The failure of either party to insist, in any one or more instances, on the performance of any of the terms, covenants or conditions of this Agreement, or to exercise any of its rights, shall not be construed as a waiver or relinquishment of such term, covenant, condition or right with respect to further performance.

14.5 TITLES AND GROUPINGS The titles given to the articles of this Agreement are for ease of reference only and shall not be relied upon or cited for any other purpose. The grouping of the articles in this Agreement and of the Owner's specifications under the various headings is solely for the purpose of convenient organization and in no event shall the grouping of provisions, the use of paragraphs or the use of headings be construed to limit or alter the meaning of any provisions.

14.6 JOINT DRAFTING The parties to this Agreement expressly agree that this Agreement was jointly drafted, and that both had opportunity to negotiate its terms and to obtain the assistance of counsel in reviewing its terms prior to execution. Therefore, this Agreement shall be construed neither against nor in favor of either party, but shall be construed in a neutral manner.

14.7 RIGHTS AND REMEDIES The parties' rights, liabilities, responsibilities and remedies with respect to this Agreement, whether in contract, tort, negligence or otherwise, shall be exclusively those expressly set forth in this Agreement.

14.8 OTHER PROVISIONS ♦

ARTICLE 15

EXISTING CONTRACT DOCUMENTS

The Contract Documents in existence at the time of execution of this Agreement are as follows: ♦

As defined in Subparagraph 2.4.1, the following Exhibits are a part of this Agreement:

EXHIBIT NO. 1 Dispute Resolution Menu, one page.

EXHIBIT NO. 2 Agreement establishing Fast-track approach and Schedule of the Work, ________________ pages. ♦

EXHIBIT NO. 3 Labor Relations provisions, ________________ pages. ♦

This Agreement is entered into as of the date entered in Article 1.

OWNER: ________________________________ ♦

ATTEST: ________________________________ ♦ BY: ________________________________ ♦

PRINT NAME: ________________________________ ♦

PRINT TITLE: ________________________________ ♦

DESIGN-BUILDER: ________________________________ ♦

ATTEST: ________________________________ ♦ BY: ________________________________ ♦

PRINT NAME: ________________________________ ♦

PRINT TITLE: ________________________________ ♦

Chapter

13

Where Do We Go from Here?

Design-build has developed a track record of delivering a project more quickly than the conventional design-bid-build process, and this ability to accelerate substantial completion appears to compensate for slightly higher initial costs. Time is money and nowhere is this more evident than in the capital facilities marketplace. So the quest to explore more rapid project delivery methods that compress both the design and construction cycles will continue.

This push may well augur a four-pronged approach:

1. Interoperability and outsourcing to obtain the most rapid, error free, and cost-effective path—from design development to contract documents.
2. Optimizing the mobility afforded by a wireless global environment.
3. Increasing productivity in the supply chain of men, materials, and equipment.
4. New advances in construction technology, materials, and equipment.

Interoperability Coupled with Outsourcing

When the interoperability cycle is fully completed, accepted, and implemented by all industry software producers and users, the rapid transfer of documents of all types will further compress communications to a nearly real-time basis. The continuing and probably accelerating case for outsourcing will further stretch the 8-hour workday, compressing 16 or even 24 hours of actual production time into a standard workday cycle. With CAD technicians in India earning $5.00 per hour and their qualified design engineers making $12,000 per year on the other side of the globe, it is hard to imagine that U.S. firms won't continue to take advantage of both costs and time by calling upon this market and others in the emerging and developing nations around the world.

With three-dimensional (3D) and four-dimensional (4D) modeling affording designers the ability to achieve a more comprehensive review of construction documents, no matter where those design consultants are located geographically, this should drastically reduce many of the nagging coordination, interference, and error and omission problems that have been of major concern to the industry.

Optimizing Mobility

No longer is the project superintendent or project manager tethered to their field office communication center by copper wire umbilical cords. The first bulky cell phones in a lunch-box sized carrying case rapidly morphed into shirt-pocket sized flip phones. Then came the handheld Palm Pilot and now the multifunctional Bluetooth devices replete with camera and Internet connections that offer the optimum (so far!) in mobile communications. Both, the project superintendent and project manager are now fully mobile, able to send and receive voice and data transmissions quickly. This instantaneous communication and the ability to move documents electronically has transformed the industry and is just a precursor of things to come.

A company in North Carolina, Field2Base in Morrisville, has developed a sort of one-device-does-all software program utilizing a PC tablet with an integral camera. The tablet has the ability to photograph objects in the field and even portions of contract drawings. The device can store a variety of standard company forms such as production forms, safety inspection forms, and daily logs in its memory. The screen accepts handwritten notes and sketches; the camera captures images; and the software program allows simultaneous distribution to whoever is designated—architect, engineer, subcontractor, vendor, or owner.

So now, a superintendent can go out in the field, take a photo of, say, a questionable ceiling repair (Fig. 13.1), and e-mail it to the architect and appropriate subcontractors for instructions. The superintendent can take a digital photograph of a portion of the contract drawings and bring it up on the tablet's screen, while standing right at that spot on the drawing. He or she can then write his or her comments directly on the screen in the appropriate area in question and e-mail it to the architect/engineer (A/E) for review and approval (Fig. 13.2). When a design change is requested from the field, or requires clarification, it can be sketched on the tablet (Fig. 13.3) and e-mailed to the architect, engineer, and selected subcontractors for their comments.

A pictorial punchlist can be developed using the tablet's camera and the screen's ability to accept handwritten notes to explain the exact nature of the item. An architect or engineer should welcome this simple method of creating what was once a laborious job and having it instantly transmitted to a general contractor who can then pass it on to the appropriate vendors and subcontractors. No explanations needed, just comply and respond.

Now that's productivity—all due to mobility and the electronic world of today.

Figure 13.1 Field2Base Tablet PC with camera transmitting a question of a ceiling repair to the architect and subcontractor. (*By permission: Field2Base, Morrisville, N.C.*)

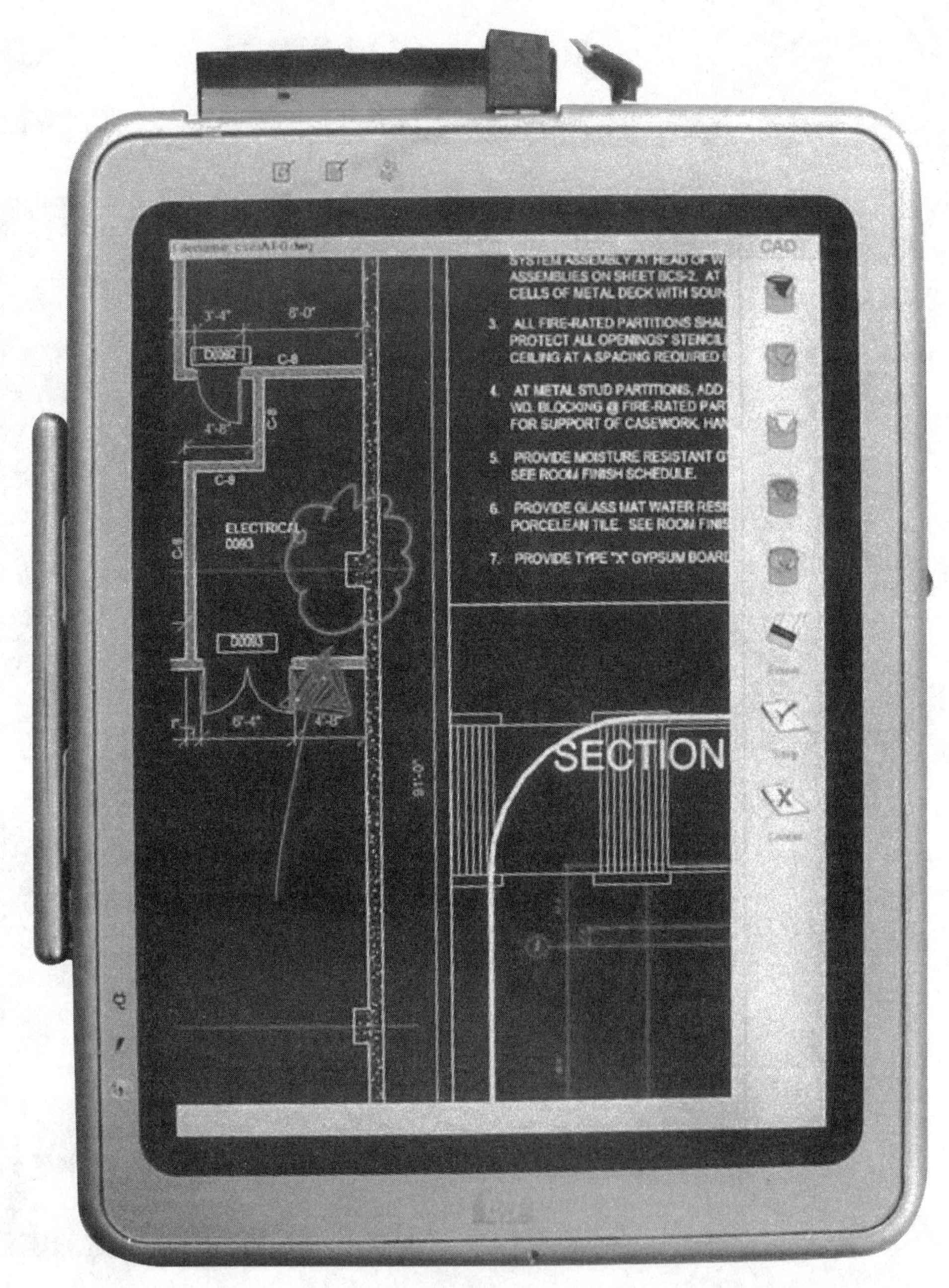

Figure 13.2 Field2Base Tablet PC transmitting portion of contract drawing to architect/engineer with questions about electrical room's wall construction. (*By permission: Field2Base, Morrisville, N.C.*)

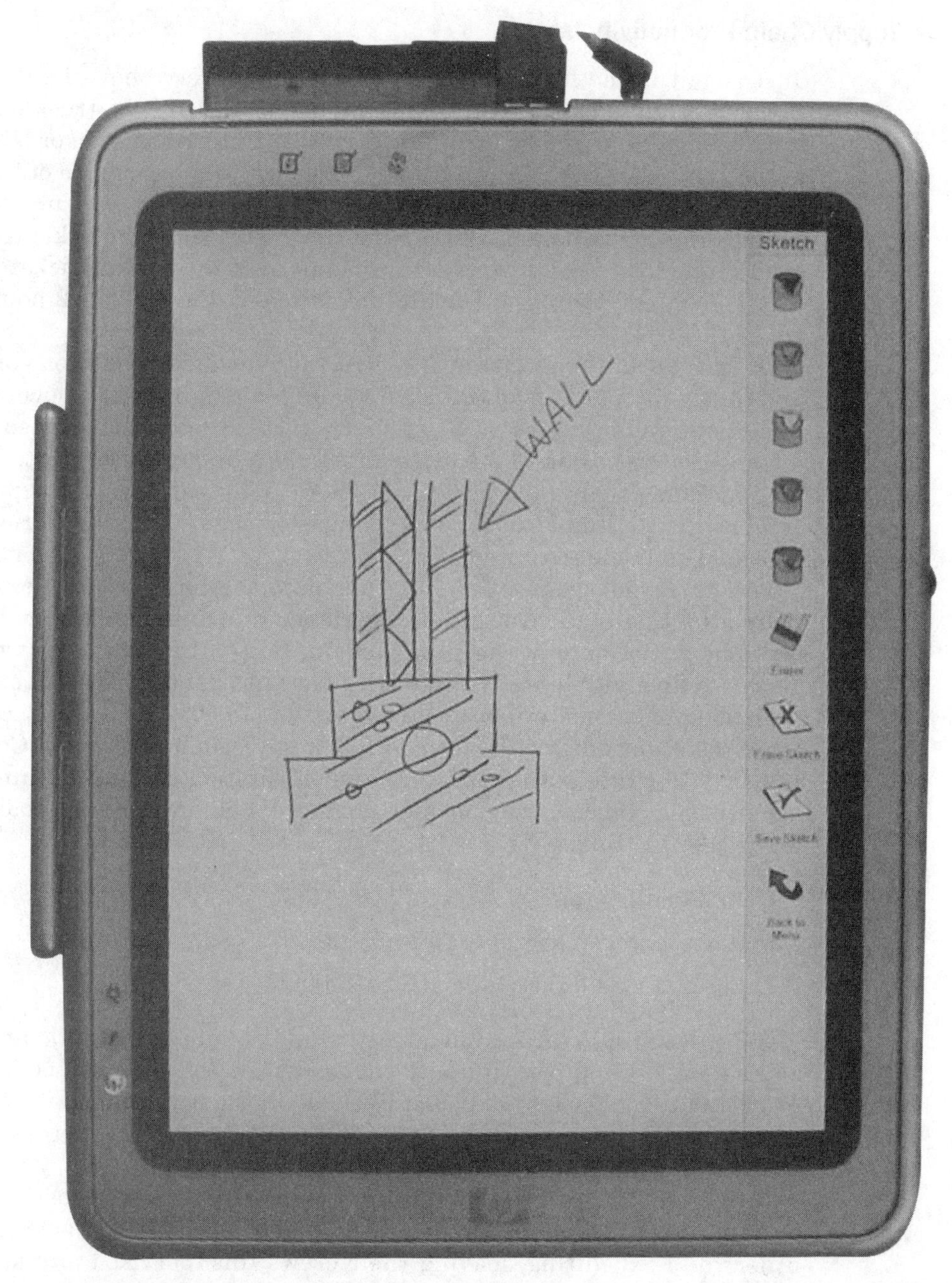

Figure 13.3 Field2Base Tablet PC screen with superintendent's sketch of proposed change to electrical room's wall. (*By permission: Field2Base, Morrisville, N.C.*)

The Supply Chain Productivity Issue

Just go out to the field on any given day and observe how the supply chain works. Perhaps yesterday, the superintendent called for a truckload of framing materials and at 8:25 a.m. they arrived on site. A foreman or laborer steps away from the assigned task, directs the truck to the east side of the building so that half the load can be unloaded, and then directs the truck to the west side to drop off the balance of the load. He, or she, signs the ticket (maybe they even count the items to verify the amount on the receiving ticket—but don't count on it) and then goes back to the assigned duties—1 or 2 hours or more wasted?

Or, how about the mechanical contractor who had a whole load of ductwork and fittings dropped off on the first floor and sent a mechanic down five flights to sort out and bring 5 pieces of 18 × 30 duct, 5 elbows of assorted sizes, and 4 transition pieces up to the 6th floor. Or, how about the carpenter searching high and low for that 3/4 inch drill? There is a great deal of lost productivity created by such simple tasks as receiving, storing, sorting, and retrieving materials and equipment.

As new technologies are developed and new monitoring systems are perfected, more attention is turned to the importance of managing field materials and equipment, not only in the supply chain, but on the jobsite as well. A 1982 Business Roundtable study, although somewhat dated, revealed that in a construction project, materials accounted for 40% to 50% of total project costs and labor costs ranged from 50% to 60%. A study conducted by the Construction Industry Institute (CII) found that implementation of a basic materials management plan on construction sites produced a 6% increase in productivity. So let us quantify this:

On a $5 million project:

Materials are, on average, 45% of total cost = $2,250,000

Labor cost would therefore be $2,750,000

A 6% increase in labor productivity would equate to $165,000 and that's no small deal. If the project initially included a 5% fee, that is $250,000, this 6% productivity increase would most likely go to the bottom line and that 5% fee would suddenly translate into $415,000—a whopping 65% increase. Does this interest you now?

A Bell and Stukhart study in 1987 found that foremen lost 20% of their working time searching for materials and another 10% tracking purchase orders and expediting, leaving their crews unsupervised and slowing productivity. Another study on power plant construction in 1980 observed that 27.7% of the craft worker's time was idle or nonproductive due to the unavailability of correct materials and tools. Electrical workers were observed to spend as much as 42% of their time handling materials and preparatory operations.

Well, Help Is on the Way in the Form of Global Positioning Satellites and Radio Frequency Identification Devices

Global positioning satellites

In a trial study using Global Positioning System (GPS) to locate materials, FIATECH, Kellogg, Brown & Root Inc. (KBR), and several equipment suppliers conducted a pilot study at a petrochemical plant in Texas, devised to locate various types of spool pieces—those short sections of pipe of various lengths used to join longer sections. The conventional method of locating a particular spool piece involved unloading a shipment of spool pieces that were sorted into small grids physically located on the site; each item was then identified with a colored tape. When needed, workers would go to the grid location, visually locate the required piece, and attach a flag to identify the spool piece for a later pick up. In this test program, a worker with a handheld computer used a focused GPS system to locate the exact position of a spool piece by inputting ID numbers on a grid system that appeared on the screen of the handheld. Under the old system it took 6 minutes and 42 seconds to recall and flag each spool; the new GPS system reduced this recall/flag time to a mere 55 seconds per spool. This same kind of tagging/grid system recovery process is easily adaptable and can find applicability in locating tools, equipment, and materials just as easily as spools. The technology is here, it is not very expensive; and all that remains is for a company to turn this process into a commercial product for contractors. Now how long do you think it would take that HVAC mechanic to locate those 14 pieces of ductwork?

Radio frequency identification devices

Radio Frequency Identification Devices (RFIDs) are tiny programmable tags that store information and transmit the stored information wirelessly when activated. These tags are composed of three parts: a chip that contains electronic data about the physical object to which it is attached, an antenna to transmit the stored information to a "reader" via radio waves, and an encasement so that both chip and antenna can be attached to the object in question.

An RFID may be similar in some respects to bar codes in that they both store information, but RFIDs are much more adaptable and can cope with more demanding tasks and environments.

These RFIDs can:

- Be read or updated without being in a line-of-sight.
- Be read in multiples.
- Cope with harsh and dirty environments.
- Be automatically tracked.
- Identify a specific product or item.

- Be programmed—electronic information can be written over several times.
- Be read even if concealed in concrete, steel, or water.

RFIDs versus bar codes. Used in somewhat similar situations, bar codes have several disadvantages, particularly when it comes to their adaptability in the construction industry. They require a line-of-sight in order to be read, can only be read one at a time and not in multiples, have limited read range, and can't be read if covered with dirt or debris. Bar-coded information cannot be updated.

RFIDs are currently manufactured in two forms—passive, those that must be energized by a powered reader and active, those that are battery powered. The passive RFIDs require a "reader" held about 1 foot away to activate them, whereas an active chip can be read up to 100 yards away, and therefore it would appear that the active chip is the one most suited to construction use at the present time. The current cost of an active RFID is now about $50.00, but this cost is expected to drop significantly as future technology and future demand drive this product's development. One manufacturer indicated that they are targeting a price of about $3.00 for active RFIDs by 2007.

RFIDs are currently being used in diverse industries such as retailing to track inventory from the manufacturer to the store shelves to the check-out counter (Wal-Mart is a leader in that field); tracking library books and beer kegs (high-value stainless steel containers!); airline baggage tracking, and truck, trailer, and container tracking in shipping yards.

Job-site equipment theft cost the construction industry between $300 million to $1 billion in 2004, according to the National Equipment Register. At Robert Bosch's Arkansas plant they are affixing RFIDs to each tool they produce in order to create barriers to these thefts. At the jobsite, when one of their tools is issued to a worker the RFID stays with the equipment and the manager can tap into a mobile computer to track that tool no matter where it goes because it is also listed on an asset-tracking database from Tool-Watch Corp.

RFIDs attached to construction equipment as it leaves the manufacturer can contain installation instructions, start-up procedures, operating and maintenance information, and, in the future, even some diagnostic tips.

Just as the automobile industry learned the value of just-in-time inventory and sophisticated retailers like Wal-Mart and Costco used inventory control to drive down costs and save consumers money, the construction industry is closely looking at how just-in-time inventory control can affect productivity.

RFIDs embedded in a product or piece of equipment can be programmed to automatically reorder itself by activating a computer terminal at the vendor's place of business, thereby cutting down the amount of dollars a builder has tied up in inventory, and avoid not having the product or equipment when needed. Equipment deliveries can be tracked so closely that unloading equipment has little or no waiting time and crews required for unloading and installation waste little downtime.

RFIDs Make Concrete Pours More Effective

These little RFIDs are "delivering" cast-in-place concrete more rapidly by being able to monitor its curing cycle electronically instead of resorting to destructive testing or calculations based on existing engineering formulae.

The more familiar method of taking random sample concrete cylinders, placing them in a cure box until the testing lab picks them up, and then waiting for the 7, 14, and 28 day breaks may become a thing of the past.

Sensors now being placed in concrete pours permit contractors to accurately monitor the state of cure in real time. Fluor Corporation, an early user of these embedded wireless sensors, reported that it allows them to remove forms earlier and load the concrete earlier. On multistoried buildings imagine what cost savings could accrue from stripping and moving forms one day earlier. Concrete testing costs will probably be more than offset by the purchase, installation, and monitoring of these in-situ devices. Fluor is also testing these chips in cold weather concrete pours, and, again, just imagine the savings in temporary heat and temporary protection when more accurate and precise cure times can be established.

These examples of technology, replacing labor-intensive and time-consuming construction procedures, with their obvious savings in time and money, are in their infancy and portend the future.

Where Are the Advances in Construction Components and Materials?

We still place one 2' × 7" brick on top of another as masons advance up the side of a building—in much the same manner we did 100 years ago. Observe concrete pumps disgorging their product on a floor where crews of workers swarm to spread the mix evenly over the surface and finishers wait with their bull floats until needed; finishers waiting in the wings for their call to action. These labor-intensive operations should be of much concern to builders. Granted, an influx of foreign workers are currently willing to perform some of the dirty, demanding tasks that the construction industry offers but our "graying workforce" may leave us with a dearth of skilled mechanics.

Some of these labor intensive and repetitive operations may be open to technological innovation that will ultimately drive down unit costs and also improve quality, much like those welding robots in the automobile industry.

On a visit to Japan to meet their leading contractors in 1989 and again in 1992, the writer was amazed to see the fervor with which their research institutes were investigating new products and new technologies. The largest contractors referred to as the Big Six were so named because they had an annual sales volumes in 1991 ranging from $9 billion to $16 billion and they had an aggregate backlog of $190 billion. These giant construction companies had developed carbon fiber reinforced architectural concrete panels weighing 40% less than conventional steel reinforced panels. They were testing thin sheets of stainless steel used as lifetime roof membrane and computer designer roof drains that created a vortex that actually pulled surface water down the rain leaders. They were into second and third generation robots to not only produce quality work and ease the drudgery of their workers but were also looking decades ahead when they expected severe shortages of skilled labor, because theirs, like ours was a "graying workforce."

The real estate bubble that followed in the late 1990s substantially reduced the net worth of many of these first-tier contractors calling a halt to further investment in these endeavors, but at least they showed the world what technology can achieve—a concrete floor finishing robot, controlled by lasers; a paint spraying robot making this messy operation much cleaner and eliminating some workforce health hazards; a wall-climbing robot that tapped its way up the side of a masonry or tile-clad building creating a log of the exact location of failed mortar joints, a rebar bending and placement robot; and a real-time compaction testing robot.

A pilot project sponsored by Big Six contractor Shimizu Corporation in 1992 called SMART successfully completed a robot-controlled structural steel erection cycle on the grounds of their research institute. Although not commercially practical at that time, this project showed what engineering feats can be achieved.

An August 5, 2005 download from the National Institute of Standards and Testing (NIST) contained a small photo of a complex-looking machine entitled "Autonomous steel beam docking using the NIST RoboCrane™." Maybe there is nothing new under the sun—rising sun, that is.

Is the A/E/C Industry Becoming Just Another IT Business?

The wireless movement and the hardware and software it spawned have really cut managers and supervisors loose from many labor intensive and grunt work functions. Preparing and transmitting RFIs, ASIs, RFQs, faxing sketches, and even e-mailing partial or complete drawings is easier and much less time-consuming. Managers can now devote more time to managing effectively, managing more complex projects, and possibly managing more projects. Will this combination of wireless mobility and sophisticated hardware and software be the driving forces behind greater productivity, higher quality, and increased profits?

Will the construction company of the mid-twenty-first century look more like an IBM Systems Program office? We may not recognize the "construction" company of 2050 if wireless technology and software advances expand exponentially.

Will robotics controlled by computer nerds with joysticks replace the operating engineers and will the work boot industry go the way of the horse and buggy? Microsoft or Cisco or IBM could be the builders of the future. What does IBM know about building? Let's first review some of the primary roles a general contractor plays in the construction process.

- Preparing a budget.
  - With a database CAD system, budget and design can be produced concurrently and with an automatic update of materials, equipment, and labor rates from plugged-in on-line sources, we won't need all those guys and gals in the estimating department. One individual can adjust the estimate for difficulty factors, local market conditions, and so forth, but can't a programmer do that? By tapping into our database to compare comparables, and into local labor stats, that one estimator won't have much to do; transfer them to the IT department.
- Preparing a preliminary schedule.
  - It is not inconceivable that a software program couldn't establish the order of precedence for each operation; after all isn't this what we do now manually and depending on a number of factors from the database, i.e, number of items, cubic yards, square feet, and figure out the length of time required for each operation. Plug in some float time for winter conditions, unforeseen site conditions, and so forth, and we've got a schedule. That part-time estimator could probably provide some human input for schedule updates.
- Distribute plans and specifications to vendors and subcontractors to obtain competitive bids. Aren't plans and specifications being stored and distributed over the Internet now? Let's outsource this grunt work to BidWare in Bangalore and have them distribute, collect, collate, and prepare a spreadsheet of all bids.
- Issue contracts/purchase orders by filling in the blanks of subcontract agreements and purchase-order forms, attaching a schedule and exhibits, and distribute the finished product to appropriate parties. Sounds like another outsourced job to BidWare.
  - Our project manager will certainly have time to oversee all these steps.
- Order and coordinate deliveries of materials and equipment. RFIDs might solve that problem by integrating them into the scheduling process. Schedule updates would concurrently update related RFIDs, thereby continuing to coordinate and regulate deliveries driven by the schedule.
- Ensure that subcontractors man the job properly, or notify them to increase their forces when required. Sounds like some human intelligence is needed here but many of these tasks can be handled automatically.
  - The current use of worker's ID cards with magnetic strips, bar codes, or RFIDs can be expanded to solve that problem. A monitor at the project's entry gate could be preprogrammed to check manpower levels. Plumbers scheduled for 10 men on site, but if only 8 showed up, it would trigger one of those computer-generated voices to call that subcontractor's office. The call would be adjusted from stern to scream level depending on the shortage of workers.
- Daily logs can be automatically generated by those magnetic card readers or RFIDs. An onsite small GPS weather station tied back to the daily log software

will record temperature. Bar-coded or RFID subcontractors will provide the daily log with work-task information as they move about the building monitored by a GPS locator tied into the 4D model.

- Payment applications, bookkeeping, and accounts payables and receivables are now being generated by in-house PCs, but for more efficiency and lower cost, this is another operation that can easily be outsourced.

Although many of these schemes may appear to be farfetched, the limits to which our ingenuity coupled with technology can achieve seem boundless and we can only guess what lies ahead for our industry.

Index